Performance Assessment of Urban Infrastructure Services

Performance Assessment of Urban Infrastructure Services

Edited by
**Enrique Cabrera Jr.
and Miguel Ángel Pardo**

cost

IWA
Publishing

Published by IWA Publishing, Alliance House, 12 Caxton Street, London SW1H 0QS, UK

Telephone: +44 (0) 20 76545500; Fax: +44 (0) 20 7654 5555; Email: publications@iwap.co.uk
Web: www.iwapublishing.com

First Published 2008, Reprinted 2015

ISBN: 9781843391913 (Hardback)
ISBN: 9781780401942 (eBook)

Table of contents

Performance Assessment

Definition and establishment of performance assessment systems

Decision support tools based on performance indicators

Managing utilities with performance indicators

Performance measures to support infrastructure asset management

Benchmarking

Metric benchmarking as a driver for improved performance

Process benchmarking as a source for new solutions

Case Studies

About COST

COST- the acronym for European **CO**operation in the field of **S**cientific and **T**echnical Research- is the oldest and widest European intergovernmental network for cooperation in research. Established by the Ministerial Conference in November 1971, COST is presently used by the scientific communities of 35 European countries to cooperate in common research projects supported by national funds.

The funds provided by COST - less than 1% of the total value of the projects- support the COST cooperation networks (COST Actions) through which, with EUR 30 million per year, more than 30.000 European scientists areinvolved in research having a total value which exceeds EUR 2 billion per. This is the financial worth of the European added value which COST achieves.

A "bottom up approach" (the initiative of launching a COST Action comes from the European scientists themselves), "à la carte participation" (only countries interested in the Action participate), "equality of access" (participation is open also to the scientific communities of countries not belonging to the European Union) and "flexible structure" (easy implementation and light management of the research initiatives) are the main characteristics of COST.

As precursor of advanced multidisciplinary research COST has a very important role for the realisation of the European Research Area (ERA) anticipating and complementing the activities of the Framework Programmes, constituting a "bridge" towards the scientific communities of emerging countries, increasing the mobility of researchers across Europe and fostering the establishment of "Networks of Excellence" in many key scientific domains such as: Biomedicine and Molecular Biosciences; Food and Agriculture; Forests, their Products and Services; Materials, Physical and Nanosciences; Chemistry and Molecular Sciences and Technologies; Earth System Science and Environmental Management; Information and Communication Technologies; Transport and Urban Development; Individuals, Societies, Cultures and Health. It covers basic and more applied research and also addresses issues of pre-normative nature or of societal importance.

www.cost.esf.org

About ESF and EU

ESF provides the COST Office through an European Comission contract

COST is supported by the European Union RTD Framework programme.

Preface

Performance assessment has been one of the hottest topics in the water industry in the past decade. The use of performance indicators and benchmarking techniques has become a common practice during this period. There are many good reasons behind this success. Water services are provided in a monopolistic environment, and in the absence of market forces it is hard to find motivation for efficiency. All the stakeholders in the business have come to realize that by assessing the performance of the services in a systematic way, utilities are driven to continually improve their performance with the consequential benefits for all those involved.

Five years ago, a COST Action on the topic was drafted and presented lo the European Union. As a result, some of the key research institutions, utilities and administrations in Europe joined the project. The goal, to evaluate the present and explore the future needs of performance assessment in urban infrastructure services. Along with drinking water and wastewater, solid waste services were included, for although at that time the topic had been less developed, their circumstances and needs were quite similar to those of water services.

COST C18 proved to be a wonderful vehicle to discuss key topics on performance assessment. The meetings allowed for the gathering of experts from around the world. Experts that would probably have not met under different circumstances. All this fruitful exchange of ideas and work will see its zenith in the final conference of COST CI8 in Valencia. A conference also promoted by the Specialist Group on Efficient Operations and Management of the International Water Association.

This publication, supported by COST, includes the papers presented in that conference. They will provide the reader with a good knowledge of the state of the art in the topic, a foresight into the future and a handful of practical experiences that are taking place Worldwide. I am confident that it will further contribute to the development and improvement of performance assessment in urban utilities.

Enrique Cabrera Jr
Chairman. Pi08 Conference

Performance Assessment

Definition and establishment of performance assessment systems

A new approach for assessment of the performance of water distribution and sewerage networks

P. Stahre[1], J. Adamsson[2] and G. Mellstrom[3]

[1] *Malmo Water, P.O. Box, S-205 80 Malmo, Sweden, peter.stahre@malmo.se*
[2] *Aquatech, Juteskarsgatan 60, S-421 59 V Frolunda, Sweden, jan.adamsson@tele2.se*
[3] *Marks kommun, Kommunalhuset, S-511 80 Kinna, Sweden, gunnar.mellstrom@mark.se*

Keywords: quality, service, environment, cost-efficiency, sustainability

 Performance Assessment of Urban Infrastructure Services
Edited by Enrique Cabrera Jr and Miguel Ángel Pardo. ISBN: 9781843391913.
Published by IWA Publishing, London, UK.

ABSTRACT

The aim of the project described in this paper was to develop an operational model for assessing the overall performance of municipal water distribution and sewerage networks. In the model three dimensions of the performance were covered: The first dimension "quality, service and environment" was assessed by evaluating a number of different aspects, all of which were quantified by means of well established performance indicators. The overall performance was quantified by the calculation of performance scores for each network. The second dimension "relative cost efficiency" was assessed with an econometric model, by which real costs was compared with normalized cost, calculated with the help of multiple regression analysis. As an alternative to the regression approach, data envelopment analysis (DEA) was tested. The third dimension "long-term sustainability" could not be assessed in an objective way. For this dimension the following three strategic areas were discussed: the rate of pipe renewal, the financing of pipe renewal and special demands on a sustainable management of the pipe network.

1 INTRODUCTION

The municipal water sector has in recent years been subject to a process of constant change. There is today a pronounced focus on how to become more efficient. As a result the interest for different types of benchmarking has increased significantly. Of special interest in this connection is how to assess the performance and cost-efficiency of utility companies and how to compare them in an objective way.

The Swedish branch organization Swedish Water has supported an initiative to develop a model for assessment of the performance of municipal water and wastewater services, Stahre et. al. (2007). An important prerequisite for this development work was that all necessary input data should be accessible for most municipalities. It was therefore decided to use only input data available in Swedish Water's national benchmarking system VASS or in other official databases of national statistics.

2 METHODS

2.1 Initial study

In an initial study to the project a survey was made of existing practices for how to evaluate quality, service and cost efficiency within the water sector in different countries, Adamsson and Stahre (2004). The study showed that many countries today consider implementing national systems for benchmarking and performance assessment of municipal water and wastewater services. The pressure to organize a more systematic follow-up and control of the quality of service and of the costs for these services is increasing. The main reason for this is that municipal water services is a natural monopoly.

The initial study, which was carried out to create a knowledge base for the subsequent development work, identified a number of interesting approaches for performance assessment. This applies to complete assessment models as well as to activities that can lead to such models. There are many elements in the identified models and activities that have served as inspiration for this project.

2.2 Project idea

The general goal with the project was to develop a pilot model for assessing the performance of municipal water services. The model was to be especially adapted to the rather scattered organizational structure of the Scandinavian water sector, where each individual municipality is responsible for the water services within their respective geographic area.

The developed pilot model is primarily intended for national comparisons of the relative performance and cost-efficiency of municipal water utilities. The comparisons should be made in such a way that areas with potential for improvements can be identified.

One problem with all performance comparisons is the availability of consistent and reliable data. The idea with the developed pilot model was that the performance assessment shall be based on a reasonable amount of data, which shall be accessible for most utilities.

To be able to evaluate the efficiency of water and wastewater organizations it is necessary to evaluate different dimensions of the water services. The following three dimensions were covered in the project: “quality, customer service and environment”, “cost” and “long-term sustainability”. A conceptual illustration of the chosen three dimensional approach is shown in figure 1.

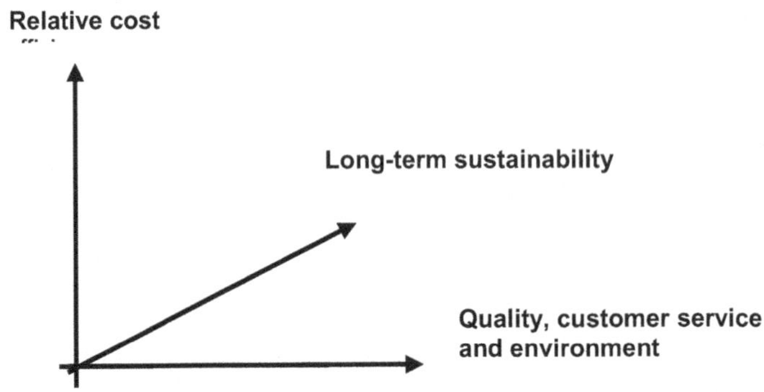

Figure 1 .CONCEPTUAL ILLUSTRATION OF THE THREE DIMENSIONAL APPROACH FOR PERFORMANCE ASSESSMENT OF MUNICIPAL WATER SERVICES.

2.3 Limitations of the project

The work with developing a model for performance assessment of water services has in this project been limited to cover only water distribution and sewerage networks. Thus production of drinking water, treatment of sewage water and administrative functions have not been covered in the project.

3 ASSESSMENT OF QUALITY, CUSTOMER SERVICE AND ENVIRONMENT

3.1 Introduction

The most extensive system for assessing quality, customer service and environment within the water sector has been developed by Ofwat - the regulatory authority for England and Wales, see Ofwat (2006). Some basic principles in the Ofwat approach have been adopted for this project. Inspiration has also come from the assessment approach used by the regulatory authority for the water services in Western Australia, see Office of Water Regulation (2001). The approach in this project has however been modified and adjusted to the special conditions prevailing in Scandinavia.

3.2 Assessment approach

To be able to assess the overall performance within the dimension quality, customer service and environment, a number of different aspects must be considered. As a base for the development work, relevant characteristics of water distribution and sewerage networks as well as different environmental factors of importance for their performance were analysed. A careful selection of performance indicators relevant to quality, customer service and environment was made. The ambition was that the indicators should be well established and easily accessible for most utilities. One set of indicators was selected for water distribution networks and another for sewerage networks.

The performance assessment is done by comparing each individual utility with the "best" and the "worst" performing utility in the compared group. For each indicator, a performance score is calculated with the following formula:

$$Score = \left[\frac{(actualPI - value) - ("worst"\, PI - value)}{("best"\, PI - value) - ("worst"\, PI - value)} \cdot 90 + 10 \right] \qquad (1)$$

The formula is set up in such a way that the number of scores for each performance indicator will have values in the interval 10 – 100. The higher score the higher relative quality and customer service. The "worst" performing utility gets the score 10 and the "best" the score 100.

Obviously the different performance indicators used in the comparison do not have the same weight. The calculated scores must therefore be multiplied by a weight factor, which is decided based on the relative importance of the different indicators from the customers' point of view. The overall performance regarding quality, customer service and environ¬ment is calculated as the sum of the weighted scores for the selected set of performance indicators.

3.3 Performance assessment of water distribution networks

Seen from the customers' perspective the quality of the drinking water, the frequency of service interruptions and the pressure in the system are the most important aspects in the evaluation of the performance of water distribution networks. As low water pressure is not a problem in Scandinavia, this aspect has not been included in the developed assessment model. When the environmental performance of a water distribution network is concerned, the most important aspect is how much water is leaking out from the network.

A list of the selected performance indicators used for performance assessment of water distribution networks is shown in table 1. For each indicator is given the weight factor that indicates its relative importance. As indicated in the table the maximum number of scores in the assessment of water distribution networks is 600.

Table 1. OVERVIEW OF THE SELECTED SET OF PERFORMANCE INDICATORS USED FOR ASSESSING THE PERFORMANCE OF QUALITY, SERVICE AND ENVIRONMENT FOR WATER DISTRIBUTION NETWORKS

Aspect	Performance indicator	Weight factor	Score interval
Drinking water quality	Microbiological water quality Proportion of the samples that are not approved or approved with some remarks	1.8	18 - 180
	Chemical water quality Proportion of the samples that are not approved or approved with some remarks	0.6	6 - 60
Customer service	Complaints on drinking water quality Number of complaints per 1000 customer	0.6	6 - 60
	Interruptions on water mains Minutes per customer and year	1,8	18 - 180
	Interruptions on service connections Minutes per customer and year	0,8	8 - 80
Environment	Water leakage m^3 per km water main and day	0.4	4 - 40
Total scores			60 - 600

The developed assessment model was tested by entering data from the branch organisation Swedish Water's national benchmarking database "VASS". Input data for the dimension quality, customer service and environment was in the database available for about 100 Swedish municipalities.

In the assessment of the dimension of quality, customer service and environment the "best" and "worst" performing utility was derived from all 100 municipalities. The evaluation of the performance was however restricted to those 30 municipalities that had complete input data also for the cost dimension.

The result of the evaluation of the 30 water distribution networks is illustrated in figure 2. Here the network performance is illustrated in a ranking diagram, in which the best performing utility got the highest score.

3.4 Performance assessment of sewerage networks

From the customers' perspective the frequency of sewer blockages and basement floodings are the most important aspects in the evaluation of the performance of sewerage networks. When the environmental performance of sewerage networks is concerned, the most important aspect is the amount of inflow/infiltration into the system. Combined sewer overflow is an environmental aspect that could be considered in this connection. As this type of pollution discharge has to do with how the sewerage network is built up (combined/ separate systems) it was in the project decided to handle this aspect in connection with the evaluation of pollution discharges from the sewage treatment plants and from other sources.

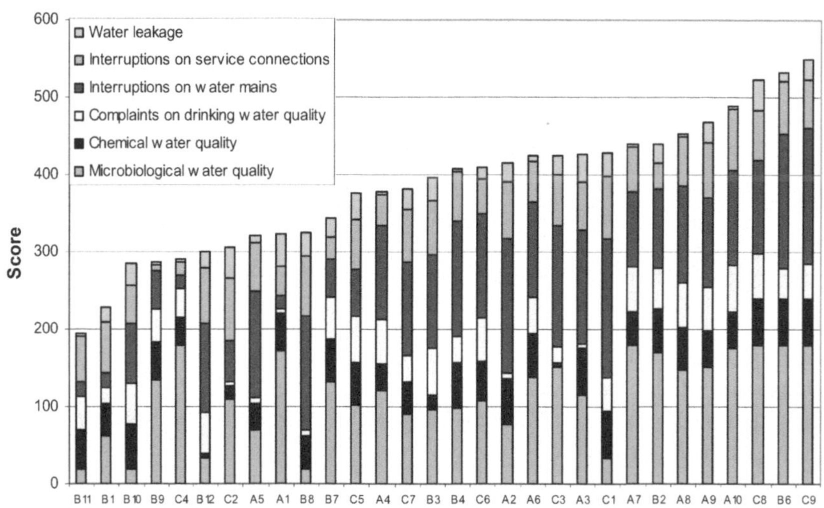

Figure 2 RANKING OF THE PERFORMANCE OF THE WATER DISTRIBUTION NETWORKS IN 30 SWEDISH MUNICIPALITIES.

A list of the selected performance indicators used for performance assessment of sewerage networks is shown in table 2. For each indicator is given the weight factor that indicates its relative importance. As indicated in the table the maximum number of scores in the assessment of sewerage networks is 400.

Table 2 OVERVIEW OF THE SELECTED SET OF PERFORMANCE INDICATORS USED FOR ASSESSING THE PERFORMANCE OF QUALITY, SERVICE AND ENVIRONMENT FOR SEWERAGE NETWORKS

Aspect	Performance indicator	Weight factor	Score interval
Customer service	Blockages in main sewers Number of blockages per km sewage pipes (sanitary+combined)	1,3	13 - 130
	Blockages in service connection Number of blockages per 1000 sewage connections (sanitary+combined)	0,5	5 - 50
	Basement floodings Number of floodings per 1000 of service connections (sanitary+combined)	1.8	18 - 180
Environment	Inflow/infiltration in the sewerage system m^3 per km sewage pipes per day	0.4	4 - 40
Total scores			40 - 400

The developed assessment model for sewerage networks was tested by entering data from the same 30 municipalities that were included in the comparison of the water distribution networks. The result of the evaluation is illustrated in the form of a ranking diagram, see figure 3. The best performing utility got the highest score

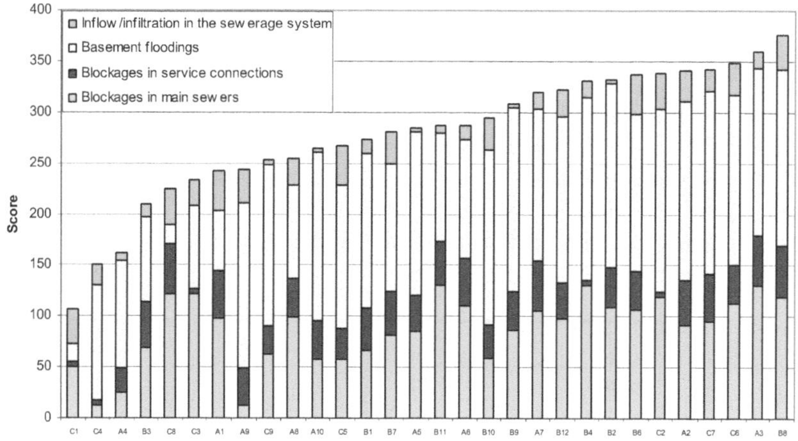

Figure 3. RANKING OF THE PERFORMANCE OF THE SEWERAGE NETWORKS IN 30 SWEDISH MUNICIPALITIES.

2.5 Collective performance ranking of water and sewerage networks

In the two previous paragraphs the results of the ranking of 30 water distribution and sewerage networks was presented. It can be of interest to look at the performance of both types of networks collectively.

In figure 4 the result of such a ranking is illustrated in a so-called matrix diagram. The number of scores have for each of the two types of network been divided in five groups: high scores (10 % of the networks), higher than the mean (25 % of the networks), around the mean (30 % of the networks), lower than the mean (25 % of the networks) and low scores (10 % of the networks). The marked quadrant in the upper right of the diagram shows municipalities which have got scores around the mean or higher for both water distribution and sewerage networks. The marked quadrant in the lower left of the diagram shows municipalities which have got scores lower than the mean for both types of network.

4 ASSESSMENT OF COST-EFFICIENCY

4.1 Introduction

To evaluate the cost-efficiency of water distribution and sewerage networks is a very complex task. The main reason for this is that the costs are dependent on a variety of local conditions. As examples can be mentioned: size of the utility, topography, demography, accounting principles etc. In the assessment of the cost-efficiency one has to find ways to handle these differences to greatest possible extent.

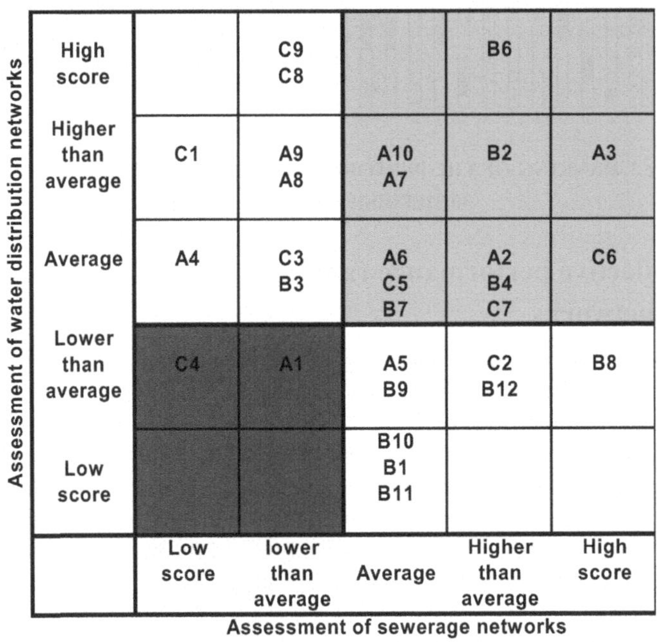

Figure 4. MATRIX DIAGRAM ILLUSTRATING A COLLECTIVE PERFORMANCE RANKING OF WATER DISTRIBUTION AND SEWERAGE NETWORKS IN 30 SWEDISH MUNICIPALITIES.

To be able to make fair comparisons of the cost-efficiency of different utility networks it is of utmost importance to distinguish between various cost components. In the developed assessment model the following three types of costs have been dealt with:

- Costs for daily operation and maintenance (including acute reparations)
- Capital costs
- Costs for pipe renewal

If these cost components are not analysed separately, it is hardly possible to draw any conclusions about the cost-efficiency of the utility.

4.2 Assessment of costs for operation and maintenance

Ofwat – the water regulator for England and Wales – has developed econometric models for evaluating the relative cost efficiency of different sectors of the water industry, see Ofwat (2005). The Ofwat approach has served as one of the sources of inspiration for the assessment model used in this project.

To get a picture of the factors effecting the costs for operation and maintenance of water distribution and sewerage networks (=cost drivers), multiple regression analysis was carried out with data from 30 municipalities in the Swedish benchmarking database VASS. More than hundred different combinations of input parameters were tested. As a result regression relations were established for the operation and maintenance costs for water distribution networks, sewerage networks as well as for both of these networks together. As example figure 5 shows the relation between calculated operation and maintenance costs and real operation and maintenance costs for water distribution networks.

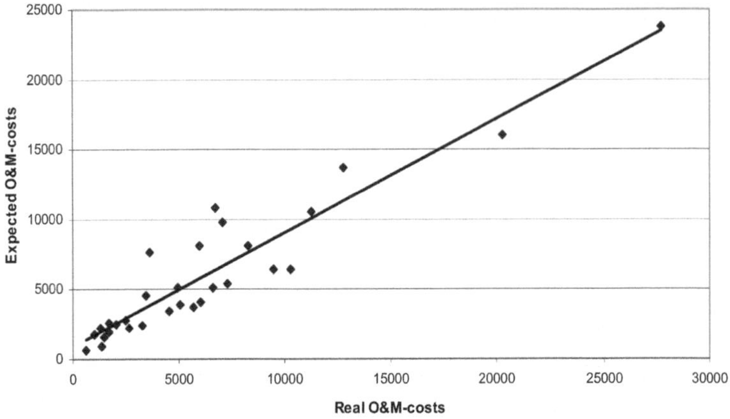

Figure 5. REGRESSION RELATION BETWEEN EXPECTED AND REAL OPERATION AND MAINTENANCE COSTS FOR WATER DISTRIBUTION NETWORKS BASED ON MULTIPLE REGRESSION ANALYSIS

For water distribution networks the best correlation in the regression analysis was achieved with the parameters "fictive" pipe length in kilometres and the number of boosters in the system per kilometre. The concept of "fictive" pipe length, which was used in the analysis, is the total length of water mains with an addition of 5 meters per each service connection.

For sewerage networks the best correlation was achieved with the parameters "fictive" pipe length in kilometres of sewer pipes carrying sanitary sewage water and the number of pumping stations per kilometre.

The relative cost-efficiency of the compared utilities was in the assessment model derived by calculating the difference in percent between the calculated and the real operation and maintenance costs. The third best performing utility was selected as benchmarking utility, against which all other utilities were compared. This "normalisation" was made to get a picture of the relative cost efficiency. The result of the efficiency assessment was used to rank the networks in a similar way as the previously described quality ranking.

4.3 Assessment of capital costs

When comparing the costs for different utility networks it is necessary to consider also the capital costs that are associated with renewal and other improvements of the existing networks. The principles for calculation of capital costs vary quite a lot between different utilities. This relates among others to depreciation time and interest rate. Therefore it is not possible to use capital cost figures from the utilities' accounting systems. The approach used in this project was to calculate "fictive" capital costs derived from the age profile of the network, a fixed depreciation time and a fixed interest rate.

As a base for the calculations of the fictive capital costs, a simplified model for determination of the construction costs for a pipe network was developed. This model takes the soil condition, the depth of the pipes and the number of service connections per kilometre into account.

4.4 Collective ranking of the costs for water distribution and sewerage networks

In the two previous paragraphs the procedure for ranking of operation and maintenance costs and of capital costs was described. To get a more complete picture of the costs it can be of interest to make a collective ranking of the two types of cost.

In figure 6 the collective ranking of the costs for operation and maintenance and capital costs for water distribution and sewerage networks has been illustrated in a matrix diagram. The operation and maintenance costs and the capital costs was divided in the following five groups: high scores (10 % of the networks), higher than the mean (25 % of the networks), around the mean (30 % of the networks), lower than the mean (25 % of the networks) and low scores (10 % of the networks). The marked quadrant in the upper right of the matrix diagram in figure 6 shows utilities which have a relative cost-efficiency around the mean or higher (= low costs for operation and maintenance) and at the same time relative fictive capital costs around the mean or lower.

Relative cost-efficiency for operation and maintenance	High	Higher than average	Average	Lower than average	Low
High			A1	A3	C2
Higher than average		A7 A8 B8	C3 C7	C9 B3	
Average	B11	A10	A5 C5 C1 B10	C6 C4	B4 B7
Lower than average		A2 B9 C8	A9 B12	B6 B1	
Low	A6 B2		A4		

Relative "fictive" capital costs

Figure 6. MATRIX DIAGRAM ILLUSTRATING THE OPERATION AND MAINTENANCE COSTS AND THE FICTIVE CAPITAL COSTS FOR 30 ANALYSED SWEDISH WATER DISTRIBUTION AND SEWERAGE NETWORKS

5 ALTERNATIVE APPROACH FOR ASSESSMENT OF COST-EFFICIENCY

As an alternative to the multiple regression analysis described in the previous section, assess¬ment of cost-efficiency can be accomplished with so called data envelopment analysis (DEA). In the DEA-approach the achievements (output) are set in relation to the utilized resources (input) for a selected process. The method is non-parametric in so far as the observation material does not need to follow any established cost function for the analysed process. Analysis with the DEA-method can be carried out either scale-dependent or scale-independent. The method is schematically illustrated in figure 7.

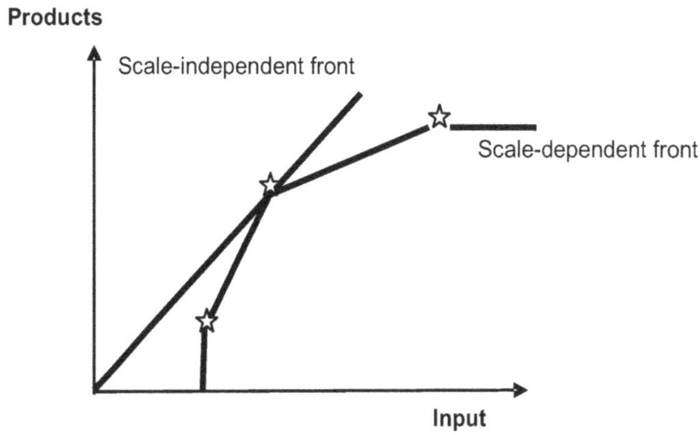

Figure 7. SCHEMATIC ILLUSTRATION OF DATA ENVELOPMENT ANALYSIS (DEA).

The evaluated cost-efficiency is relative and reflects the potential for improvements in relation to other utilities in the comparison. The relative efficiency is expressed in scores from 0 to 100 where the score 100 represents the most efficient utility.

The DEA-method was tested on the same 30 utilities that were included in the multiple regression approach. Both the scale-dependant model and the scale-independent model were tested for a number of different combinations of input and output variables.

In the DEA-simulations the costs for operation and maintenance was used as input variable. Output variables were fictive pipe length and number of boosters and pumping stations. Figure 8 shows the results of the efficiency ranking of the water distribution and sewerage networks with a scale independent DEA-model. As indicated in the figure two utilities are in the front with 100 scores. The least efficient utility got 30 scores and the median utility 59 scores.

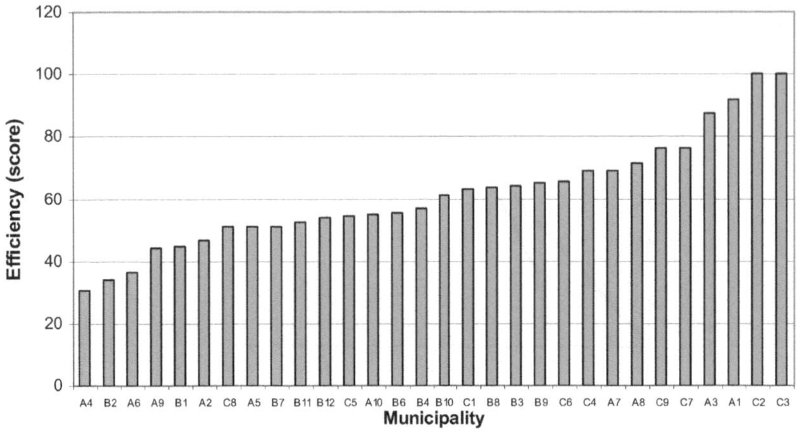

Figure 8. EFFICIENCY RANKING OF 30 SWEDISH WATER DISTRIBUTION AND SEWERAGE NETWORKS WITH A SCALE INDEPENDENT DEA MODEL.

Efficiency simulations were also carried out with a scale dependent DEA-model. In these simulations eight utilities ended up in the front with 100 scores. The least efficient utility got 42 scores. The limited number of utilities gave in the scale dependent DEA-model the result that too many benchmarking networks were created, which equalized the differences. The scale dependent DEA-model is however of big interest if a larger number of utilities are included in the analysis.

An interesting question is how the result of the efficiency ranking with DEA-modelling corresponds to the result of the efficiency ranking with the multiple regression approach. The correlation between the two methods is illustrated in figure 9. Even if the efficiency ranking was carried out on only 30 utilities the results indicate that DEA-modelling is an interesting tool in efficiency ranking. There is every reason to carry out further tests on efficiency ranking with DEA-modelling for a larger number of utilities.

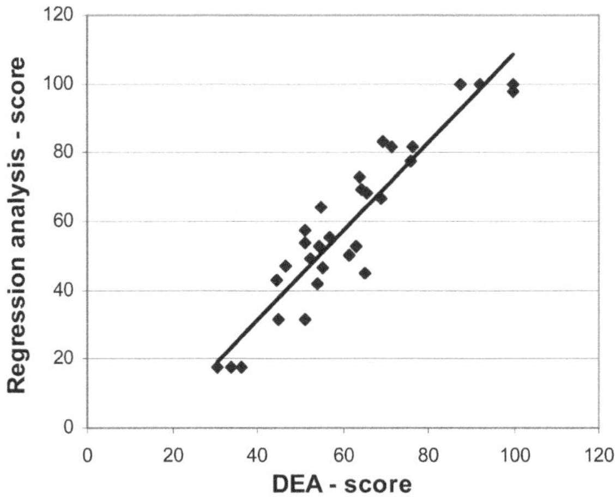

Figure 9. CORRELATION BETWEEN THE RESULT OF EFFICIENCY RANKING WITH MULTIPLE REGRESSION ANALYSIS AND DEA MODELLING.

6 ASSESSMENT OF LONG-TERM SUSTAINABILITY

To get a more complete picture of the performance it is necessary to analyse also the sustainability of the management of the networks. To assess the long-term sustainability is a very complex task. As the practices vary so much between different municipalities it is not possible to make an objective comparison. It was in the project decided to limit the assessment of the sustainability dimension to a discussion of the following three areas, which were considered most important in this connection:

- Level of pipe renewal
- Financing of pipe renewal
- Strategic management and planning of utility networks

6.1 Level of pipe renewal

A strategic question for all utility managers is to decide to which extent an existing utility network shall be renewed in order to maintain an acceptable level of service for the customers and at the same time avoid an accelerated deterioration of the pipe network. The problem in this connection is that the deterioration process is so slow and that the negative effects of the deterioration become apparent first after many years. In the project a conceptual illustration of the level of pipe renewal was introduced, see figure 10. It was anticipated that there is a relation between the actual level of renewal and the frequency of functional disturbances occurring on the pipe network and further that all utility networks should be subject to a certain base level of renewal.

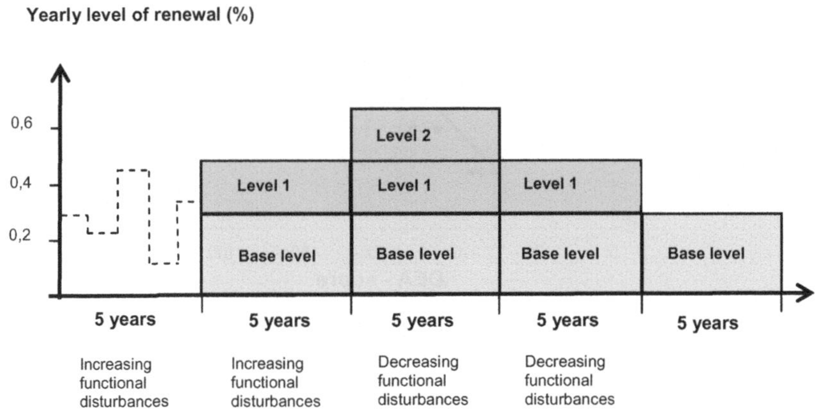

Figure 10. CONCEPTUAL ILLUSTRATION OF THE LEVEL OF LONG-TERM PIPE RENEWAL

The base level has in figure 10 been set at 0,3 % of the total length of the network. This level of pipe renewal is only representative for the situation in Sweden. The base level can of course vary from country to country. If the frequency of functional disturbances over a five year period is increasing, the level of renewal is to be increased by another 0,2 %, e.g. to a total of 0,5 % over the next five year period. For the following 5 year period the level of renewal is increased or reduced with 0,2 % depending on the development of the functional disturbances. The renewal level shall however never be lower than the base level 0,3 %.

The frequency of functional disturbances can for example be measured as the number of scores for the dimension quality, service and environment.

6.2 Financing of pipe renewal

In Sweden the practice for financing pipe renewal differs from municipality to municipality. In some municipalities the renewal of deteriorated pipes are financed over the investment budget while in others the renewal is directly financed over the operation budget. There are many arguments for both methods. Seen from a long-term perspective the most cost effective is to finance the renewal over the operation budget.

6.3 Strategic management and planning of utility networks

A strategic management and planning is an important prerequisite for an efficient network operation. To evaluate to which extent strategic management and planning is applies, cannot be done in an objective way. There is simply not enough data to make such an evaluation when several utilities are compared. In the project it is suggested that the practices in the following areas are analysed:

- Management by objectives (clear goals)
- Management systems for continuous improvements (quality, environment, working environment, security)
- Leadership and development of competence
- Strategic network planning
- Monitoring and analysis of achieved results (benchmarking)
- Information systems for a rational handling of data
- Customer orientation (communication and public relations)
- Organisational structure

For each one of these areas checklists have been developed to get an indication whether the utility focus on strategic management and planning or not. The developed checklists are intended for self control within the organisation and as a source of inspiration for improvements.

7 CONCLUSIONS

In Scandinavia there is today no accepted method for assessment of the performance of the municipal water sector. Benchmarking is carried out on a national scale but the results of these efforts have so far not been used for evaluating the overall performance. In this paper is described how the overall performance of water distribution and sewerage networks can be assessed. The characteristic feature of the new approach is that the amount of data in the evaluation is quite limited.

The developed approach for the performance assessment is primarily intended for national comparisons of a large number of utilities. The comparisons can be applied by branch organisations or regulator authorities to get a general overview of differences among the utility operators. The results of the suggested performance assessment can be of great help for the utility operators in the communication with their politicians.

The assessment approach described in this paper has been limited to cover only water distribution and sewerage networks. The intention is to in the future develop assessment models also for the other parts of the water sector (water production and sewage treatment)

8 REFERENCES

Adamsson J and Stahre P. (2004). Performance Assessment of Water and Wastewater Services - Survey of best practice. Swedish Water: VA-Forsk report B 2004-101

Office of Water Regulation (2001). Statistical profile and performance benchmarking of water supply services in 32 major western Australian towns 1999/2000. Perth. Australia

Ofwat (2005). Water and sewerage service unit costs and relative efficiency, 2003 – 2004 report. Birmingham, England

Ofwat (2006). Levels of service for the water industry in England and Wales, 2004 – 2005 report. Birmingham, England.

Stahre P, Mellstrom G and Adamsson J (2007). Assessment of the performance of water and wastewater networks. Swedish Water: SVU report 2007 (in Swedish)

Cross-link performance indicators for sustainability evaluation of supply and disposal systems at mountain refuges

N. Weissenbacher[1], E. Mayr[1], T. Ertl[1], G. Steinbacher[2], P. Weber[3], R. Haberl[1]

[1] *Univ. of Natural Resources and Applied Life Sciences Vienna, Institute for Sanitary Engineering and Water Pollution Control, Muthgasse 18, 1190 Vienna, Austria; norbert.weissenbacher@boku.ac.at; +43 1 36006 5806*
[2] *Steinbacher+Steinbacher Civil Eng. Inc., Breitwies 16, 5303 Thalgau, Austria*
[3] *DAV- Deutscher Alpenverein e.V., Bundesgeschäftstelle, Von-Kahr-Str. 2-4, 80997 Munich, Germany*

Edited by Enrique Cabrera Jr and Miguel Ángel Pardo. ISBN: 9781843391913.
Published by IWA Publishing, London, UK.

Keywords: mountain refuges, infrastructure evaluation, performance indicators, supply, disposal

ABSTRACT

Mountain refuges comprise systems for energy and water supply, wastewater and waste disposal operating under unfavourable conditions. To identify optimization potential an integrated evaluation of existing systems and experiences is carried out on international basis (data base: 100 mountain refuges). Therefore, an adapted performance indicator approach was elaborated as component of the overall evaluation. Whereas sector-PIs describe the use of resources for a particular supply or disposal chain, cross-link-PIs reflect interactions and interdependencies between them.

1 INTRODUCTION

Mountain refuges represent public stand-alone infrastructure comprising systems for energy and water supply, wastewater and waste disposal operating under unfavourable conditions like low temperature or far transport distances (Grinzinger, 1999; Wett et al., 2002 and 2003). The sensitive ecology of alpine regions, climate change and increasing number of visitors require sustainable, resource oriented solutions. Sustainability has been defined in various ways comprising different principles like environmental accounting and eco-efficiency (Glavic and Lukmann, 2007; Alfsen and Greaker, 2007). Corporate responsibility for social and environmental sustainability is a matter of increasing importance (CEC, 2002). Alpine Associations have been dealing with these issues for decades providing service for the broad public - always balancing financial pressures and environmental considerations.

To identify optimization potential in Austria, Germany, Italy, Switzerland, Slovenia and the Czech Republic, the project 'Integral evaluation of supply and disposal systems of mountain refuges' was launched in 2007. With the aim to 'learn from experience', data and operation results shall be evaluated in order to elaborate guidelines for planning, construction and operation. The project results shall foster sustainability on an international level based on experiences made by all stakeholders involved in the project (Alpine Associations, Legal Authorities, Planners, Operators and Researchers).

Compared to large scale service providers usually involved in benchmarking (IBNET, 2006; Larsson et al., 2002), the organisation of mountain refuges differs significantly. Alpine Associations are responsible for investments and lease the refuges to private operators responsible for operation. The project does not aim at an overall benchmarking but defines a 'sustainable model refuge' that aims to fulfil the following criteria weighted dependent on the surrounding conditions (e.g. freshwater availability, transport):

- Low resources consumption per capita (capita = daytime or overnight visitor)
- Low emission production per capita (wastewater, waste)
- Low maintenance efforts (costs and operator + voluntary labour)
- Low investment costs
- Provision of adequate and undisturbed operation of the mountain refuge (operation reliability)
- Fulfilling all legal requirements (e.g. effluent quality, waste disposal)

The objective of this paper is to present an adapted performance indicator approach as component for the sustainability evaluation of interacting small scale supply and disposal systems present at mountain refuges.

2 MATERIAL AND METHODS

2.1 System analyses

The procedure is based on the IWA performance evaluation for water supply services (Allegre et al., 2006) and wastewater services (Matos et al., 2003). An adapted performance indicator (PI) system was elaborated to cover the four systems (energy supply, water supply, wastewater disposal, waste disposal) separately (sector-PIs) and to cover connections or influences between the systems and to the surrounding conditions (cross-link-PIs). This procedure comprises metric benchmarking methodology (to make optimisation potentials evident) as well as process benchmarking methodology (to clearly confine each process as well as the sub processes) looking at process connections within the mountain refuge. Grouping is based on three categories: Criteria with reservation of supply resources, geographical position and regulatory framework, criteria with reservation of capacity and operation and criteria with reservation of different operational systems (e.g. biological wastewater treatment). To fulfil the requirements of an integrated approach, data analyses were carried out in two steps:

Performance assessment of urban infrastructure services

1. Sector analyses: energy, water, wastewater, waste systems
2. Cross-link analyses: interdependencies/interactions

According to the criteria for a 'sustainable model refuge' defined in the introduction, the PI framework presented in Matos et al. (2003) was adapted and implemented as first part (sector analyses) of the two step procedure (Table 1). The data to be surveyed was selected to cover all possible technical and operational options present at the mountain refuges and to answer all aspects of the system analyses addressed by the adapted PI framework.

Table 1. ADAPTATION OF THE PI FRAMEWORK.

Matos et al. (2003) Indicators	Adapted Indicators	Adapted Short description
Environmental	Environmental	Impacts on receiving environment: e.g. wastewater discharge, solid waste disposal, sewage sludge disposal, compliance to legal requirements. Use of resources: e.g. water consumption, consumption of fossil fuel, solid waste production and prevention, water and energy saving measures.
Personnel Operational	Operational	Operation and maintenance activities including training, qualification and labour of the undertaking personnel (operator), energy consumption (including transport of e.g. fuel or wastes) and failures.
Physical	Physical	Capacities of the systems to fulfil the service targets: e.g. water storage capacities, water, wastewater and sewage sludge treatment capacities, thermal and electric energy coverage; automation and control.
Quality of service	Quality of service*	Level of service provided to the operator: e.g. comfortable control, reliable operation, low labour efforts, stable service coverage, start up and shut down efforts, external maintenance needs, system coordination.
Economic/ financial	Financial	Effectiveness and efficiency of financial resources: running cost and investment costs carried by the operator and the Alpine Associations, respectively.

* The purpose of mountain refuges is to provide only a limited level of services like food and shelter to hikers and mountaineers. Therefore, the operator of the refuge was chosen to be addressed as user.

2.2 Data collection

100 mountain refuges have been selected for data collection. Altitudes range from 700 to 3200 m a.s.l. (Fig. 1). All main types of technical systems (e.g. sanitation, hybrid energy, treatment, reuse), operation as well as the surrounding conditions (e.g. altitude, hydrogeology) are covered. The data set of variables and context information was collected in two parts. Baseline data on investments, planning and legal framework was gained from the Alpine Association responsible for the single refuge. Operation data was gained from detailed on-site interviews with the operator including a survey of all systems and devices. Sampling and other measurements were included if available. The amount of data collected is shown in Table 2.

Table 2. OVERVIEW ON THE COLLECTED DATA PER MOUNTAIN REFUGE.

# Data sets		Sector			
		Supply		Disposal	
Type	General	Energy	Water	Wastewater	Waste
Context Inf.	13	55	35	58	15
Variables	17	51	14	50	16
Total	30	106	49	108	31

Accuracy of data has been defined according to the suggestions of Allegre et al. (2006) using accuracy and reliability bands. As the ongoing data collection showed, many data sets are based on the subjective experience of the system operator leading to a high share of estimated values. Further, different management structures of the Alpine Associations lead to different levels of documentation of planning, construction and operation.

3 RESULTS AND DISCUSSION

3.1 Grouping

Due to the great variety of institutional, operational and surrounding conditions, mountain refuges are difficult to compare. Nevertheless, grouping shall be the basis for comparison considering the main criteria influencing system set up and performance. Criteria with reservation of supply resources, geographical position and regulatory framework:

- *Altitude:* Temperature effects, weather conditions (Fig.1).
- *Hydrogeology:* Sensitivity of the receiving environment.
- *Water source:* Available water quality and quantity and treatment required (rainwater harvesting, springs).
- *Emission standards:* Sludge disposal and treatment required. Required effluent quality of wastewater treatment.
- *Transport:* Transport of goods and wastes by road, cable car or helicopter.

Criteria and indicators with reservation of refuge capacity and operation:

- *Season:* Operation in summer, winter only or during the whole year.
- *Capacity:* 'Size' of the mountain refuge: Number of beds.
- *Service level:* Degree of comfort offered to the visitor (e.g. showers, menu).
- *Visitation characteristics:* Ratio of daytime visitors to overnight visitors (visitor ratio, Fig. 2).

Criteria with reservation of supply and disposal systems:

- *Energy supply:* Connection to public energy supply or stand alone.
- *Wastewater treatment:* Type of treatment system installed (primary, secondary, tertiary treatment; Fig. 1).
- *Reuse:* Fraction of treated wastewater reused.

The distribution of altitude and wastewater treatment of 70 refuges (number of datasets available so far) is shown in Fig.1. The grouping for visitor based performance indicators is given in Fig. 2.

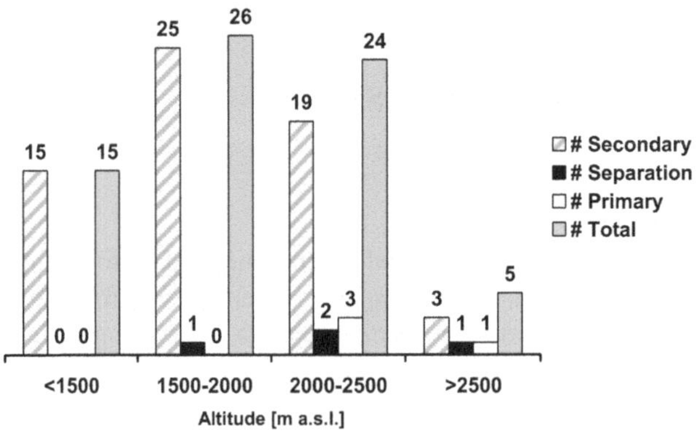

Figure 11. DISTRIBUTION OF ALITUDE AND WASTEWATER TREATMENT.

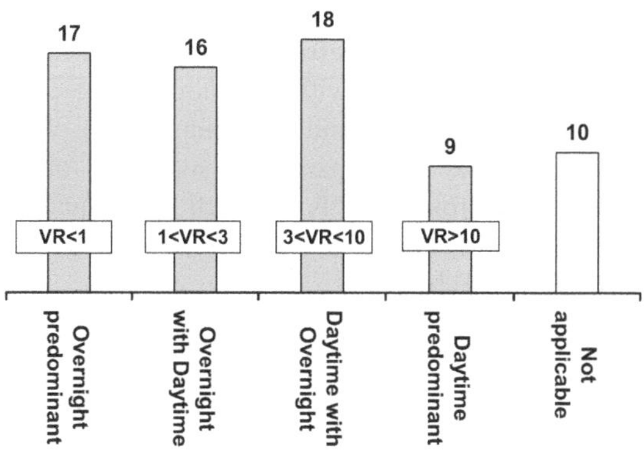

Figure 12. GROUPING: DAYTIME/OVERNIGHT VISITOR RATIO (VR).

3.2 Sector-PIs

With respect to the PIs presented in the IWA best-practice series, a set of similar applicable PIs for mountain refuges has been elaborated for sector analyses. The single visitor (daytime and overnight) has been defined as a reference base for the PI calculations. PIs are serving as basis for sector system description, cross-link dependencies as well as for grouping (Fig. 2).

Table 3. EXAMPLES OF ENVIRONMENTAL PIs FOR SECTOR ANALYSES.

Sector	Indicator *Unit*	**Concept** Processing rule *-Grouping* *#Datasets| Range| Median*
Waste	Waste production *L/Day*	**Daily average production of residual waste to be transported.** Volume of average residual waste per transport divided by the number of transports per day. *-Capacity* *70 | 1-400 L/Day| 18 L/Day*
Wastewater	Wastewater production *L/Visitor/Day*	**Daily average wastewater production per visitor (daytime and overnight).** Daily average water consumption divided by the sum of daytime and overnight visitors. *-Visitor ratio/Reuse* *70 |2-111 L/V./Day| 32 L/V./Day*

3.3 Cross-link-PIs

Cross-link PIs shall show the interdependencies of the single systems emphasising on the concept of eco-efficiency which means to provide the service (mountain refuge operation) with the minimum use of resources like fossil fuel, water and labour (Glavic and Lukmann, 2007). Whereas sector PIs describe the use of resources for a particular supply or disposal chain in a quantitative way (Table 3), cross-link PIs shall reflect the qualitative adjustment between them. To illustrate this approach, three simple demonstrative examples are shown in the following (Table 4).

Table 4. EXAMPLE PIs FOR CROSS-LINK ANALYSES.

Indicator *Unit*	**Question-Concept-Comment** Processing rule *-Grouping*	*#Datasets| Range| Median*
Water and fossil energy *%*	**Are the systems for water supply and wastewater treatment drivers towards production of electricity from fossil fuel? – Comparison of the system consumption to the production from fossil fuel. Low values indicate that fossil fuel consumption is not linked to the water and wastewater sector.** Sum of the daily average energy consumption of the sectors water and wastewater as ratio of the daily electricity production from fossil fuel. *-Energy supply*	*70| 0-11 %| 1.3 %*
Energy for toilets *Wh/Flushing*	**Is it reasonable to use water for toilet flushing? - Energy consumption per flushing; not applicable for dry toilets. Sufficient water availability presumed, low values would justify the use for flushing.** Sum of the average energy consumption of the sectors water and wastewater per unit volume divided by the volume of a standard toilet flush. *-Wastewater treatment* *-Water source*	*70| 0-114 Wh/F.| 4.0 Wh/F.*
Maintenance for visitors *min/Visitor*	**How is the complexity of the overall system for the operator? – Maintenance efforts for a single visitor. High values indicate a too high workload for the technical systems in comparison to the customer support.** Sum of the daily maintenance of the sectors divided by the sum of daytime and overnight visitors. *-Visitor ratio*	*70| 0.01-1.46 min/V.| 0.29 min/V.*

Positive results for cross-link PIs may originate from integrated planning, experienced operators and comprehensive technical support by Alpine Associations. Nevertheless, a variety of factors may be hindering the successful integration of systems like gradual construction/ renewal/ upgrade of systems over decades, change of operators, low level of training or high investment costs (at different levels of public subsidising).

4 CONCLUSIONS

The data collection and the first analyses showed that the integral evaluation is most important for the overall operation assessment. Sector PIs are applicable for the system description, where quantitative data is available. They are also useful for grouping purposes and plausibility testing of data. As subsequent step, cross-link PIs are capable of describing interdependencies and interaction based on the sector description. This procedure is an important step towards the identification of sustainable solutions and of optimisation potential present beside the analyses of experiences of the operators and the Alpine Associations involved.

5 ACKNOWLEDGEMENTS

The authors thank all partners (Alpine Associations, Legal authorities and others) for their contribution to the project work. Beside the financial contribution of the partners, this project was funded by the Deutsche Bundesstiftung Umwelt (DBU) and the Austrian Federal Ministry of Agriculture, Forestry, Environment and Water Management (BMLFUW).

6 REFERENCES

Alegre H., Baptista J.M., Cabrera E. Jr, Cubillo F., Duarte P., Hirner W., Merkel W., Parena R. (2006): Performance Indicators for Water Supply Services - Second Edition. Manual of Best Practice, IWA Publishing, London, UK. ISBN 1843390515

Alfsen K.H., Greaker M. (2007) From natural resources and environmental accounting to construction of indicators for sustainable development. *Ecological Economics*, Vol **61**, 600-610

CEC (2002) Communication from the Commission concerning Corporate Social Responsibility: A business contribution to Sustainable Development. COM/2002/0347 final.

Glavic P., Lukman R. (2007) Review of sustainability terms and their definitions. *J. of Cleaner Production*, Vol **15**, 1875-1885

Grinzinger U.(1999) Die umweltgerechte Schutzhütte (The environmentally sound mountain refuge). VAVÖ, Vienna, AT. ISBN 3-9500087-3-X [in German].

IBNET (2006): The International Benchmarking Network for Water and Sanitation Utilities; http://www.ib-net.org/ (15th of August 2006)

Larsson M., Parena R., Smeets E. and Troquet I. (2002). Process benchmarking in the water industry. - IWA-Publishing, London, ISBN 1-84339-010-8.

Matos R., Cardoso A., Ashley R., Duarte P., Molinari A., Schulz A. (2003) Performance Indicators For Wastewater Services, Manual of Best Practice, IWA Operations & Maintenance Specialist Group , IWA Publishing, London,

Wett B., Becker W. and Ingerle K. (2002) Sludge management regulations and their applicability in mountainous regions, *Water Science & Technology*, Vol **46** No 10, pp 305–310

Wett B., Eladawy A. and Becker W. (2003) Carbonate addition – an effective remedy against poor activated sludge settling properties and alkalinity conditions in small wastewater treatment plants, *Water Science & Technology*, Vol **48** No 11–12, pp 411–417

Definition of performance indicators for urban water distribution systems in drought conditions

C. M. Fontanazza[1], *G. Freni* [1], *G. La Loggia* [1] *and V. Notaro*[1]

[1] *Dipartimento di Ingegneria Idraulica ed Applicazioni Ambientali, Università degli Studi di Palermo. notaro@idra.unipa.it, Viale delle Scienze 90128 Palermo (Italy), Tlf +39 0916657720, Fax: +39 0916657749*

Keywords: Drought condition, equitable water supply, intermittent water service, urban water distribution systems

 Performance Assessment of Urban Infrastructure Services
Edited by Enrique Cabrera Jr and Miguel Ángel Pardo. ISBN: 9781843391913.
Published by IWA Publishing, London, UK.

ABSTRACT

The present study proposes a procedure for analysing urban water distribution system characterised by intermittent supply service under drought conditions. Namely, ten new performance indicators have been defined able to identify inequality in water supply both considering effects on users, whose consumption reduces, and effects on network, whose hydraulic behaviour during the service cycle greatly varies. The procedure has been applied to analyse the performance of an existing supply system in Palermo (Italy). The network and users' behaviour in intermittent distribution conditions have been monitored and a mathematical model has been arranged and calibrated in order to evaluate the impact of water scarcity events on population and consequently suggest actions for improving resources distribution equity.

1 INTRODUCTION

Water shortage is, nowadays, one of the main problems for social and economic growth in the world. Many programmatic reports (among others Intergovernmental Panel on Climate Change, 2007) recently foresee a drastic reduction of the water resources available for human purposes due to climate change. Although all these studies show some uncertainty in quantifying the severity of this reduction and declares that the unbalance between the water demand growing and the resources available decrease is a real problem that should not be neglected. Like any unbalance between water resources demand and offer, water shortage can be linked both to natural and climatic phenomena both to atrophic activities (such as incorrect management of supply networks, inefficient use of water resources, etc...). Thus, severity and duration of water shortage negative effects for people and economic activities depend on the climatic drought characteristics and on the atrophic systems adaptability to the new environmental conditions.

In such context, distribution networks are usually the weak point of the whole water supply system. During water shortage period, networks operating conditions are very far from the usual design ones: discontinuous water distribution and water resources rationing are often used as the main measure to cope with water scarcity. This approach is widely adopted not only in developing countries (McIntosh *et al.*, 1997; Hardoy *et al.*, 2001; Vairavamoorthy *et al.*, 2001) but also in developed ones for solving short-term scarcity conditions, which can be caused by unexpected or unpredicted drought periods (Cubillo Gonzales & Ibanez Carranza, 2003). When a continuous system is managed as an intermittent one, pressure distribution is often inadequate to provide a sufficient service level and water supply results to be inequitable and not homogenous in space and time (Fontanazza *et al.*, 2007). Moreover, intermittent water service has a relevant impact on population causing modification on usual social and economic activities of users, above all on those located far from the supplying nodes or at higher elevation in the network. The main cause of such behaviour can be researched in the traditional methodologies adopted to design supply network that fail when supplied flows are lower than design user demand. Usually network is aimed to satisfy users demand (residential, commercial, industrial) on the base of an optimistic prevision of the available water volumes. In such complex situation, network behaviour and user accessibility to water resources are determined by pressure levels and advanced tools are needed for analysing and representing system ability to meet users needs. The analysis of such operational conditions has to be tacked by the mean of head-driven mathematical models. Such models are widely discussed in literature (Reddy & Elango, 1989; Chandapillai, 1991; Jowitt & Xu, 1993; Gupta & Bhave, 1996) even if some questions are still open especially on the definition of relationships between supplied water volumes and network pressure.

System behaviour can be synthetically assessed by Performance Indicators (PIs) that are defined in literature for characterizing the system in continuous supply conditions, and, for this reason, they may not be suitable to represent the status of the network or to evaluate drought mitigation measures effectiveness.

Therefore, the aim of the present study is the definition of new PIs focused to analyse supply systems under water scarcity and intermittent operational conditions. This new set of PIs can help to identify the network sectors where the detrimental effects of intermittent distribution are more relevant and to guide the proposition of operational approaches that can reduce the impact on users.

2 PERFORMANCE INDICATORS FOR URBAN WATER SUPPLY SYSTEMS IN SHORTAGE CONDITIONS

In the last years, many definitions have been proposed to explain the concept of performance of water supply system and many methodologies have been implemented to quantify performance in the most objective and general way (La Loggia & Mazzola, 1989; Alegre *et al.*, 2006). This aim is usually pursued by the definition of Performance Indicators (PIs) synthetically representing system behaviour.

Generally, urban water systems PIs are grouped in homogeneous categories according to the different relevant aspects for water distribution management (Alegre *et al.*, 2006): Water Resources (WR), Personnel (Pe), Physical (Ph), Operational (Op), Quality of Service (QS), Economic and Financial (Fi).

Comparing intermittent and continuous water supply systems, Pe, Fi and Ph indicators do not require substantial changes, even if some of them may assume "unusual" values. The other categories need the definition of new PIs describing the risk that water supply may not be equitable and some groups of users may not be able to have sufficient access to water resources.

According to this aim, ten new performance indicators have been proposed in the present study. Namely, the adoption of the proposed PIs is focused to identify the peculiarly situations of the intermittent operating mode and to take into account these ones during the global system performance assessing.

With regard to WR category, two performance indices have been proposed (table 1): WV1 is defined as the ratio between the average daily water volumes supplied at each node and those should be distributed to users in a service cycle according to water resources availability; WV2, instead, is defined as the ratio between hourly water supply in intermittent and continuous conditions.

Table 1. WATER RESOURCES AVAILABILITY PIs (WR)

$WV_1 = \frac{V_{\sup}}{V_d}$	V_{int}	Water volume supplied to users in a service cycle (the time lag between two subsequent activations of water services in intermittent distribution conditions)
	V_d	Water volumes that should be distributed, in a service cycle, to users during shortage conditions according to water availability
$WV_2 = \frac{Q_{\text{int},i}}{Q_{cont,i}}$	$Q_{\text{int},i}$	Average water flow supplied to users at hour i in intermittent distribution conditions
	$Q_{cont,i}$	Average water flow supplied to users at hour i in continuous distribution conditions

With regard to the Op category, seven indices have been proposed (table 2).

The first two indices SD1 and SD2 are related to the intermittent distribution service duration. Namely, SD2 highlights the major problems caused by an intermittent service adopted during the day instead of during the night.

The last five indices, instead, are linked to the system resilience:

- *SR1* shows the distribution network resilience. High values of this index, due to great storage volumes in the system tanks (public and private), point out a good adaptability of the network to an unexpected supplying interruption caused by a short-term scarcity condition, while low values show high system vulnerability.
- *SR2* points out the supply system resilience. Low values of this index, due to high storage volumes in the reservoir, identify good system adaptability to a reduction of the extractable water from its supply sources caused by a long-term scarcity condition.
- *SR3* represents the frequency of the supply service interruptions.

- *SR4* represents the total duration of service interruptions in a year.
- *SR5* evaluates the gap between the average residential water demand in the analysed area and in the country.

Table 2. OPERATIONAL PIs (OP)

$SD_1 = \frac{ST}{24}$	ST	Duration of the water supply service [hour]
$SD_2 = \frac{ST_{8-20}}{ST}$	ST_{8-20}	Hours of water supply service in the period of the day in which population is active (from 8:00 to 20:00)
$SR_1 = \frac{V_s}{V_d}$	V_s	Volume of reservoirs available in the network (both considering public and private tanks) [cubic meters per capita]
	V_d	Average user demand [cubic meters per capita per day]
$SR_2 = \frac{Q_{mg}}{\sum_{i=1}^{N} Q_{avail,i}}$	Q_{mg}	Daily average water flow request by the urban centre
	$Q_{avail,i}$	Water flow derivable by the ith water source (considering all available water resources in the area)
$SR_3 = \frac{N_{inter}}{365}$	N_{inter}	Number of supply service interruptions in one year
$SR_4 = \frac{Turn}{365}$	$Turn$	Total duration of service interruptions in a year
$SR_5 = \frac{D_{urb\_centre}}{\overline{D}_{region}}$	D_{urb_centre}	Average residential water demand in the analysed area
	$\overline{D}_{region}$	Average residential water demand in the region

The first two indices SD1 and SD2 are related to the intermittent distribution service duration. Namely, SD2 highlights the major problems caused by an intermittent service adopted during the day instead of during the night.

The last five indices, instead, are linked to the system resilience:

- *SR1* shows the distribution network resilience. High values of this index, due to great storage volumes in the system tanks (public and private), point out a good adaptability of the network to an unexpected supplying interruption caused by a short-term scarcity condition, while low values show high system vulnerability.
- *SR2* points out the supply system resilience. Low values of this index, due to high storage volumes in the reservoir, identify a good system

adaptability to a reduction of the extractable water from its supply sources caused by a long-term scarcity condition.

- *SR3* represents the frequency of the supply service interruptions.
- *SR4* represents the total duration of service interruptions in a year.
- *SR5* evaluates the gap between the average residential water demand in the analysed area and in the country.

The reliability of the proposed PIs to analyse network behaviour in water shortage conditions, can be quickly verified carrying out a performance analysis of an existing supply system. According to this aim can be applied the performance analysis procedure firstly proposed by Coelho (1997) where system variables is compared with threshold values of user-defined curves (penalty curves) allowing for calculating PIs performance score and assess system performance. For each PI, a suitable penalty curve is defined in order to evaluate the performance associated to every network element. The system performance is assimilated to the level of service, varying between a "void service" and an "optimum service" condition, and the curve is built in penalising the behaviour far from "optimum service" conditions. Figure 1 shows penalty curves related to indices WV1 and WV2.

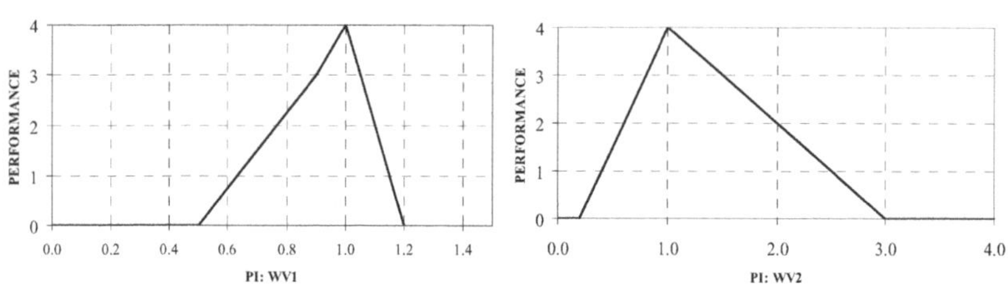

Figure 1. EXAMPLES OF SOME PENALTY CURVES ADOPTED IN THE STUDY

3 PI APPLICATION AND RESULTS DISCUSSION

3.1 The case study

The performance analysis has been carried out using as case study one of the 17 distribution networks of Palermo (Fontanazza *et al.*, 2007).

Figure 2 shows the distribution network adopted in the present study.

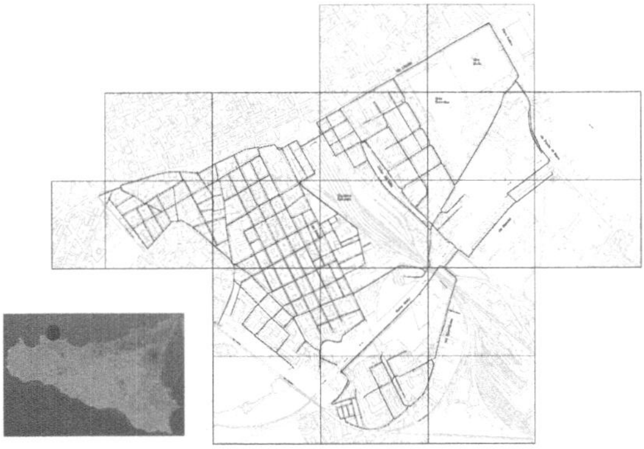

Figure 2 . CASE STUDY NETWORK SCHEME (Fontanazza *et al.*,. 2007).

Users elevation range is between 47 m and 3 m above the sea level. The distribution system is fed by two tanks at different levels that can store up about 40.000 m^3 per day, and supply around 35.000 inhabitants. The network is long about 40 km and pipes are made of polyethylene with diameters ranging from 110 to 225 mm. The design user water demand is about 400 L/capita/d but the actual mean consumption is about 280 l/capita/d. As consequence, in ordinary conditions, the network is characterised by low water velocities and correspondently high pressures causing in the past high leakages. This condition together with the recurrent lack of water resources has not permitted to maintain continuous distribution in the last 5 years (at least in summer period) and intermittent distribution on daily basis has been introduced as a common practice convincing the users to build up local private reservoirs.

3.2 The intermittent distribution network model

The network has been analysed by the mean of an intermittent water distribution mathematical model schematising both user and manager dependent regulation structures (pumps, private reservoirs, etc.), typical of the systems where water distribution is periodically provided on intermittent basis (Fontanazza *et al.*, 2007). Namely, a private reservoir under the roof and a pump have been associated with each supply node (figure 3).

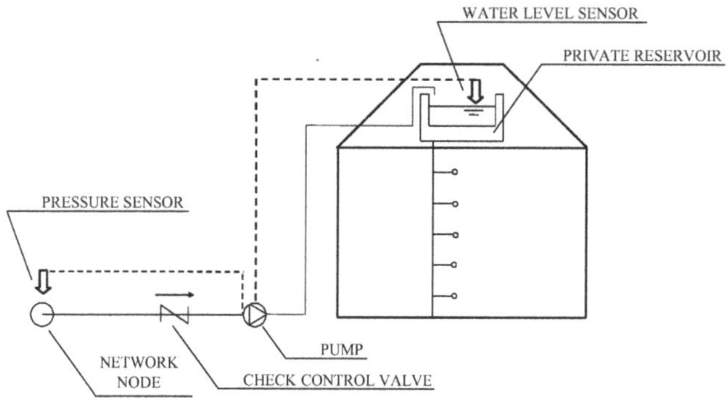

Figure 3. DISTRIBUTION NODE NUMERICAL SCHEME

The reservoir has been designed according to nodal daily water demand and a pump has been chosen being able to fill the reservoir in 4 or 5 hours. The pump is turned on if the reservoir is empty and turned off if the reservoir is full or the pressure on the network is negative.

To take into account that pumps permit to provide private reservoirs even if nodal pressure is lower than minimum required to have outflow at the node, the user demand has been simulated introducing a pressure-consumption relationship in the range of node head lower than the minimum value. Specifically, the actual outflow at node j q_j^{avl} has been defined by the following equation:

$$q_j^{avl} = k \cdot H_j{}^p \tag{1}$$

where Hj is the nodal head and k and p are calibration coefficients.

This algorithm has been readily implemented into an existing hydraulic network solver, EPANET 2 (Rossman, 2000).

The model has been calibrated by the mean of a wide set given by six months of continuous water flow data at the two network inflow nodes and correspondent pressure and flow data in 6 nodes distributed inside the network. The data has been collected on hourly basis. The only parameter that has been considered in the calibration process was the pipe roughness and the calibration has been performed by the least square method fitting the simulated pressures and flows in the network internal nodes with the measurements.

3.3 The network performance analysis

Three possible operative scenarios have been simulated with water resources availability respectively equal to the 95%, 85% and 75% of users demand.

System performance analysis has been carried out using all new proposed PIs. Namely, for each simulated scenario and network node index *WV1* has been computed in the whole simulation period, while index *WV2* has been computed for each analysis time step. The related nodal performance scores have been assigned using the penalty curves shown in figure 1. For each operational (OP) index, instead, has been computed a single global system value (table 3).

Table 3. OPERATIONAL PIS (OP) VALUES

NETWORK PARAMETERS						PIs (OP)	
						SD1	1.00
ST	[h]	24	$Q_{avail,1}$	[m^3/h]	43.2	*SD2*	0.50
$ST_{8\text{-}20}$		12	$Q_{avail,2}$	[m^3/h]	435.6	*SR1*	5.96
$V_{s\_private}$	[m^3]	17500	$Q_{avail,3}$	[m^3/h]	399.5	*SR2*	0.46
V_{s_public}	[m^3]	40000	D_{region}	[m^3/h]	0.2	*SR3*	0.25
$Vs=V_{s\_private}+V_{s\_public}$	[m^3]	57500	$D_{urban\ centre}$	[mc*ab/d]	0.3	*SR4*	0.49
Q_{mg}	[m^3/h]	401.9	Turn	days	180	*SR5*	1.33

With regard to *WV1* index the analysis of figures 4a, 4b and 4c shows inequity in water volumes distributed to users. Namely, in the first scenario (fig. 4a), corresponding to a mild shortage conditions, low performances are present in the lower part of the network, due to the fact that users collect more water than the needed, while, in the higher part of the network, users demand is not satisfied due to the low pressures. Analysing the other two scenarios (figg. 4b and 4c), corresponding to water volumes availabilities equal to 85% and 75% of demand, average network performance reduces slightly while inequalities among users are more evident with a large part of the network affected by water supply deficiencies and few users almost unaffected by water scarcity effects.

With regard to the first scenario the analysis of index *WV2* has demonstrated that, in the first part of the service day (just after the restart of the water service), all nodal performances are zero: the lower zone of the network drains the most part of the available resources for filling their local reservoirs showing values of *WV2* near to 320%; instead, due to low pressure, the high part of the network cannot drain resource showing *WV2* near to 0%. During the service period, lower nodes are filling and resource becomes progressively available for higher nodes (fig. 5a). At the end of service cycle, when lower reservoirs are filled (fig. 5b), *WV2* values in the higher part of the network grow while local reservoirs are filling.

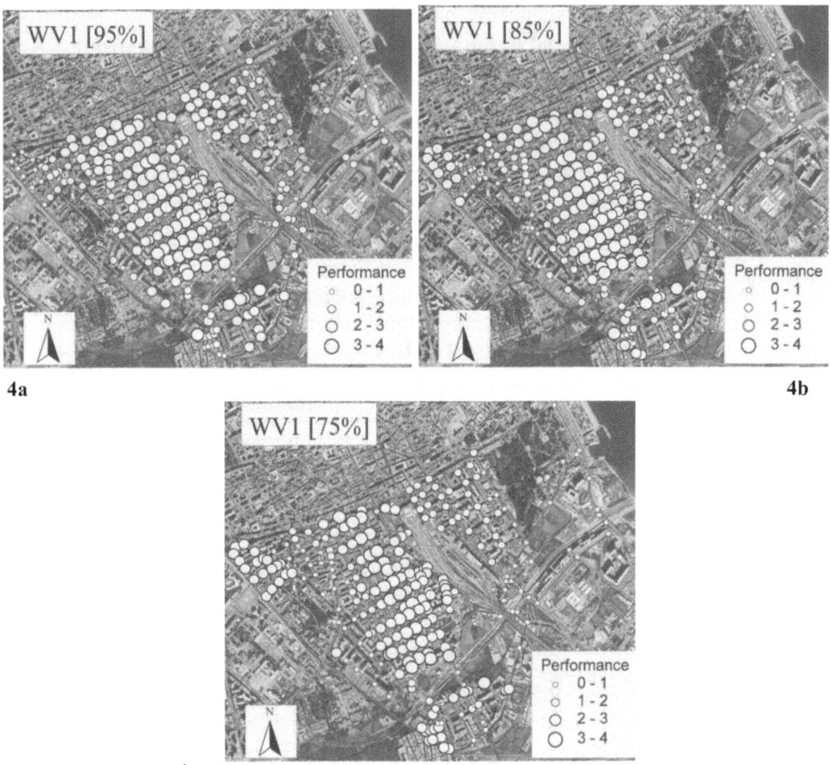

4a **4b** **4c**

Figure 4. PERFORMANCE RELATED TO WV1 INDEX: a) 95%, b) 85% AND c) 75% SCENARIOS

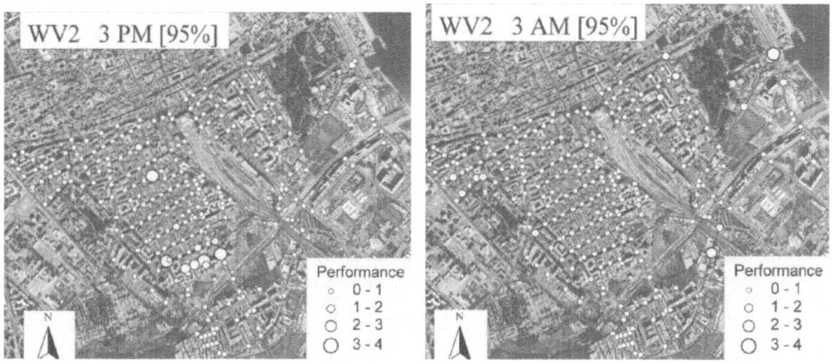

5a **Figure 5. PERFORMANCE RELATED TO WV2 INDEX: a) 3 PM, b) 3 AM** 5b

In intermittent distribution under water scarcity conditions, *WV2* indices are highly variable demonstrating that the network subjected to intermittent distribution will work in conditions that are quite far from the design ones and also from the operating conditions that take part in continuous distribution.

4 CONCLUSIONS

The present study has proposed a procedure for analysing a distribution network under intermittent distribution condition. The methodology has defined ten performance indicators able to identify inequality in water supply both considering long and short term effects.

The network and users' behaviour in intermittent distribution conditions have been monitored and a mathematical model has been arranged and calibrated.

The analysis has been allowed for evaluating network performance in different water scarcity scenarios and for highlighting inequitable water distribution. Intermittent supply can greatly affect both water availability for the users and the network behaviour. In the case study, even in presence of a low reduction of available water resources to fulfil user demand (95% scenario), intermittent distribution produces inequalities by reducing water supply to high elevated nodes and increasing supply to lower nodes.

The intermittent distribution has thus a great impact on users and networks. Users change their water supply patterns with higher peaks (connected with the filling of local reservoirs) and large periods with very low discharges. They can also receive higher or lower water volumes depending on their elevation and position in the network. Intermittent service changes radically the behaviour of the network with parts of it interested by flows that are much higher than the design ones and parts interested by almost null discharges.

5 REFERENCES

Alegre H., Baptista J. M., Cabrera E. Jr., Cubillo F., Duarte P., Hirner W., Merkel W. and Parena R. (2006). Performance indicators for water supply services. IWA Publishing, Alliance House, 12 Caxton Street, London SW1H0QS, UK.

Chandapillai, J. (1991). Realistic simulation of water distribution system. Journal of Transportation Engineering, **117**(2), 258–263.

Coelho S.T. (1997). Performance in water distribution - a system approach. Research Studies Press Ltd.. Tauton, Somerset, England.

Cubillo Gonzales F. L. and Ibanez Carranza J. C. (2003). Manual de abastecimento del Canal de Isabel II. Graficas Fanny, Madrid.

Gupta R., Bhave P. R.. (1996). Comparison of methods for predicting deficient-network performance. Journal of Water Resources Planning and Management, **122**(3), pp. 214-217.

Hardoy J. E., Mitlin, D. and Satterthwaite D (2001). Environmental Problems in a Urbanizing World: Finding Solutions for Cities in Africa, Asia and Latin America. Earthscan, London.

Intergovernmental Panel on Climate Change, (2007), Climate Change 2007: the Physical Science Basis. Summary for Policymakers. Contribution of Working Group I to Fourth Assessment Report of the Intergovernmental Panel on Climate Change.

Jowitt P. W., Xu C.. (1993). Predicting pipe failure effects in water distribution networks. Journal of Water Resources Planning and Management, **119**(1), pp. 18-31.

Fontanazza C. M., Freni G. and La Loggia G. (2007). Analysis of intermittent supply systems in water scarcity conditions and evaluation of the resource distribution equity indices. Proceedings of 4th International Conference on Water Resources Management. Kos, Greece, May 21-23.

La Loggia G., Mazzola M.R., (1989). Syntetic performance indices in a real decision process, Closing the gap between theory and practise, Proceedings of the Baltimore Symposium, May 1989, IAHS Plubl. N°**180**.

McIntosh A. C. and Yñiguez C. E. (1997). Second Water Utilities Data Book: Asia and Pacific Region. Asian Development Bank, Manila.

Reddy L. S., Elango K., Analysis of water distribution networks with head dependent outlets. Civ. Eng. Syst., **6**(3), pp. 102-110, 1989.

Rossman, L.A., Epanet (2000): User's Manual. Environmental Protection Agency, EPA: Reston, Virginia.

Vairavamoorthy K., Akinpelu E., Lin Z. anf Ali M. (2001) Design of sustainable system in developing countries. Proceedings of the World Water and Environmental Resources Challenges, Environmental and Water Resources Institute of ASCE, Orlando, Florida, May 20-24.

Development and implementation of a water quality performance assessment and management system for local government in South Africa

F. Stevens[1]*, A. Wensley, G.S. Mackintosh and E. Delport*

[1] *Institute of Municipal Engineering of Southern Africa, 3 Prior Rd, Durban, South Africa, 4000 (E-mail: frankst@dmws.durban.gov.za)*

 Performance Assessment of Urban Infrastructure Services
Edited by Enrique Cabrera Jr and Miguel Ángel Pardo. ISBN: 9781843391913.
Published by IWA Publishing, London, UK.

Keywords: water quality, performance assessment and management system; eWQMS

1 INTRODUCTION

The provision of safe drinking water and effective sanitation are considered the most important determinants of public health. Both are fundamental municipal engineering services which, if substandard, can significantly raise the health risks and hazards to which South Africans are exposed. Many municipalities have inadequate systems in place as regards these key Water Quality Management (WQM) aspects. As the water services sector leader and national regulator, DWAF is proactively deploying a range of initiatives to support municipal WQM, raise awareness of the importance thereof, and to regulate municipal WQM performance. The Institution of Municipal Engineering of Southern Africa (IMESA), a voluntary professional association with specialist expertise in the municipal engineering sector, has been proactively partnering with DWAF regarding municipal WQM since 2004. An important component of this joint initiative is the deployment and support of a comprehensive municipal engineering oriented electronic Water Quality Management System (eWQMS) at all municipalities across South Africa.

This paper provides feedback on the successful joint DWAF/IMESA municipal Water Quality Management initiative. Attention is drawn to:

- The multi-faceted challenge of effective and sustainable municipal Water Quality Management;
- Progress in the DWAF/IMESA municipal Water Quality Management Initiative;
- Key contributors to achievements to date; and
- Challenges Ahead.

2 CHALLENGES AND CONSTRAINTS FACED BY MUNICIPALITIES IN IMPLEMENTATION OF MUNICIPAL WATER QUALITY MANAGEMENT

Following on from considerable progress with the expansion of municipal water service provision since 1994, South Africa has over the last three years embarked on important initiatives regarding the development of municipal operations and management. An important component thereof has been the goal of ensuring that all South Africans have access to sufficient and safe drinking water and sanitation. The provision of effective and safe water supply and sanitation services requires municipalities to design, implement and operate holistic risk based WQM programmes, as shown in figure 1 overleaf.

The challenges relating to the establishment of effective WQM processes across all municipalities is not insignificant. In many instances these WQM processes have historically not existed or were wholly inadequate, and even in those municipalities where relatively good WQM practices were in place these practices tended to be monitoring oriented and not preventative risk management based (as shown in Figure 1 overleaf). As such the national engagement with municipalities to move towards effective WQM comprises considerable challenges as a change management process. As in any business environment, municipal engineering processes only work where they are supported by compatible structures, culture, and systems (see Figure 2 overleaf). The manner in which the municipality is organised, the systems that are in use to support processes, and the prevailing culture of the municipality will all have an influence on how well any process works. This is particularly important where the function of meeting target service quality goals is historically weak as is the case with municipal WQM; for example, a November 2006 DWAF survey has found that only 66% of municipalities believe that they are adhering to DWQ standards - up from 55% in March 2006 (DWAF, 2006).

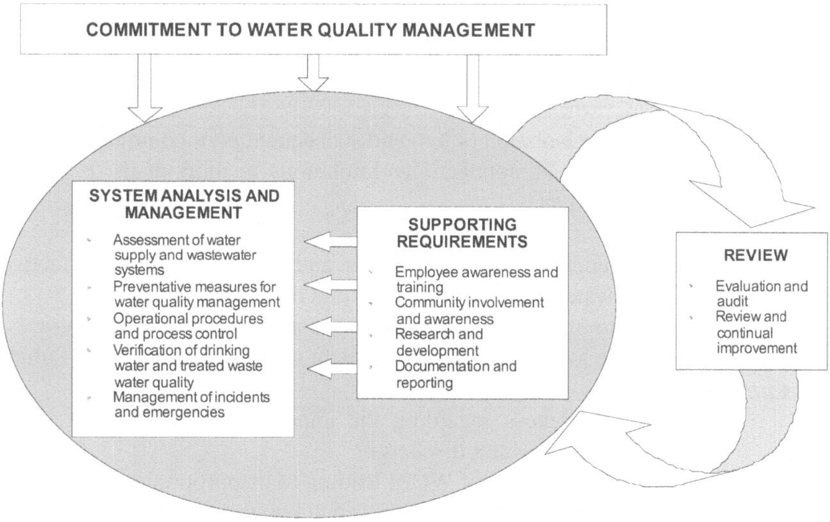

Figure 1: FRAMEWORK FOR MUNICIPAL WATER QUALITY MANAGEMENT

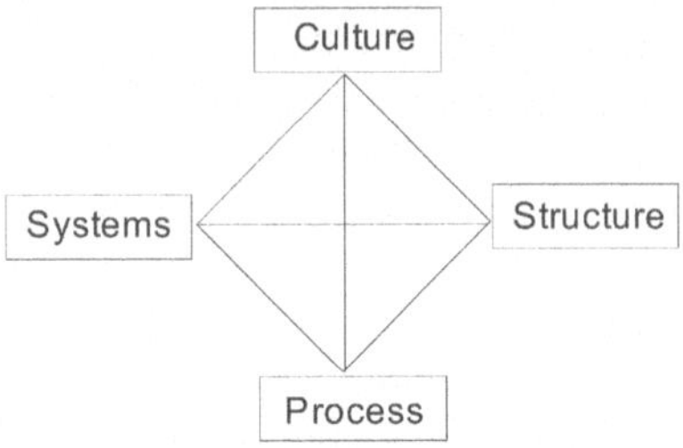

Figure 2: **THE MUNICIPAL ORGANISATIONAL CONTEXT FOR WATER QUALITY MANAGEMENT PROCESSES**

With reference to the organisational context for WQM processes, the following key constraints and challenges faced by municipalities in the introduction and operation of effective WQM are noted:

- ***Systems***
  - Financial administration burden on poorly resourced municipalities
  - Lack of IT connectivity at small, remote municipalities
  - Lack of appropriate staff mentoring, appraisal and feedback
- ***Structure***
  - Inability to employee adequate staff (numbers and skills)
  - Inability to fill all required posts on organogram (budget or organization constraints)
  - Loss of staff (to private sector, disease)
  - Constant restructuring
  - Operation in isolation (silos) and subsequent poor communication
  - Imposition of IT connectivity limitations to individuals (e.g. only top management)
- ***Processes***
  - Lack of understanding the needs and requirements for effective drinking-water quality monitoring and management
  - Inability to implement interventions to address issues of concern
  - Inadequate infrastructure management resulting in premature failure
- ***Culture***
  - Lack of awareness regarding the importance safe drinking-water and treated wastewater discharge
  - Lack of prioritisation of WQM leading to inappropriate budgeting
  - Inadequate institutional awareness and know-how regarding WQM
  - Constant imposition of budget cuts on essential technical inputs (operations and maintenance)

3 KEY AREAS OF PROGESS WITH THE DWAF/IMESA WQM INITIATIVE

Taking note of the ultimate objective of the DWAF/IMESA initiative, i.e. effective Municipal WQM, and the recognised challenges and constraints identified, the following areas of progress are highlighted.

3.1 Deployment and Use of an electronic Water Quality Management Tool by Municipalities

The eWQMS is a novel Open Source Software based management system accessible via the internet (www.wqms.co.za). eWQMS has been developed in a "bottom up" iterative approach with municipalities, IMESA and DWAF. eWQMS provides easy access to useful water quality tools and information. Data can be loaded by municipalities onto eWQMS via the internet, spreadsheet or specific import scripts. Water Quality Management outputs features of eWQMS include (a) Management Dashboard, (b) Overview, (c) Data Analysis, (d) Reports and (e) Automatic Monthly Summary Reports. ***NOTE:*** Details of the above features/functions are contained within a paper presented at this conference (De Souza et al, 2008)

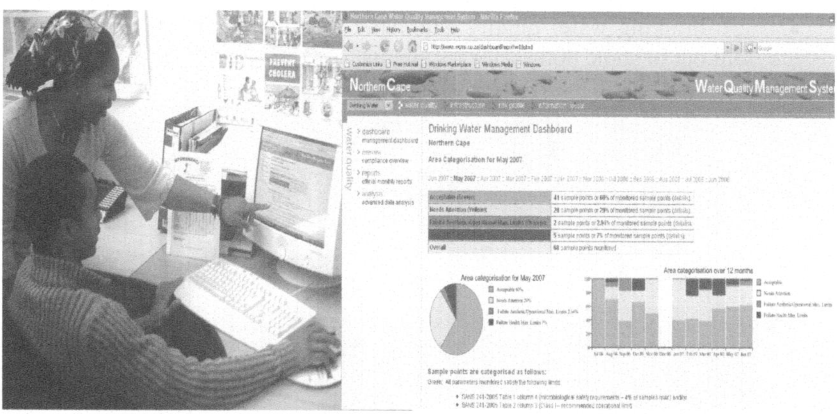

Figure 3: EWQMS USERS EXAMINE THE MANAGEMENT DASHBOARD SCREEN, BOTH HIGHLIGHTING COMPLIANCE AND DRAWING ATTENTION TO HEALTH THREATS

The capturing of WQM data from municipalities onto eWQMS has been as follows:

- Where municipalities have an existing WQM system, eWQMS is linked via an import script whereby WQM data can be seamlessly captured and passed on to DWAF National Information System for regulatory purposes.
- Where municipalities have no existing WQM system, the web-based eWQMS system is provided.
- Where water quality data can be obtained directly from a service provider (e.g. laboratory, Water Services Provider or Water Board), the linkage is established directly with the service provider.

A key yardstick as to progress with the initiative is the extent to which WQM data flows onto eWQMS. The figure overleaf shows the current participation rate as regards monthly data submission by municipalities (May to October 2007). In particular, for October 2007 the following can be noted:

- 147 out of 169 (87%) municipalities in South Africa submitted data to the eWQMS before the cut-off date for reporting purposes (i.e. before the end of the 14th November 2007).
- The data provided covered 2 883 samples sites, and 4 555 sets of data were obtained (i.e. at some sample points, more than one sample was collected in the month).
- 156 drinking water quality automatic monthly summary reports were sent out to some 800 sector recipients.

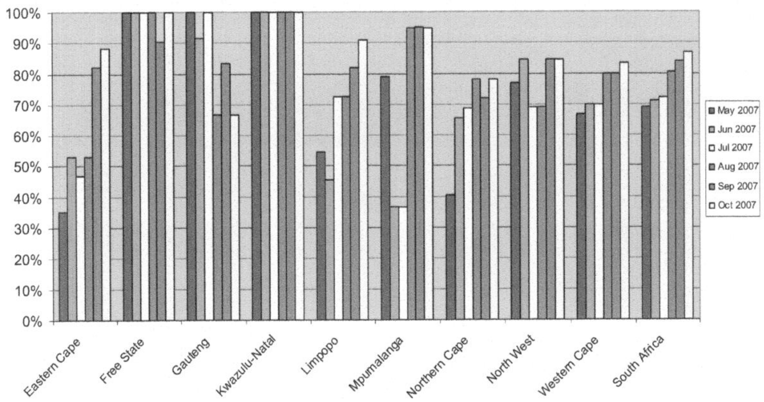

Figure 4: MONTHLY SUBMISSION OF DRINKING WATER QUALITY DATA IN SOUTH AFRICA FOR THE LAST 6 MONTHS (MAY TO OCTOBER 2007)

Figure 4 above shows that, whereas before a fairly consistent data submission (at about 70% participation) was noted, significant improvement has been noted in recent months. Whilst Figure 4 shows a very impressive national participation rate for a sector wide process that was initiated in April 2006, and which only really gathered steam by late 2006, the participation rate below 100% highlights the fact that all municipalities in South Africa are either (i) not monitoring drinking-water quality on a monthly basis, or (ii) not submitting drinking-water quality data to the eWQMS.

3.2 Other Related Water Quality Initiatives

Through the eWQMS roll-out process, additional water quality related tools have been introduced to municipalities. These include:

- Water Quality Management Decision Support Tool - Assesses in detail the six legs of sustainable and effective DWQM (Wensley et al, 2008)
- Consultative Audit – Assess municipal performance versus minimum legislatively compliant DWQM requirements (de Souza et al, 2008)
- Other Municipality Orientated Tools – various useful tools have been provided to municipalities to allow for effective and sustainable drinking-water quality management (e.g. Water Supply System Assessment).

4 DEVELOPING AN EFFECTIVE MUNICIPAL ENGAGEMENT MODEL

It is clear from the progress reflected above that the initiative has made remarkable progress with engaging with municipalities, and thus contributing significantly towards a foundation for effective and sustainable DWQM. The engagement model that has been utilised can be represented as an arch made of five blocks (see figure below). Each part of the arch is an aspect that can help or hinder the achievement of sustainable municipal WQM.

- **National Leadership: Providing Clarity**

The first foundation pillar of the arch relates to the clear and effective leadership, guidance and support from the Water Services sector leader, DWAF. This includes both strategic and practical inputs from DWAF National and DWAF Regional offices as regards the provision of appropriate legislation, guiding policies, and proactive regulation.

Figure 5: EFFECTIVE MUNICIPAL ENGAGEMENT ARCH

- **Municipal Appropriate Solutions: Ensuring That the Solution Fits**

Sometimes in a developing world environment, highly desirable and ambitious programmes are launched in a "top-down" manner only to falter in implementation. The involvement in an iterative manner of municipal representatives, and in particular the South African Local Government Association (SALGA) and IMESA, has assisted in assuring that municipalities capabilities are considered via "bottom-up" inputs, are timeously captured and that a functional progress improvement approach is supported.

- **Support Existing Good Practice: Every Municipality is Unique**

If you have a hammer, every problem looks like a nail! That there are many similarities between WQM problems does not mean that apparently similar environments have similar solutions. Many ingenious and good existing practices are in place, and these should be positively engaged with and developed towards full legislative compliance.

- **IT Supports Business (Business does not exist to support IT)**

The provision of any IT based tool, in this case the eWQMS tool which facilitates improved WQM and relays WQ data to DWAFs National Information System, must provide positive value to municipal users. Value in that it introduces efficiencies, is reliable and cost effective, is supportive of existing good practice and yet able to introduce and guide new good practice. Positive in that it is municipal appropriate and is customizable to be supportive of variable local conditions

- **Prioritisation of Municipal Water Quality Management**

The second foundation pillar of the arch relates to the need for prioritizing of municipal WQM. In the high demand world of municipal services provision, municipal management, municipal leaders and communities are often forced to prioritize areas of expenditure and effort with limited understanding of the primary importance of safe WQM.

5 CHALLENGES AHEAD

Despite the exceptional progress to date (as is most readily evidenced by municipal data flow via eWQMS), a number of key challenges will need to be addressed nationally in the near future to ensure ongoing progress towards the target of effective municipal WQM across South Africa. These include:

- **Appropriate Sample Collection**

  Both the appropriate number of samples that need to be collected and the appropriateness of sample point locations must be considered. The South African National Standards (SANS) 241 Drinking-Water Specification (SANS 2006) recommends a minimum number of samples that should be collected by municipalities (based on population served by a supply system). Considering population figures, it can be estimated that a total of 5 293 samples should be collected throughout South Africa as a minimum on a monthly basis. By also factoring in the need for sample point appropriateness (samples collected along the water supply chain, at strategic areas of high risk, etc) it could be anticipated that the actual number of samples required to effectively manage drinking water quality may be at least 50% more than the population based minimum of 5 293 samples/month – i.e. approximately 8000 samples/month for South Africa. Comparing this to the data sets received for October 2007 (4 555 data sets – see Section 3), the data submitted in October 2007 only represents 57% of the likely required number of samples (***NOTE:*** This figure was only 43% in June 2007). This provides a yard stick indication of the need for additional sampling (i.e. either increase frequency of sampling or establish additional sampling points).

- **Data Integrity**

  Monitoring for appropriate water quality determinants, the use of appropriate analysis methods and ensuring associated quality control must be considered. As regards drinking water, in many instances municipalities are not monitoring the minimum determinants required to accurately assess the status of their drinking water quality. Although it is acknowledged that bacteriological monitoring (*E.coli*, faecal coliforms and/or total coliforms) is of critical importance, many municipalities do not monitor basic physical (e.g. pH, electrical conductivity, turbidity) or chemical determinants. In particular, it is important that the residual concentration of any chemical added to the water during treatment is monitored to ensure that the concentrations in the final treated water are within acceptable levels (e.g. free chlorine residual, aluminium if aluminium sulphate coagulant is used, etc). In addition, source water specific chemicals (e.g. nitrates/nitrites and fluoride) should also be monitored.

Linked to the above is the need to use appropriate monitoring equipment with suitable accuracy and to ensure that laboratories utilized have at least implemented a basic quality control system. The above will ensure that issues identified are in fact "real" issues, thus minimizing waste of already limited resources. In particular, use of internationally recognized monitoring equipment, frequent equipment maintenance and calibration, inter-laboratory studies and proficiency testing and use of accredited laboratories is recommended.

- **Development and Implementation of Water Safety and Security Plans**

The majority of municipalities do not have structured management plans for WQM. In line with the South African National Drinking Water Quality Management Framework, and international best practice, it is desirable that each municipality develop a comprehensive set of water quality management plans. These should be used to direct WQM practises within the particular municipality, and as a means to ensure sufficient staffing and financial capacity, and by way of example should at least include (i) a relatively brief high level Water Safety and Security Plan (WSSP), (ii) more detailed system specific WSSPs and (iii) an Operations Programme, including operational procedures, compliance monitoring, and incident and emergency response procedures.

- **DWQM in Communities Not Yet Served with Potable Water**

A proportion of people in South Africa still exist without access to safe drinking water (85% of population has access to a safe water supply, and 94% to an improved water supply). The greatest proportion of these are located in rural areas where infrastructure is limited.

- **Treated Effluent Discharge Management**

As South Africa has limited water resources, it is essential that available resources be protected and managed effectively. Considering this, the threat posed by failing wastewater treatment facilities on drinking-water quality cannot be ignored. By way of example, on-going monitoring and evaluation within Free State (a province in South Africa) indicates a 36% failure with regards to faecal coliforms and a 48% failure with regards to chemical oxygen demand (COD) at wastewater treatment facilities (using DWAF General Authorisation General Limit values). Of importance to note is that similar circumstance exists in other provinces of South Africa. In order to ensure a safe drinking-water, it is therefore essential that the entire water cycle in a municipality is effectively monitored and managed.

Although the above list of challenges is not exhaustive, it highlights key areas where municipalities in South Africa need to focus to ensure that consumers within their area of jurisdiction receive a safe drinking water. In particular, addressing the above challenges will require:

- Acknowledgement by municipalities that the above gaps exist and a commitment to close the gaps.
- Commitment by municipality staff (both political and technical) to continually strive to improve water services to consumers
- Continued close co-operation and on-going communication between national, provincial and local government
- On-going DWAF support and guidance to municipalities regarding DWQM

6 CONCLUSIONS

The DWAF/IMESA roll-out of the electronic water quality data capture, information dissemination, and management tool (eWQMS) to municipalities in South Africa has made very good progress since initiation in April 2006. The proactive and supportive involvement by local, provincial and national government to ensure that municipal Water Quality Management (WQM) is adequately resourced and capacitated, can be expected to lead to a considerable improvement in WQM by municipalities across South Africa in the near future.

7 REFERENCES

De Souza PF, Ramba MH, Wensley A and Delport E (2006) Implementation Of An Internet Accessible Water Quality Management System For Ensuring The Quality Of Water Services In South Africa. Paper Presented at WISA Conference, Durban, South Africa, 2006.

DWAF. (2007). A Drinking Water Quality Framework for South Africa. Department of Water Affairs and Forestry, Pretoria, South Africa

DWAF. (2006). Strategic Overview of the Water Sector in South Africa-2006. Department of Water Affairs and Forestry, Pretoria, South Africa.

Mackintosh G, Ramba M, Wensley A, Manus L, Stevens F and Delport E (2007) Towards Effective Municipal Water Quality Management – Feedback on Key Success Factors and Future Challenges Identified during the DWAF/IMESA Initiative. Paper presented at IMESA Conference, Durban, South Africa.

Mackintosh G, Ramba M, Wensley A, Manus L, Stevens F, and Delport E (2006). Improving the Quality of Drinking Water Services Delivery by Local Government – Feedback on a Joint DWAF/IMESA Capacity Building Initiative assisting Municipal Engineers. WISA Conference, Durban, South Africa.

Mackintosh GS, De Souza PF, Wensley A, and Delport E (2005) Multidisciplinary Considerations For Satisfying Water Quality Related Services Delivery Requirements - A Practical Tool For Municipal Engineers. Paper Presented At IMESA Conference, Port Elizabeth, South Africa, 2005.

SANS: South African National Standards. SANS 241-2006: Drinking-Water, Edition 6, (2006).

Performance assessment as a management tool for water utilities in developing countries. Case study at EMCALI-Colombia

F. Garzón-Contreras[1] *and A. Agredo-Perdomo*[2]

[1] *Pontificia Universidad Javeriana. fgarzon@puj.edu.co, Calle 18 No. 118-250, Cali, Colombia, Tlf +572 3218200*
[2] *Empresas Públicas de Cali – EMCALI EICS ESP. algredo@emcali.com.co, Av. 2N # 7N-45 P7, Cali, Colombia, Tlf +572 8996222*

Keywords: Public Water Utilities in Developing Countries, Performance Indicators System, Performance Indicators Implementation

Edited by Enrique Cabrera Jr and Miguel Ángel Pardo. ISBN: 9781843391913.
Published by IWA Publishing, London, UK.

ABSTRACT

Empresas Municipales de Cali (Emcali) provides public services, including telecommunications, electric power, public lighting and waterworks for Cali, Colombia. During the last years, the company has been in a very hard financial situation that almost brought it to liquidation. Right now, EMCALI is under the Superintendent of Public Services and a board of banks delegates control that had agreed to restructure the company's debt.

This change in the director's board brings to EMCALI a new business perspective, focused in the development of modern management tools for each business unit. This paper presents the process that was initiated by EMCALI's Water Supply and Waste Water Business Unit (WSWW BU), to define and implement a performance indicators system that allows a better understanding of current situation of the WSWW BU.

Is important to note that the system the WSWW BU attends is the third biggest system in Colombia. This system has more than 480 000 active customers, and 2 560 km of pipes. Another fact to be considered is that the Water Undertakings have to report only a very limited set of PIs that are later reviewed by the Superintendent of Public Services, which focus on financial aspects of the industry; consequently, the use of more detailed or industry specific PI system is not usual in Colombia.

In this paper, the authors present the methodology used to develop and implement a PI system based on the IWA PI framework that accounts facts like: lack of information, limited human resources, low data reliability and accuracy. Additionally, it discusses the difficulties and lessons learned during the process and the first results interpretation.

1 INTRODUCTION

Santiago de Cali, often shortened to Cali, is the main city, municipality and capital of Valle del Cauca, a department in western Colombia. Cali is also the third largest city in Colombia. It is located on the Cauca river valley, on the west side of the Cauca River and to the east of the Western Mountain Range near the Farallones de Cali. The city is at approximately 1,000 meters (3,280 ft.) above sea level and the topography is fairly flat.

Empresas Municipales de Cali (EMCALI, the municipal enterprises of the city of Cali) is a public undertaking that provides energy, telecommunications, and water and wastewater services to Cali and Yumbo, a nearby city. The company traces its origin back to 1930, becoming one of the strongest public service companies in the country. However, during the last decade, EMCALI has faced a financial crisis mostly due to bad management practices influenced by obscure political interests that forced significant capital investments to build a wastewater treatment plant (7.2 m^3/s) and a termic power plant leading to substantial debt that EMCALI was unable to pay.

Due to this situation, on April 2000 EMCALI was put under the control of the Superintendence for Residential Public Services (SSPD for its initials in Spanish), a multi-sector regulatory agency that is in charge of the control and prosecution of Colombian public service companies in. At the end of 2003 year, the central government and private banks concluded a long negotiation to restructure EMCALI debts that resulted in a final agreement signed in May 2004. This agreement maintains the EMCALI public ownership and defines the level of investments and performance indicators that the Utility must fulfil in order to avoid the company liquidation.

EMCALI is governed by an elected General Director and a Board of Directors. The organizational structure of EMCALI is divided in seven functional areas: human resources, financial, information technology, customer service, water and wastewater service, energy service, and telecommunication service. The first four areas are common for all the company, while the other three attend all activities relates with the provision of each service.

One of performance indicators assess by creditors is the UfW that now is around 40 %. But for this year the goal, imposed by the agreement, is 36%. EMCALI has tried to reduce this PI, during the last couple of years, using NRW reduction strategies focused on the control of apparent losses. Unfortunately, the results were not the expected due to a 2% increase in the UfW during this period.

This situation forced EMCALI to treat the problem of water losses in more holistic approach taking in to account the latest IWA developments for leakage management. In this way, the Research Group in Advanced Urban Water Management (GIGAAU for its initials in Spanish) at Javeriana University was assigned the task to design a water loss management strategy appropriate to EMCALI local conditions and constrains.

The latest experiences in this kind of NRW projects showed that one of the key elements for successful strategy implementation is a sound project management based on an appropriate Management Information System (MIS) that will provide all the useful data including progress and performance indicators.

To do this, the first option was the development of a set of Water Loss Performance Indicators, but later and considering the actual transformation of EMCALI to be more competitive, the objectives of this task were changed to provide a general and representative overview on the efficiency of EMCALI in relation to other companies that allow them to identify their own deficiencies to initiate and control measures for improvement. This paper presents details of the development of EMCALI MIS based on IWA-PI system, problems during the implementation and lessons learned.

2 SYSTEM DESCRIPTION

The water distribution system operated by EMCALI serves the cities of Cali, Yumbo, Palmira and Candelaria, with almost 2.5 million inhabitants (455 thousand users). This system is divided into three main areas. The biggest is the *Red Baja* area, located in flood plain of the Cauca River, with flat topography. This area is supplied by the Rio Cauca and Puerto Mallarino treatment plants, which take water directly from the Cauca River.

The Rio Cauca WTP has been in operation since 1958. The water is treated using conventional means, and supplies the city of Cali through a 7 horizontal axial pumps, with a capacity of 0.35 m^3/s each. The total installed pumping capacity is 2.5 m^3/s. The contribution of this plant to the system is, on average, 21% of the total production, using a 64% of the installed capacity.

On the other hand, the *Puerto Mallarino* treatment plant has been operating since 1978. It has an installed capacity of 6.6 m^3/s of water treated by conventional means, which are delivered to the city through a 9 horizontal axial pumps, each pump with a capacity of 1.10 m^3/s, connected to a system of three transmition pipes with a diameter of 1000 mm. This plant contributes, on average, with a 59% of the total production using the 58% of its installed capacity.

The pumping of these plants is transported to 8 storage tanks: *Nápoles* (30.000 m^3), *Siloé* (24.000 m^3), *La Normal* (30.000 m^3) and *La Campiña* (32.500 m^3) from where the distribution network is fed. Additionally, the *Nápoles* pumping station, with a capacity of 0.4 m^3/s, feeds the *Ciudad Jardin* tanks (8,500 m^3).

The other distribution system area is the *Red Alta*, which is fed by the *Rio Cali* treatment plant. This plant uses water from the Cali River. This plant serves more than 400 thousand users in the old part of the city (15% of total production) and has an installed capacity of 1.8 m^3/s and is working at a 78% of its installed capacity. However, in summertime, the capacity decreases to a 45%, which makes it necessary to provide the missing capacity using WTP at *Rio*

Cauca and *Puerto Mallarino*. The plant has five gravity flow outputs: *Refuerzos Norte, Sur, Centro, Nacional, La Normal*, and a pumping output called *Bellavista*. Additionally, this area has 25 storage tanks with a storage capacity of up to 51,850 m^3 of treated water and 9 pumping stations with capacities between 11 and 117 l/s.

The last area to mention here is the network of *La Reforma*, which is fed by the waters of Melendez River and is treated at the *La Reforma* treatment plant, whose capacity is 1.0 m3 / s, but operates at 40% of the installed capacity. This area serves about 120 thousand users (5% of total production) located in the hillside area of the city. The plant feeds two main independent networks connected to the *Napoles* and *Siloe* storage tanks.

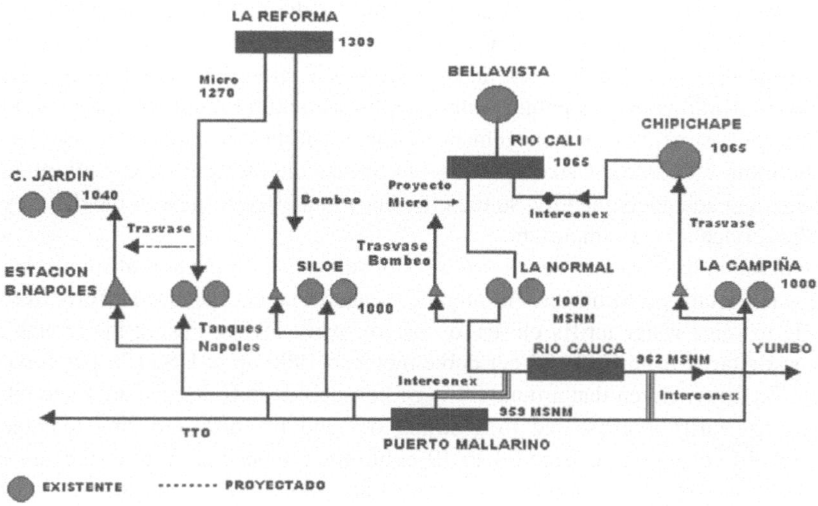

Figure 1. EMCALI WATER DISTRIBUTION SYSTEM DIAGRAM.

On the other hand, Cali's distribution network is composed of approximately 2,743 km of pipes, where 46.4% are uPVC, 36.2% Asbestos Cement, and 11.2% Cast Iron. The pipe diameters vary between 2 and 56 inches, with a distribution of 4-inch (40.3%), followed by 3 (21.8%), 6 (14%) and 8-inch (9.4%).

Additionally, the system has almost 14 thousand valves, of which 258 have a diameter greater than 12 inches. It also has 24 pressure regulation valves, all fixed-point, in pipelines from 2 to 6 inches. And the number of fire hydrants is 3010.

3 PREVIOUS PI SYSTEMS IN EMCALI

In Colombia the SSPD is in charge of the performance assessment for the water utilities. It is for this reason that it created, back in 2003, a set of PI arranged in the following groups: financial, commercial, and technical. Unfortunately, these PIs do not have clear definitions not a concise meaning, and do not provide context information, so this makes them difficult to interpret. Although, this information is put in a special public website called SUI, the SSPD results of the metric benchmarking initiative aren't public, so the utilities are not forced to compare themselves with others so the need to improve their performance disappears. SSPD only used this data to identify what utilities area experienced financial problems and then intervene them to protect the customer's interests.

In general, the water undertakings in Colombia aren't used to assess their performance. This situation may be explained by the fact that the management depends highly on the local political scenario and the organizational structure is so hierarchical that the decision making process is controlled only by very few persons. Additionally, as public water supply is a natural monopoly there is no utility interest neither its management to implement best management practices to be more efficiency. Usually, is enough to provide all time water with good quality and adequate quantity, so the efficiency and effectiveness of the delivery of the service it is not important.

Colombia has recently developed a set of reforms with the aim of improving the water sector performance. Among the measures taken, sustained efforts were made to raise water tariffs closer to cost recovery levels, raising the average domestic tariff from US$0.33 per cubic meter in 1990 up to US$0.78 per cubic meter in 2001. Given that around 88 % of households in Colombia are metered, this large tariff increase led to a strong demand reaction, reducing average household consumption from 34 to 19 cubic meters per month over the same period. Due to this situation the consumers considering the relative high costs of the Water and Wastewater Service (WWS) are claimed for a more efficient provision of the service.

In the case of EMCALI's WWS the internal use of PIs was limited to a set of 7 PIs assessed in a monthly base: UfW, service coverage, chemical costs, number of employees by 1000 consumers, actual level of investment, actual tariff structure, and volume of wastewater treated. It is clear that this set of PIs is not intended to make decisions.

On the other hand, the agreement with creditors brought another set of PIs assessed yearly, but reported monthly: number of stopped meters changed, number of new connections, number of cancelled connections, and so on. The objective of these PIs is to control the fulfilment of goals established by the agreement but not to improve the performance of the undertaking.

The general public pressure for a more efficient service provision, the aging and deterioration of infrastructure, and the challenged environment (legal, economic, social, and environmental factors) faced by Colombian water utilities, especially in big cities, as is the case of EMCALI, ask for modern practices in systems management. Business practices, customer satisfaction, and economic planning are all as important as engineering design and can no longer be disregarded by water service providers.

4 THE DEVELOPMENT OF EMCALI'S NEW PI SYSTEM

Considering the above situation, the defined management objectives of this new PI system for EMCALI are:

- to provide an appropriate level of service to consumers, while complying with national and regional policies and meeting creditors agreement and other obligations;
- to obtain the highest possible productivity from human resources;
- to protect and assure a sustainable use of water and other natural resources;
- to achieve the most efficient use of financial resources;
- To plan, construct, maintain and operate the undertaking's physical assets as efficiently and effectively as possible.

The organizational structure of EMCALI is distinguished by the concurrence of more than one stakeholder for the service provision. That is the case in which WWS does not have any participation on costumer management. Additionally, the senior staff of WWS is comprised completely by sanitary and civil engineers with no graduate and management formal education. This situation created the need for a training workshop to explain the concepts and principles of the use of PIs as a management tool.

The workshop had 28 delegates from each business area of EMCALI related to WWS. They were the senior staff who is going to use the PI information, and representatives of the departments and business units who need to provide the PI data.

It is worth to note that the GIGAAU decided in this case to use the IWA-PI system as a base for PI system for EMCALI, because this system had proved to be an universal, robust, consistent and coherent set of PI aimed to the improvement of the WWS, that allowed a truly metric benchmarking of the EMCALI performance compared with other international undertakings. The former was really important because EMCALI wanted to determine the areas where improvements have to be implemented.

During the workshop the participants classified the IWA-PI, depending on their importance for the performance assessment, as high, medium and low. Later, the variables of PIs set with high and medium classification were evaluated for their availability and quality, and a set of supporting explanatory factors was developed for them. Finally, the combination of high significance, and availability and quality of data needed to calculate the PI defined an initial set of PIs and CI.

The second step in the implementation process was the development of data requirements and guidance notes for data collection. This process was followed by the collection and entry of EMCALI data to SIGMA Pro software for performance indicator evaluation developed by Technological Water Institute (ITA for its initials in Spanish).

The results from this pilot testing at the moment of the writing of this paper are under analysis in order to evaluate the initial set of PIs and explanatory factors to confirm the relevance of the selected PI and the feasibility of assessment. For this reason, the final set of PIs and CI is not at the final stage.

In other hand, the success of implementing a PI system depends on the effective use of the results as a decision-making tool. The objective is to interpret the results (i.e. to decide whether a given result is satisfactory or not), and if the results aren't satisfactory, to establish a diagnosis and identify what improvement measures need to take place.

The process of result interpretation requires their comparison to a reference value or a reference threshold, taking into account the context information, and analysing at each time a group of related indicators and not single indicators individually.

In the case of EMCALI the internal comparison is not feasible yet because there are not historical data to compare. The comparison with other undertakings or with publicly available reference values is therefore the only possible way in this case. Consequently, the success of the PI system in EMCALI is by now under question until the results showed the benefits of this management tool for the utility.

4 MAIN PROBLEMS FACED

The main problems faced to implement the PI system were:

- The previous history of the utility marked this kind of strategies as a temporal initiative highly dependent of who is in charge of the business unit. So, the PI system is considered a waste of time and efforts by the senior staff.
- The senior staff experienced some kind of threat because their performance never was assessed, and as they are near retirement, they are not interested in the adoption of any new way to carry out their duties.
- Difficulties inherent to the company surrounded defining responsibilities and arranging suitably reliable delegation of data collecting procedures.
- Specific difficulties to obtain data concerning certain variables. The company's data have water and wastewater combined information, and in some cases it was very difficult to isolate one from the other.
- Some discrepancies between the measuring units proposed by the system and those used by EMCALI;
- The motivation of the team to collect the needed information. Their regular work did not liberate them enough time to gather all the required information, and the general feeling was that they had another difficult task to do with no practical results.
- Certain deficiencies for the use of the SIGMA Pro program associated with the modernization of IT platform in EMCALI.

5 CONCLUSIONS

The IWA-PI system has been a very good basis for the establishment of an indicators system tailored to the needs of EMCALI. The feedback from the participants of this project were entirely positive and generated interesting discussions and relevant conclusions about the EMCALI procedures for service provision that allowed the establishment of the critical success factors for the WWS Business Unit that in a short-term could improve the performance of the utility.

Although the project faced some problems, it is our belief that from this on-going project a clear gain may result for the service users, for the sector generally speaking and for the country at last.

6 ACNOWLEDGEMENTS

The authors express their acknowledgements to all people from EMCALI for their encouragement, support and credit provided during this study.

An agent-based model for simulating stakeholders interactions in water services through PIs

A. Ancarani[1]

[1] *Department of Civil and Environmental Engineering, University of Catania – Viale A. Doria, 6 - 95125 Catania (ITALY) – Ph. +39 095 7382718 – Fax +39 095 7382748 – aancaran@dica.unict.it*

Keywords: Agent-based model, value for customer, customer satisfaction

Edited by Enrique Cabrera Jr and Miguel Ángel Pardo. ISBN: 9781843391913.
Published by IWA Publishing, London, UK.

ABSTRACT

A methodology for evaluating the supplier's performance in water services is presented. The basic concept is that the quality of the service should be determined in terms of supplier's performances. Such performances can be evaluated in terms of technical and functional quality. It can be expressed in terms of prevalent model of value for customer in the interaction between customer and supplier. The interaction in the case-study has been simulated through and agent-based model.

1 INTRODUCTION

The on going reforms in water services started all over the world with the separation between the public authority that owns the assets and the supplier (a public, private, or public-private organisation) to which the service provision is contracted out. With reference to water service provision, the customer of water services become the pivot around which provider organizations have, and will have, to build new strategies of development, based on principles of efficiency, effectiveness, and economicity of management with the aim of assuring suitable performance to customer's requirements, both in terms of the interaction between customer and provider and in terms of the quality of service.

In the water services it is important to consider that the roles of buyer (Local Authority), client (customer/citizen) and supplier (external organisation, public or private) have been clearly differentiated. The three stakeholders interact in different ways. In particular, in figure 1 are reported the main interactions.

The triangulation interactions reruire a modified approach to the evaluation of the quality of service provided. While, up to a few year ago, it was requested to the supplier to provide value for money, nowadays there is an increasing request for value for customer (VC) (Bolton, 1998; Bolton and Drew, 1991) for obtaining public service improvement (Boyne, 2003).

Value for customer is a broader concept then customer satisfaction that since its early conceptualizations (Gronroos 1982; Parasuraman et al., 1985) has been based on the disconfirmation paradigm employed in the physical goods literature (Churchill and Surprenant, 1982; Oliver, 1977). In such an approach, the service quality depends on the capability to minimise the gap between expected and perceived quality by customers.

The service quality cannot be constrained within customer satisfaction concept (customer's expectations) since it is strictly referred to a specific transaction (Grönroos, 1984). Water service provision is a long term recursive service whose characteristics have to be identified by Local Authorities for satisfying citizens' needs. It seems clear that this identification is complex due to the several aspects of the service to be considered. Hence, there could be a gap between the performance required by Local Authority and performance expected by customers/citizens.

A recent model by Woodall (2003) identifies five primary forms of value:

- net VC, in terms of balance of benefits and sacrifices;
- derived VC, in terms of use/experience outcomes;
- marketing VC, in terms of perceived product attributes;
- sale VC, in terms of value as a reduction in sacrifice or cost;
- rational VC, in terms of assessment of fairness in the benefit–sacrifice relative comparison.

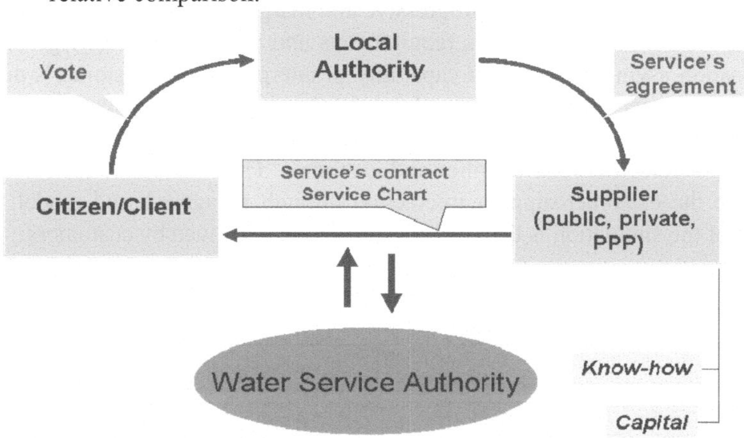

Figure 1. **RELATIONSHIPS AMONG ACTORS IN PUBLIC SERVICES PROVISION.**

Even if some limitations may be outlined, however, the model can be considered appropriate for the quality in water services. The final goal is to evaluate whether the value provided by the supplier match the value required by both buyer and customers. Since the buyer's institutional duty is to satisfy the customer's needs, two kind of analysis can be carried out. The first aiming at comparing the value provided by supplier with the one required by customer. The second aiming at determining the gap between customer expectations and buyer perceptions of that perceptions.

2 METHODOLOGICAL APPROACH

The methodology here proposed is based on the definition of service quality as a compendium of functional and technical quality of service (Grönroos, 1984), (fig. 2). The methodology builds on a previous framework (Ancarani and Capaldo, 2001) aiming at assessing the performance of public service suppliers. The methodology applies a model of value for customer including progressive gains of value along both the experience cycle with the service and the interaction cycle with the service provider. It requires the identification of a set of performance indicators representative of the service level to be guaranted. The set of indicators provides progressive and marginal values that the customer obtains. Sometimes the model's requirements and the indicators diverge due to information asymmetry, lack of capability in interpreting the customer's needs, and contract's incompleteness. Finally, a set of supplier's behavioural rules have been defined, along with a set of customer's behavioural rules and the interaction rules between supplier and customer. The rules have been used to simulate the customer/supplier interaction through an agent-based model. The output of the simulation is the level of satisfaction perceived by customers.

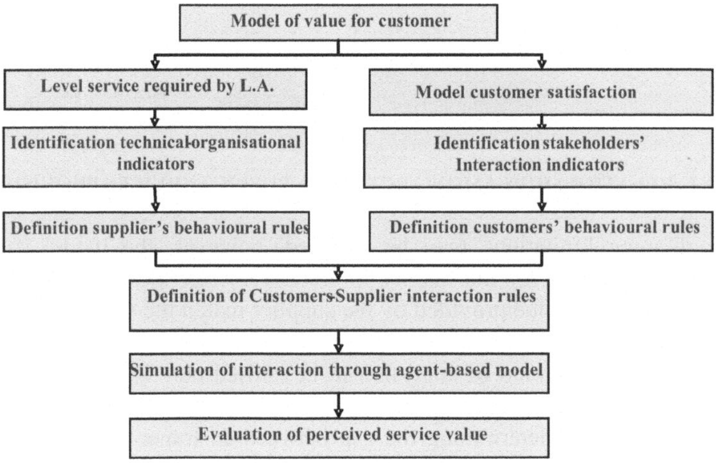

Figure 2. METHODOLOGICAL FRAMEWORK FOR SUPPLIER'S EVALUATION

3 A MODEL OF VALUE FOR CUSTOMER

The Woodall's model acts in the present paper as reference model in terms of measurement of service value as an aggregated amount along the different time phases of interaction, experience cycle with the service and experience cycle with the provider. In water services time duration and relevance of the phases are different for the two type of customers above mentioned (Table. 1).

Customer-Supplier interaction and Customer-Service interactions influence value perception, as they influence the aggregated value the customer perceives. The model for customer can be applied by considering aggregated value cube projections on planes corresponding to the time phases (Figure 4).

Table 1. TIME AND EXPERIENCE PHASES ALONG THE PUBLIC SERVICE PROVISION

Time phases	Experience phases	Local Authority	Citizen
1) Ex ante (pre-purchase)	*Collecting information*	*Extremely Relevant*	*Not relevant*
2) Transaction (at the point of contract/ during experience)	*Purchase*	*Relevant/ N.A* (*)	*Normal/ Relevant*
3) Routine operations (normal operation)	*Learning*	*Normal*	*Relevant*
3 bis) Failure/Recovery from failures	*Emergency*	*Normal*	*Extremely relevant*
4) Re-agreement or Disposition	*Purchase/ Disposition*	*Relevant*	*N.A.* (**)

(*) The L.A. do not pay the tariff (**) It rarely happens for citizens

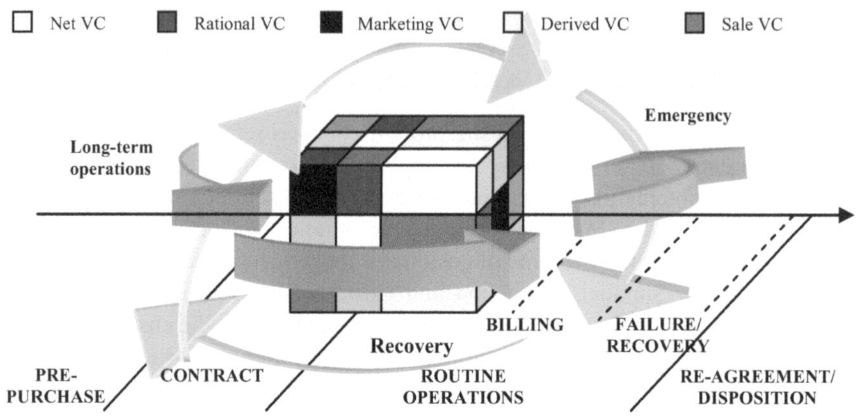

Figure 3. MODEL OF AGGREGATED VALUE FOR CUSTOMER WATER SERVICES

Pre-Purchase	Contract	Routine Operations	Re-Agreement/ Disposition
	Marketing VC Sale VC	Marketing VC Sale VC Derived VC **Failure/Recovery**: Net VC, Marketing VC, Sale VC, Derived VC **Billing**: Net VC, Marketing VC, Rational VC	

Figure 4. APPLICATION OF THE MODEL OF VALUE FOR CITIZEN

4 QUALITY MODEL ADAPTED TO WATER SERVICES

In the present paper the PZB model (Parasuraman et al., 1985) has been adapted to the water services conditions by taking into account the three main actors. The Local Authority should monitor the gap between the quality required by contract (Q_g) with the quality supplied (Q_s): $\Delta Q_C = Q_g - Q_s$ (1)

The quality gap the client/customer requires to overcome can be defined as follows: $\Delta QCUS = Qe - Qp$ (2)

Qg can differs from Qe due to the different perspectives and goals by Local Authority and customer: $\Delta QG\text{-}E = Qg - Qe$ (3)

Qg depends on the supplier's market power (image, financial availability, expertise) and by informative asymmetry between buyer and supplier.

Qs might differ from Qp due to a lack of effective communication (ΔQCOM) between the supplier and the client/customers, as already stated by PZB model:

$$\Delta QCOM = Qs - Qp \quad (4)$$

Q_s is the total quality level as a weighted mean between technical and functional qualities. Q_g is defined according to the legislative standards.

5 PERFORMANCE INDICATORS

The International Water Association (Alegre et al., 2006) has defined a set of performance indicators (PIs) able to provide information about efficiency and effectiveness of service management and provision. The model of value was used as a filter to split the identified subset of PIs into two classes: the first referred to the interaction customer/supplier (PI1), the second referred to technical and organizational supplier efficiency (PI2). PI1 and PI2 represent respectively the customer's state variables and the supplier's technical-organisational parameters (Table. 2).

In phase 1, the value for customer depends on the existence of information and on the capability by front-end people to quickly reply to customers answers. In phase 2, the value depends on the amount of sacrifices required in terms of organizational time (front-end office interactions) and "technical" time (time for supply activation). Finally, in phase 3, the value for customer depends on the quality of water supplied, the precision of meter reading (efficiency in billing), the time for interacting with the supplier (wait time on the phone, on window, time between complaints forward and reply by the supplier, technical and organizational capabilities of supplier to guarantee continuity of service, to notice users when an interruption occurs (planned maintenance), to play a proper role when a failure occurs (unplanned maintenance), that can bring any kind of risk for customer wellness in general.

Table 2. PERFORMANCE INDICATORS PI1 AND PI2

Codex	Description PI1	Codex	Description PI2
G1gen	Time to have an estimate made	**G8gen**	Routine use information systems
T3acq	Per capita supplied water	**G9gen**	Automation and control
G3gen	Waiting time in office	**G22acq**	Quality system adoption
G4gen	Waiting time at telephone	**G23acq**	Internal laboratory existence
G11acq	Connection time	**G24acq**	Mains inspections
G12acq	Time for water service activation	**G25acq**	Leakage detection
G5gen	Service complaints	**G26acq**	Pumping inspection
G6gen	Response to written complaints	**G27acq**	Storage tank cleaning
G7gen	Time for emergency operations	**G28acq**	System flowmeters
G13acq	Customer reading efficiency	**G29acq**	Pressure meters
G14acq	In advance time communication planned failures	**G30acq**	Network inspections
G15acq	Time for planned service connection failures	**G31acq**	Replaced/renewed mains
G16acq	Interruptions for planned failures	**G32acq**	Mains/service connection failures
G18acq	Service connection recurring failures	**G33acq**	Pumps failures
G19acq	Duration service connection non-planned failures	**G34acq**	Power failures
G20acq	Service connection non-planned failures	**T1acq**	Population coverage
G21acq	Emergency service availability	**T4acq**	Total losses
T17acq	Quality of supplied water	**T5acq**	Non-revenue water by volume
[-]	Billing payment alternatives	**T6acq**	Total cost coverage ratio
[-]	Billing efficiency	**T14acq**	Water storage capacity
[-]	Pressure of supply adequacy	**T16acq**	Water quality at sources
[-]	Guaranteed standards scheme existence	**T22acq**	Water sources alternatives

6 AGENT-BASED SIMULATION

6.1 General concepts

The general structure of an agent-based model follows the ERA scheme (Environment-Rules-Agents) (Terna, 2000), that considers four different layers to build the model and the agents: 1) the environment in which the elements interact (agents and rules); 2) the agents built as element of one or more classes; 3) the mechanisms through which agents define their behaviour (*Rule Master*); 4) the definition of the rules (*Rule Maker*).

The ERA model can be single-agent based or multi-agent based (MAS). The interactions in a MAS can be cooperative, conflicting or neutral. A MAS at an high aggregate level can show characteristics that do not belong to any agent part of it; this kind of synergy allow moreover to study MAS like a greater and unique agent.

Wooldridge and Jennings (1995) classified agents into three groups, according to cognitive nature: *reactive*, *deliberative* and *hybrid*. Reactive agents can only have pre-defined behaviours to certain inputs, without taking into account where the whole system is going. Deliberative agents are thinking agents that elaborate complex plans of action combining many simple behavioural schemes. They take into account their internal state and try to predict effects of their actions before they act.

A MAS acts as a dynamic multi-criteria system for decision making. By coding each decisional scenario, choosing an adequate cognitive nature for agents, modelling environment, and selecting proper coordination mechanisms, it is possible to unify different perspectives and try to reach an optimal solution for the whole system and not only for one or some parts. Such advantages allow to consider both the different aspects of service evaluation (functional and technical service quality) and the different perspectives of customers and Local Authorities. Moreover, it allows the non-conflicting co-existence and mix of different value models in an aggregated measure.

6.2 Building an agent-based model: a theoretical case

The model has been built as a structured environment, where the autonomous agents interact, able to transfer information on performances (from supplier to customers) and complaints (from customer to supplier).

The classes of autonomous agents are two. The class of supplier is always made of a single element. The number of subclasses of users depends on the local conditions. Each class of agents is characterized also by cognitive nature, that influences the behaviour, and state variables (tab. 3).

Table 3. COGNITIVE NATURE AND BEHAVIOUR OF THE AGENTS

Classes	Cognitive nature	Objectives	Algorithm of action
Customer	Hybrid	Complaint if something "does not work"	1. Control of state variable 2. Classification of contingency (gradually) 3. Update of internal state function 4. Complaint or not
Supplier	Hybrid	Reply to complaints "technically and organizationally" to minimize them and maximize customer satisfaction	1. Control of budget internal state function 2. Control of user's complaints 3. Classification of contingency 4. Resolving conflict situations (corrective actions according to the priority rate k) 5. Implementation of corrective actions through budget consumption 6. Update of budget internal state function

Customer agent is hybrid because reacts to any misfit occurs during service provision (reactive). However, it gradually recognizes any kind of contingency and fits his level of satisfaction to real conditions of supply (deliberative). Supplier agent is hybrid because knows the corrective actions necessary to improve the level of service (reactive), but elaborates plans of action by classifying contingencies, through a priority rate (deliberative).

A simplified water system scheme has been hypothesised considering two customers, representative of two conditions of supply characterized by different elevations bringing to different service availability. Only interactions during the third phase has been simulated. ATO is considered an observer.

Each cycle "j" of the simulation starts with the monitoring the customer satisfaction; it depends on the value assumed by the state variables. Each customer compares its internal state, component by component (CS_i), with the standard value fixed by regulations ($CSobj_i$ = minimal level of service). From this comparison, customer computes a gap, representative of their lack of satisfaction for each variable: $\Delta(CS_i)_j \equiv \left(\frac{CS_i - CSobj_i}{CSobj_i}\right)_j$

All the gaps are then compared to the values reported in table 4.

At the end of each cycle "j" users update their global internal satisfaction state taking into account the value of whole $\Delta(CS_i)_j$. The global customer satisfaction at cycle "j" is at maximum level only if $\Delta(CS_i)_j \geq \text{Bound}_1 \ \forall i$.

Table 4. HYPOTHESISED BOUND THRESHOLD AND COMPLAINT RULES

Threshold	Rule
No violation $CSobj_i$	$-0.2 \leq \Delta(CS_i)_j \leq 0 \Rightarrow$ customers forward an informal complaint
20% violation $CSobj_i$	$-0.4 \leq \Delta(CS_i)_j \leq 0.2 \Rightarrow$ customers communicate the misfit suffered and wait for a reply; if at the following cycle the service does not recover at the optimal value, customers will forward a formal complaint
40% violation $CSobj_i$	If $\Delta(CS_i)_j \leq -0.4 \Rightarrow$ customers forward a written complaint about variable CS_i

The supplier, when receives a complaint, controls its financial wellness in terms of budget. If its overall value is greater then zero it decides to act or not depending on priority rate "k". Priority rate k is computed by the *Rule Master*, a simple neural network, as follows:

1. at time "j" any type of misfit is additively computed;
2. trial performance indicators are built on these measurements;

3. trial PI are compared to complaints register of the supplier;
4. each state variable has high priority if both correlated failures and complaints are high;
5. the supplier controls costs and times of corrective actions, from higher to lower priority and decides to act optimizing its budget function;
6. the Rule maker updates the state variables in terms of priority rates (connections weights).

At the beginning of each simulation the components of customer satisfaction (CSi)j are equal to CS_{obj} fixed by contract for each customer. For each technical-organizational variables (TO_k), the probability of failure is computed through a negative exponential distribution formula: $f(t)=e^{-\lambda t}$. For each TO_k a random number is created and a failure occurs if $rnd \in [f(t);1]$, while no failure occurs in $[0; f(t)[$. The linear dependence between CS_i and TO_k have been calculated in terms of impacts of the correspondent performance indicators (PI1/PI2) (tab. 5).

Table 5. RELATIONSHIPS AMONG DEPENDENT AND INDEPENDENT VARIABLES

$CS_1 = f(TO_3;TO_1)$	$CS_2 = f(TO_3)$
$CS_3 == f(TO_3)$	$CS_4 = f(TO_3;TO_{14})$
$CS_5 = f(TO_3;TO_1)$	$CS_6 = f(TO_3)$
$CS_7 = f(TO_3)$	$CS_8 = f(TO_1;TO_2;TO_3)$
$CS_9 = f(TO_1;TO_2)$	$CS_{10} = f(TO_3)$
$CS_{11} = f(TO_3)$	$CS_{12} = f(TO_{14};TO_5;TO_{22};TO_{23};TO_6)$
$CS_{13} = f(TO_{14};TO_{18};TO_{22})$	$CS_{14} = f(TO_{12};TO_{13})$
$CS_{15} = f(TO_{14};TO_{15};TO_{16};TO_{17};TO_9;TO_{10};TO_{11})$	$CS_{16} = f(TO_1;TO_2;TO_3)$
$CS_{17} = f(TO_{19};TO_{20};TO_3;TO_4)$	$CS_{18} = f(TO_1)$
$CS_{19} = f(TO_1;TO_2;TO_7;TO_8)$	$CS_{20} = f(TO_{21})$
$CS_{21} = f(TO_8;TO_{18};TO_{12};TO_{13})$	$CS_{22} = f(TO_7;TO_{18};TO_{12};TO_{13})$

The simulated value CS_i is optimal ($CSobj_i$) if the related PI2 variables are in their optimal ranges. Otherwise $CS_i = (CS_i)_{j-1}[1-\Delta(CS_i)]$ with $\Delta(CS_i)$ equal to the percentage loss on CS_i. In table 6 how the independent variables changes with respect to the fixed standard is reported.

The scheme above described has been simulated on a daily base for one year. The output of the model and the parameters that can change along the simulation period are reported in figure 5. The dependent variable change with the independent variables according to the relationships fixed inTable 6.

The simulation presents a period of transition, when the supplier try to solve immediately and in any case to minimise all the problems, a period of stabilisation when the customer satisfaction is at an intermediate level between the minimum and the maximum of the previous period, and a final decrement of the customer satisfaction due to the lack of financial resources available. Even if the applied rules can be considered reasonable in the analysed service, they should be carefully evaluated when applied to a real system.

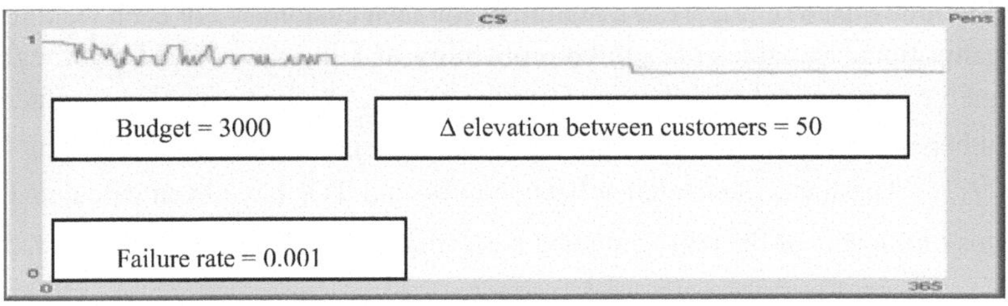

Figure 5. EXAMPLE OF OUTPUT OF THE SIMULATION MODEL IMPLEMENTED.

Table 6. CHANGES OF INDEPENDENT VARIABLES

Simulated independent variable (PI2)	Optimal value	Range to stay optimal	Loss (%)	Updated value
TO_k	$\overline{TO_k}$	$]TO_1^0;TO_1^1]$	Δ_k	$(\overline{TO_k})_j = (\overline{TO_k})_j(1-\Delta_k)$

7 CONCLUSIONS

In the present paper a methodology for evaluating the supplier performance in water service provision is presented. The evaluation of the water services requires to take into account both the technical parameters (resulting from the interaction buyer/supplier) and the customer satisfaction (resulting from the interactions supplier/customer).

The basic concept is that the quality of the service depends on the quality of the supplier's performances. Such performances can be determined in terms of technical and functional quality as linked with the customer satisfaction. If the customer satisfaction can be expressed in terms of prevalent value model for the users, the quality of the interaction between customer and supplier is function of the quality of the service provided.

The interaction in the case-study has been simulated through and agent-based model. The model shows an interesting bivalence. It allows ATO to select from different potential suppliers the one having the best investment plan to maximize customer satisfaction (forecast simulation), but it allows also to monitoring the supplier and assessing its performances (customer satisfaction) with reference to contingencies and corrective actions implemented.

Several limitations must be recognised. The results depend on the scheme implemented in terms of technical characteristics and interaction rules applied. In particular, the rules are very stringent as both the supplier aims at satisfying the complaints and the customer complaints immediately trusting in supplier's ability to improve the system. It is not considered the case, indeed frequent, that the customer does not complain at all. The PI reference values have been defined according to the minimum level of service.

The analysis of a real case is ongoing in order to test the model applicability. Moreover, the customer satisfaction simulated through the model should be compared with the result of a field survey aiming at verify whether it actually matches the customer perceptions.

8 REFERENCES

Alegre, H. et al., 2006. Performance indicators for water supply services, International Water Association Publication. London (UK).

Ancarani, A., Capaldo, G., 2001. Management of standardised public services: a comprehensive approach to quality assessment. Managing Service Quality 11 (5), 331-341.

Bolton R. 1998. A dynamic model of the duration of the customer's relationship with a continuous service provider: the role of satisfaction, Marketing Science 17 (1), 45-65.

Bolton, R., Drew, J., 1991. A multistage model of customer assessment of service quality and value, The Journal of Consumer Research 17 (4), 375-384.

Boyne, G.A., 2003. Sources of public service improvement: a critical review and research agenda. Journal of Public Administration Research and Theory 13 (3), 367-394.

Churchill, G.A., Surprenant, C., 1982. An investigation into the determinants of customer satisfaction. Journal of Marketing Research 19 (November), 491-504.

Grönroos C., 1982. An Applied Service Marketing Theory. European Journal of Marketing 16 (7), 30-41.

Grönroos, C., 1984. A service quality model and its marketing implications, European Journal of Marketing 18 (4), 36-44.

Oliver, R.L., 1977. Effect of expectation and disconfirmation on post-expense product evaluations: an alternative interpretation. Journal of Applied Psychology 62 (4), 480-486.

Parasuraman A., Zeithaml V.A., Berry L.L. 1985. A conceptual model of service quality and its implications for future research, Journal of Marketing 49 (4), 41-50.

Parasuraman, A., Zeithaml, V.A., Berry, L.L., 1991. Refinement and reassessment of the SERVQUAL scale. Journal of Retailing 67 (4), 420-450.

Terna, P., Boero, R., Morini, M., Sonnessa, M., 2006. Models for complexity, (in Italian), Il Mulino, Milano.

Woodall, T., 2003. Conceptualising vValue for the customer: an attributional, structural and dispositional analysis, Academy of Marketing Science 12, 1-42.

Wooldridge, M., Jennings, N.R., 1995. Intelligent agents: theory and practice, Knowledge Engineering Review 10 (2), 115-152.

Decision support tools based on performance indicators

PI and Municipal Solid Waste service quality evaluation: experiences, needs and perspectives

A. Ancarani, A. Santamaria and F.G.A. Vagliasindi

Department of Civil and Environmental Engineering, University of Catania, aancaran@dica.unict.it, viale A. Doria 6, 95125 Catania Italy, Tlf +39 0957382715 Fax: +39 7382748

Keywords: Decision support system, municipal solid waste service, holistic solid waste management

Edited by Enrique Cabrera Jr and Miguel Ángel Pardo. ISBN: 9781843391913.
Published by IWA Publishing, London, UK.

ABSTRACT

Performance Indicators (PI) may be used to measure the performance assessment of urban infrastructure services, and in particular of integrated solid waste services. Based on recent studies performed to identify appropriate sets of performance indicators for the solid waste sector, in order to implement an effective performance evaluation system, a Decision Support System (DSS), based on PI and different performance measures, is here proposed, aiming at identifying sustainable solutions in quality assessment of integrated Municipal Solid Waste (MSW) management. The DSS could represent a tool too assess and to improve the performance of provided services, taking into account aims and objectives of different stakeholders (service providers, regulators and proactive customers).

1 INTRODUCTION

Performance assessment of urban infrastructure services, aiming at rating the global efficiency in integrated and quantifiable terms, is becoming a major issue worldwide, and particularly in Europe.

The role of performance indicators is becoming more important in order to verify the quality of public services in terms of efficiency, effectiveness, economy. The PI may be used to measure progress towards achieving corporate objectives and targets, promote accountability to the public and other stakeholders, to study performance development for internal improvement, compare performance to identify opportunities for improvement in comparison with other organisations (benchmarking) or promote service improvement by publishing service levels. PI are also very useful to measure improvement after effecting actions, preferably after using PI to identify problems or improvement areas.

This is particularly significant in the solid waste sector, which is characterized by fast technological innovation, high and frequent interaction between public administration and private companies, and an active role of citizens. Specifically, in the solid waste service performance indicators could take into account various aspects for making clear to the public Authority the most relevant and critical issues of the service provided. Moreover, they should suggest also improvement measures.

The study and the analysis of performance measures in integrated solid waste service management based on PI had a major development just recently (Cost Action C18, 2004-2008).

Due to the particular characteristics of the municipal solid waste sector, in terms of both complexity and required participation of the citizens to the service provision, it was clear that there was a great need for improving the use of performance indicators both in itself, as a policy tool, and in conjunction with appropriate methods and models for performance assessment able to support effective and sustainable policies in Municipal Solid Waste (MSW) sector.

The objective of the present work is to develop a Decision Support System (DSS) based on performance indicators and performance measures as a tool for supporting stakeholder involved in the management of municipal solid waste systems. Specifically, in order to overcome numerous and recurrent management difficulties often due to an overestimation of the importance of technical issues, the DSS has been envisioned and developed on the basis of a holistic approach which includes, in a comprehensive and multifaceted approach, all essential fields and dimensions of the waste management operation (technical, social, economical, environmental, etc.). The general framework and features of the DSS are presented here.

3 PI USE IN MSW SERVICE: EXPERIENCES

Current EU waste policy is based on a concept known as the waste hierarchy. This means that, ideally, waste should be prevented and what cannot be prevented should be re-used, recycled and recovered as much as feasible, with landfill being used as little as possible. Landfill is the worst option for the environment as it signifies a loss of resources and could turn into a future environmental liability. The waste hierarchy should not be seen as a hard-and-fast rule, particularly since different waste treatment methods can have different environmental impacts. However, the aim of moving towards a recycling and recovery society means moving up the hierarchy, away from landfill.

At present in the EU municipal waste is disposed of through landfill (49%), incineration (18%), recycling and composting (33%). In the new Member States, where major efforts and investments have been made to align with the EU position, the situation is evolving rapidly but still dominated by landfill. There are wide discrepancies between Member States, ranging from those which recycle least (90% landfill, 10% recycling and energy recovery) to those which are more environmentally friendly (10% landfill, 25% energy recovery and 65% recycling) (Source: Eurostat http://epp.eurostat.ec.europa.eu).

Despite the strategies for integrated waste management (i.e., the hierarchy: waste reduction, recycling and reuse, energy recovery from waste, and landfill disposal), issues related to solid waste management became in recent year more significant due to numerous and recurrent emergencies experienced in assuring appropriate treatment and disposal solutions.

The urban solid waste services, due to its direct impact on the urban environment and health, to the resulting costs on the citizens, and to the competing private companies, still needs significant improvement with regard to efficiency, effectiveness, and resources (human, financial and structural) allocation.

Indicators are often used in conjunction with other types of information (system characteristics, environmental factors, etc.) to present the complexity of the system in a simple and direct way, namely performance measures.

In the solid waste sectors PI may be used to measure progress towards achieving corporate objectives and targets, promote accountability to the public and other stakeholders, to study performance development for internal improvement, compare performance to identify opportunities for improvement in comparison with other organisations (benchmarking) or promote service improvement by publishing service levels. PI are also very useful to measure improvement after effecting actions, preferably after using PI to identify problems or improvement areas.

In Europe, the study and the analysis of performance measures in integrated solid waste service management had a major development just recently (Cost Action C18, 2004-2008). The main objective of the Action C18 has been to increase the knowledge and to promote the use of effective, scientifically robust and well devised methodologies for decision-making based on the use of performance indicators (PI) for urban infrastructure services, able to attract utilities to use them as routine management tools. Based on the use of the existing PI systems for water supply and wastewater services, a specific objective is the establishment of effective, scientifically robust and well devised methodologies for decision-making. Another specific objective is the definition of research priorities regarding the integration of performance measures into other engineering tools, as referred in the previous section. The third specific objective is to promote the use of performance indicators as a routine management tool in other areas of urban infrastructures, using the sector of solid waste as case study.

In Italy, specifically efforts were made to identify the role of performance indicators in the integrated waste management sector to evaluate and verify the quality of services, and to identify specific sets of indicators (APAT, 2003 and 2004; Ancarani et al., 2004; Ancarani et al., 2005).

A set of 84 indicators was developed at DICA (Department of Civil and Environmental Engineering, University of Catania), including:

- context indicators: information of the characteristics of the served area and typology of solid waste;
- service level required: definition of technological equipment required and frequency of service to be provided;

- service quality indicators: financial and economical management, customer relationship management, applied fares, treatment plants characteristics.

Technical and organisational indicators should include different aspects of the municipal solid waste management like characteristics of the area under investigation and assets conditions and management. The indicators proposed aim at evaluating the efficiency and effectiveness of the service provided and the impact on the environment. Moreover, some indicators have been added in order to support managers in planning and verifying the adequacy of the investments.

This set was applied to study the solid waste management in a city of 350.000 people, in order to evaluate and compare services provided by public and private companies operating in different neighbourhoods of the city.

Differently, the Italian national agency for environmental protection and technical services (APAT, 2003) developed a set of indicators base on the DIPSIR (drivers, impacts, pressures, state, and responses), an evolution of the model PSR developed by OCSE, which included indicators on social-economics (3), urban waste production and management (18), and special waste production and management (21).

APAT (2004) has then defined a set of 32 environmental indicators related to solid waste management in urban areas which were applied in different cities. Specifically, seven domains of general satisfaction (not individual satisfaction) were proposed, including: Waste minimization and reuse; Technical standards in collection and urban hygiene services; Planning and energy; Treatment and disposal of waste; Companies strategies for the management of urban waste Actions to modify waste production, and recycling directed to the citizens; Compliance with regulations.

3 NEEDS AND PERSPECTIVES

Based on the experiences gained, the following could be highlighted:

- developed indicators often do not really constitute a coherent set of performance indicators and there is no benchmarking between the different local authorities;
- several indicators are valuable to analyze the performance at national and regional level, other directly refer to the performance of the local companies entrusted with solid waste services provision;
- further efforts are necessary in order to develop a standard set of performance indicators able to support policy makers at different level (national, regional, local);

- the definition, monitoring, and calculation of indicators are not sufficient for implementing an effective performance evaluation system; an assessment methodology able to measure the quality level attained and the impact on the socio-economic systems is necessary;
- for the assessment of waste management schemes a holistic approach should be applied including all essential fields and dimensions of the waste management operation;
- it is necessary to develop appropriate strategies and policies, supported by adequate tools, for performance assessment of MSW management systems.

3.1 DSS for sustainable MSW management solutions

Based on the above considerations, it is evident the usefulness of a tool to make, based on performance indicators and performance measures, a performance assessment of solid waste services and capable of identifying sustainable solutions. The general framework of this tool, envisioned as a DSS, is presented in Figure 1.

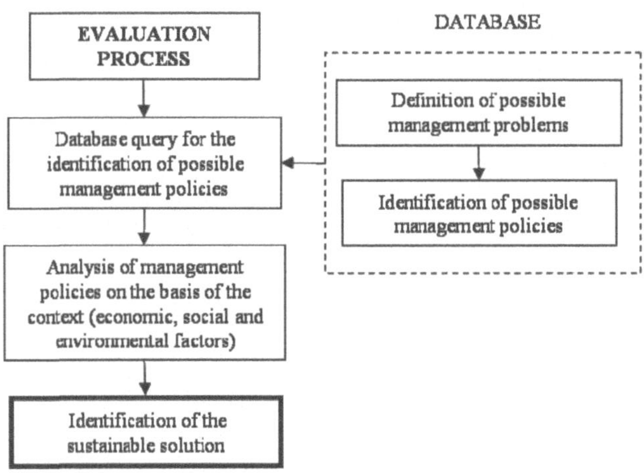

Figure 1 .DSS STRUCTURE.

Specifically, the DSS includes a methodology for the monitoring and assessing the performance of solid waste services which involve the whole assessment process from the analysis of the information and strategies to the identification of priorities for improving the quality level of the service.

The evaluation process identify spatial characteristics of the MSW service and, in particular, areas under critical service conditions for each management problem. On the basis of these results, following a database query by the

Autority (Autorità d'Ambito), who supervises the activities of the company entrusted with the services provision in a given Optimal Territorial Unit (ATO, ambito territoriale ottimale) and monitors the performance and verifies the customer satisfaction, for the identification of possible management policies to attain better quality of the service.

The database is structured to propose management policies in order to solve management problems independently from specific territorial characteristics. Then, based on economic, social and environmental context, the sustainable solution is identified.

3.2 Methodological framework for evaluating quality of MSW service

The methodology has been based on the definition of service quality as a compendium of functional and technical quality of service (Grönroos, 1990). From one side this means that the service quality is proportional to the quality of the performances of the service provider and to his efficiency and effectiveness. On the other side, it is possible to interface this definition with the customer satisfaction in terms of perceived quality. In literature seems clear that customers and service providers interact in such a manner that judgement on the quality is influenced both by the service delivered (i.e., technical quality) and how it is delivered (i.e., functional quality).

Starting from the methodology for the evaluation of the quality of services (Ancarani and Capaldo, 2001), the proposed methodological framework for performance assessment of the provider of the solid waste service has been developed into ten main steps (Figure 2).

The first step regards the analysis of regulations for the identification of minimum standard of service that must be guaranteed. The study of stakeholders involved in the management and supply of the service, and the study of informational flows between them allow also to identify and define institutional roles and objectives of the stakeholders of the service-system. The study of service-system and all processes referred to have also been considered.

The output of this first step introduces to the evaluation process and allows, in the second step, for the definition of a set of performance indicators. This set is representative of the level of service that provider must guarantee to different stakeholders of service-system and it is helpful to evaluate the service performance in a simple and direct way. The evaluation of the service performances can be carried out either weighting each indicator or verifying the compatibility with existing standards.

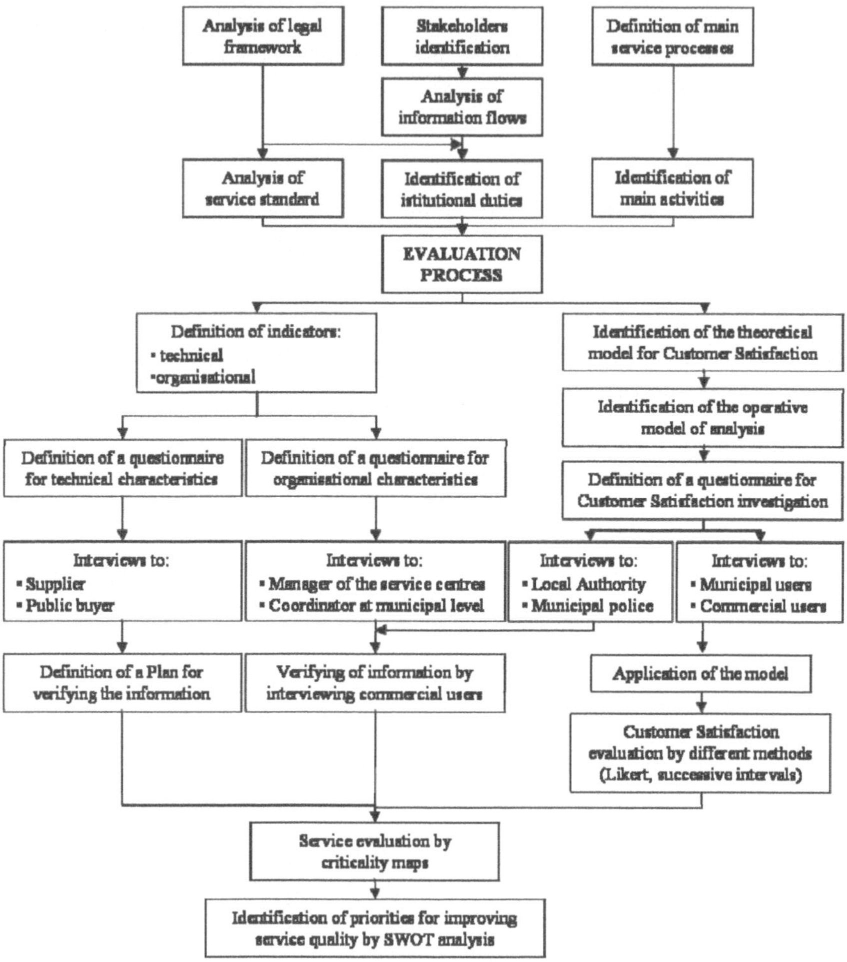

Figure 2. METHODOLOGICAL FRAMEWORK FOR EVALUATING QUALITY OF MSW SERVICE.

The developed methodology also involves the preparation of specific questionnaires, to obtain the necessary information for the constant monitoring of the service provided and for the calculation of the indicators. The questionnaires have been designed and administered to different stakeholders (customers, service provider, public authorities).

The next step involves the definition of a protocol to verify the technical and structural characteristics and to acquire information from selected customers to verify operational aspects.

Another step involves the identification of the model for the description of the service quality based on customer satisfaction. The theoretical GAP model (Parasuraman et al., 1985; 1991) and the operative scheme SERVPERF (Cronin and Taylor, 1992) are used. A specific questionnaire can be defined and it has to be administered to public authorities and customers in order to collect the information for the application of the model.

Eventually, the level of customer satisfaction evaluation can be carried out by psicometric analysis using Likert method and the successive intervals method (Franceschini, 2001).

A comprehensive evaluation is made by preparing maps showing areas under critical service conditions. These tools can provide useful information to the Authority and to the provider to identify priorities and to implement specific actions to improve non-adequate service conditions.

All the information collected are used to identify, based on SWOT analysis, the necessary actions to attain better quality of the service. The SWOT analysis provides information helpful in the identification of the priorities for improving service quality. SWOT is based on scan of the internal (Strenghts and Weaknesses) and external (Opportunities and Threats) environments.

The methodology has been implemented and applied to an Optimal Territorial Unit in Italy in order to verify and refine the developed work.

4 CONCLUSIONS

Based on recent studies on PI and on the analysis of performance measures in integrated solid waste service management, a DSS was proposed aiming at identifying sustainable solutions in quality assessment of integrated MSW systems. The proposed DSS includes a methodology for assessing the level of service provided which is integrated with the definition of appropriate reference standards and performance indicators able to describe the level of MSW service provision. The DSS has been envisioned and developed on the basis of a holistic approach which includes, in a comprehensive and multifaceted approach, all essential fields and dimensions of the waste management operation (technical, social, economical, environmental, etc.).

The proposed tool could help stakeholders and, specifically, managers of public organizations and regulators, in identifying critical and relevant issues and appropriate measures to improve the system performances.

6 REFERENCES

Ancarani, A., Capaldo, G., (2001), Management of standardised public services: a comprehensive approach to quality assessment, Managing Service Quality, vol. 11 (3).

Ancarani, A., Vagliasindi, F.G.A., Leotta, S., (2004), Definizione di indicatori di qualità del servizio di raccolta e smaltimento dei rifiuti solidi urbani. (in Italian), Atti del Simposio Internazionale di Ingegneria Sanitaria Ambientale, Taormina 23-26 giugno.

Ancarani A., Santamaria A. and Vagliasindi F.G.A., (2005). Identification of a set of indicators for analysing performances in solid waste sector, *Proceedings of the 9th International Conference on Environmental Science and Technology*, Rhodes island, Greece, 1-3 September 2005, vol. (A), pp. 84-92.

APAT, ONR, (2003), Rapporto annuale rifiuti, (in Italian), Estratto.

APAT, ONR, (2004), Rapporto annuale rifiuti, (in Italian), Estratto.

Cronin J. and Taylor S. (1992) Measuring Service Quality: A Reexamination and Extension. Journal of Marketing, 56 (July), pp. 55-66.

Franceschini (2001) From products to services. The new approaches to quality measurement. UTET, (in Italian).

Grőnroos, C (1990) Service Management and Marketing-Managing the Moments of Truth in Service Competition (Massachusetts, Maxwell MacMillan).

Parasuraman A., Zeithaml V., and Berry L. (1985), A Conceptual Model of Service Quality and Its Implications for Future Research. Journal of Marketing, 49 (Fall), 41-50.

Parasuraman,A.;Berry,Leonard L.;Zeithaml,Valarie A., (1991) "Refinement and Reassessment of the SERVQUAL Scale", Journal of Retailing, 1991, 67, 4, 420-450.

Performance Indicators (PIs) as a Control Mechanism for the Effective Use of Donor Funds

K. Rohrhofer[1], K. Leroch[2] and F.M. Cate[3]

[1] *OEKOREAL, Research and Realisation, Operation and Maintenance of Water and Wastewater Systems; Email: karl@rohrhofer.at Carl Reichert-Gasse 28, 1170 Vienna, Austria Tel.: +4314805010 Fax.: +4314805010-99*
[2] *GWCC - General Water Consult Corporation, Vienna, Austria; email: Klaus.Leroch@gwcc.at*
[3] *INTERIVAL - Integrated River Basin and Agricultural Land Use Management; Vienna, Austria; email: frederick.cate@interival.at*

Keywords: Performance indicators, donor funds, asset management, water and wastewater, benchmarking

Edited by Enrique Cabrera Jr and Miguel Ángel Pardo. ISBN: 9781843391913.
Published by IWA Publishing, London, UK.

ABSTRACT

The use of Performance Indicators for the evaluation of existing assets and for a sustainable and flexible allocation of donor funds within a multi stakeholder approach in "rich" and "non-rich" countries is being discussed in this paper.

1 INTRODUCTION

This paper is based on the experience of the authors of more than 10 years in "non-rich" and "rich" countries, where donor funds were and are being spent to assist water and wastewater utilities when constructing or rehabilitating their assets (Rohrhofer 2002). It was identified that in many cases donor organisations do not put sufficient stress on checking whether the assets for water supply and wastewater systems, which were funded by them, are subsequently used correctly and maintained properly after finalization of the implementation. A sort of "after sale service" including handover to previously trained key staff is often severely neglected (Rohrhofer et al. 2008, submitted).

In "non-rich countries", in some cases it also seems that different donor organisations do not exchange information among each other about their intentions to provide funds for certain projects. In addition, the decisions about which physical assets to be constructed/rehabilitated and about the amount of the money to be made available is sometimes made without serious investigations on site, and often even without assessing the condition of already existing assets.

The authors firmly believe that such investigations should always be carried out together with the local experts who are operating the existing systems, as thereby not only the true needs can be filtered out from "romantic development ideas", but also the local level of expertise (human resources, soft assets) can be evaluated at the same time.

Such shortcomings could be avoided by establishing and using specific Performance Indicators (PIs) which refer to the existing condition of the assets and the present way of using available resources and operating/maintaining the assets. A selection of PIs which also include medium- and long-term perspectives from both the donor's and the beneficiary's point of view are the missing tools to safeguard a control mechanism for the effective use of the donor funds.

PIs as a control mechanism for the effective use of donor funds

1.1 State of the Art

Definition: *A performance indicator (PI) is a quantitative measure of a particular aspect of the undertaking's performance or standard of service. It assists in the monitoring and evaluation of the efficiency and effectiveness of the undertaking, thus simplifying an otherwise complex evaluation (*Alegre *et al.* 2000).

Worldwide many experts for water supply and wastewater infrastructure know about PIs, which shall be used e.g. to measure the efficiency and effectiveness of a utility in achieving its objectives (Cabrera 2006). In 2003 Matos et al. defined a comprehensive set of Performance Indicators for Wastewater Services. Performance Indicators for Water Supply Services were defined, e.g. by Alegre et al. (2000) and Alegre et al. (2006). These publications reflect a generally accepted definition of PIs where a set of values is defined and used to measure the results of an implemented action and the quality of its services against a set target. A PI system can be considered as a key assessment tool for the achievement of targets, by applying a coherent set of indicators. PIs must be well interpretable and clearly defined by auxiliary context information. Specific targets for each indicator must be established, routinely monitored, tracked and adjusted when needed (ISO TC 224).

1.2 ISO Series 24500 (ISO 24510, 24511, 24512)

The ISO standard series 24500 represents a framework for the definition and measurement of services relating to water supply and wastewater systems. This Series, elaborated by ISO TC 224, provides guidelines for the assessment of services related to the management of water supply and wastewater systems. In addition, it should aid all stakeholders to continuously improve the level of service provided to the user / customer at optimal cost. The ISO Series 24500 is for voluntary application and does not include targets, limits or thresholds for the suggested PIs. It can be used by all stakeholders, regardless of their legal status.

2 USING PIS

2.1 General

The systematic use of performance indicators, when preceded by adequate planning and implementation and followed by an in-depth comparative analysis of the results and by the improvements derived there from, can have a major leverage effect on the increase of effectiveness, efficiency and productivity of an organization (Alegre, 2002).

When talking about the water supply and wastewater sector, a number of possibilities for the application of PIs can be identified:

1. Assessment of existing assets and the efficiency of its operation (=Asset Management) (Rohrhofer et al. 2007)
2. Planning of infrastructure measures and supervision of its implementation
3. Parameters for Decision Support Systems
4. Benchmarking

The focus of this paper is on the first 3 options when analysing whether PIs could be used for supporting the decisions to be made by donor organisations for an optimal use of donor funds for projects in both, "rich" and "non-rich" countries. PIs in the context of benchmarking are dealt with in a separate section of this paper.

For any of the above objectives, the use of specific PIs will most efficiently safeguard the correct and efficient use of donor funds only when all project partners are included. Especially in "non-rich" countries this means that, additionally to the donor and the beneficiary, not only ministries and other high ranking organisations shall be involved, but also local partners like communal and regional councils, and local waterworks and associations. There should always be mutual consent between all stakeholders about the objectives of the project to be implemented. This is usually not such a big issue in most "rich" countries, where infrastructure investments are approached within a multi stakeholder approach, but especially in "non-rich" countries, where the initiative and the funding for the implementation of infrastructure projects comes in many cases from abroad, either from national or international donor organizations.

All stakeholders should already be involved in the first planning phases of the intended project. However, the greatest input must be provided at least from two sides, the donor side and the beneficiary side, in order to allow for a flexible reallocation of donor funds within the project, if necessary, based on the evaluation of interim results by means of PIs.

2.2 Step by step approach

To reach the above goal, several steps will have to be followed before, during, and after the implementation phase:

Before the project implementation:

- Ensure that development goals correspond realistically to available donor funds.
- Assess the condition of existing assets and their functionality. Compile PIs which allow for the assessment of the achievement of medium- and long-term objectives. This should, if applicable, include the maintenance of the new/rehabilitated system and the training of the operation personnel.
- Setup of a realistic time target
- Compile specific PIs
- Clearly define and divide the duties of donor, consultant, beneficiary and local authorities.
- Provide clear and simple templates for reporting procedures.
- Compile PIs which allow for a monitoring of the implementation success during and after the implementation phase

During the implementation phase:

- Using PIs, check if the chosen way of implementation is suitable to reach the objectives. Allow for adjustment of objectives or reallocation of donor funds if necessary.
- Release donor funds in a step by step procedure only after achievement of interim objectives.

After the implementation phase:

- Assessment of achievement of the previously set targets by use of PIs.
- Release donor funds for potential further project phases only after achievement of interim objectives.

2.3 Use of PIs as prerequisites for granting donor funds

The granting of public as well as private money for the purpose of investment into the infrastructure, or into improved operational procedures, both in "rich" and "non-rich" countries, shall be bound to the fulfilment of targets defined by means of PIs. The donor and the beneficiary shall together work out key PIs which then shall be used to evaluate even during the implementation whether the donor funds are used in the foreseen way and if it is likely that the intended impact can be reached.

Also the granting of soft loans should be linked to the fulfilment of targets defined by means of PIs.

In further project phases the progress of the project can be constantly monitored. In case of unsatisfactory performance, donor funds can be rededicated or strategies of implementation can be altered. This should not only allow the donors to control the progress of a project by evaluating the work of consultants and the cooperation of local authorities, but also by controlling their own agenda and long-term strategies.

Therefore PI should be used by the donor organisations not only to control the use of its funds by the recipients, but also for a sustainable self-control of the donor's use of public and private funds.

The implementation of this type of management tool requires the active participation and engagement of the respective managers in every stage of the process. A key factor for a successful application is that responsibilities must be clear and that the working group / steering committee is led by a senior expert.

2.4 Requirements for PIs

The selection of a PI system and of the specific PI's to adopt is another step of major importance. The use of standardized PI systems is recommended as a better alternative to ad-hoc systems developed expressly for a specific objective (Matos et al. 2003).

A sub-set of relevant PIs, which reflect the objectives of the project, and which also take into account the availability of data, should be selected either from the generic lists, published for example in the respective IWA reports for water supply and wastewater, or based on the ISO series 24500.

In any case, "Push-Pull" effects must be avoided when selecting a small number of PIs. It is thus important that the chosen PIs are independent from each other in a statistical sense, as especially negatively correlated PIs may lead to false conclusions and disagreement among the project partners, especially if the kind of correlation between the PIs is not obvious. This may occur where improvements in one area consequentially cause deterioration in performance in another area (Matos et al. 2003).

3 DATA AVAILABILITY AND QUALITY

The use of PIs requires input data with sufficient accuracy. Before the final decision is made about the set of PIs to be used, the data availability and quality must be checked. The use of inadequate and/or inaccurate data obviously can have serious consequences. The availability and quality of the data will affect not only the choice of individual PIs, but also the character and composition of the set of PIs proposed (see clause 2.2).

A confidence-grading scheme is necessary to quantify the reliability and accuracy of the available data. IWA (Matos et al. 2003, Alegre at al. 2000) recommends to use a confidence grading scheme which was developed in England and Wales by the Office of Water Services (OFWAT, 2002), and which is now being successfully applied in a number of different countries worldwide.

In any case, confidence grades should reflect the current status of the data, not the future status that it may be intended to achieve.

It has to be checked thoroughly whether using confidence grading as suggested by the IWA (e.g. Matos et al. 2003) is applicable for the respective project, or if rather a simple confidence grading procedure should be applied, simply qualifying data quality into "good", "average", and "bad".

When implementing a PI system with an initially unknown ability for data collection it might furthermore be reasonable to establish a priority ranking for the considered PIs. Table 1 shows an example of guidelines, for discussion when deciding for PI and the level of information Sjøvold (2005).

4 PIs IN THE CONTEXT OF BENCHMARKING

Benchmarking in the context of performance indicators means the comparison with others, doing a similar or comparable job. In a first step indicators are calculated, then the values are compared with former periods, other departments or other similar enterprises. For real comparisons a PI-system is needed, which is described exactly and which takes into account all relevant influence, e.g. the IWA-PI-system presented herein. The result of the comparison is a ranking (Matos et al. 2003).

While benchmarking by means of PIs may be an appropriate means of increasing the efficiency and effectiveness of well established water supply and wastewater undertakings, the authors of this paper do not believe that benchmarking is a primary priority, at least not for development projects.

Table 1. EXAMPLE DESCRIBING DIFFERENT AMBITIONS AND DIFFERENT USE OF PI SYSTEMS (3 LEVELS), SJØVOLD, F. (2005)

Level	No. of PI	Information	Questions to answer	Users (ranked)
Obligatory	e.g. 20 (high level)	Economy Environment / Health Physical PI	"Do we fulfil the basic requirements regarding cost" Regulation basic service level	Regulators Politicians Public Media Managers Operators
Voluntary 1	e.g. 50 (more detailed level)	Operation Quality of service Sustainability	"How is the level of service? Are we efficient?"	Managers Operators Regulators Politicians
Voluntary 2	e.g. 100 (detailed level)	Same as above with more detail and PI lnked to specific issues	"Where are the bottlenecks? Where are we doing well? How can we improve?"	Operators Managers (Politicians)

Nevertheless, we agree that a PI system should be independent of the state of development and internal organisational structure of an undertaking which is assessed by means of PIs. The possibility to use the same set of PIs on a long-term basis will not only enable the beneficiary to easily get involved into benchmarking procedures in future phases of a project, but it will also allow for a long-term monitoring of the performance of the new assets which again can help the donor to adopt its development strategies on a long-term basis.

5 SUMMARY

The use of PIs, which are to be established as described above, will allow public as well as private donor organisations to quantify the grade of achievement of their own objectives. At the same time the donor organisations will be forced to put stress not only on the provision of infrastructure but also on its long term maintenance and sustainable operation.

6 REFERENCES

Activities relating to drinking water and wastewater – Guidelines for the improvement and for the assessment of the services to users (ISO 24510), 2007, International Organisation for Standardisation, Geneva

Activities relating to drinking water and wastewater – Guidelines for the management of wastewater utilities and for the assessment of wastewater (ISO 24511), 2007, International Organisation for Standardisation, Geneva

Activities relating to drinking water and wastewater – Guidelines for the management of drinking water utilities and for the assessment of drinking water services (ISO 24512), 2007, International Organisation for Standardisation, Geneva

Alegre H., Hirner W., Baptista, J.M., Parena R. (2000) Performance indicators for water supply services. *Manual of Best Practice series, IWA Publishing, London, ISBN 1 900222 27 2 (160pp.)*

Alegre H. (2002) Performance indicators as a management support tool. in Mays, L.W., *Urban Water Supply Handbook, MacGraw-Hill, ISBN 0-07-137160-5.*

Alegre H., Baptista J.M., Cabrera E., Cubillo Jr.F., Duarte P., Hirner W., Merkel W., Parena R. (2006) Performance indicators for water supply services. *Second edition 2006, IWA Handbook*

Cabrera E. (2006) Understanding of the IWA Performance Indicator System. Structure and Implementation. *IWA World Water Congress, Session Workshop on Strategic Asset Management, Benchmarking and Performance Indicators, September 11, Beijing, China*

Matos R., Cardoso A, Ashley R, Duarte P., Molinari A., Schulz A. (2003) Performance Indicators for wastewater services *IWA Handbook*

Rohrhofer, K. (2002): Lean technologies in Shkodra, Albania. Best Practices Newsletter, 12; 3/2002; p. 22 ff

Rohrhofer, K, Cabrera, E., Cate, F., Derntl, Th., Gortan, P., Jovani, A., Olšinská, M (2007) Research Needs for Strategic Asset Management; IWA Conference "Leading Edge Asset Management", 2007 Lisbon, Portugal

Rohrhofer K., Leroch K., Gjondrekaj E., Derntl, T. (2008, submitted) The correct use of donor Funds for Water Supply and Wastewater Disposal in Non-prosperous countries. *IWA World Water Congress 2008 Vienna, Austria*

Sjøvold F., Røstum J., Sand K., Fagerhaug T. (2005) Målestokkonkurranse for VA-sektoren, Trondheim, Norway *STF50 A05 217www.odin.dep.no*

A web-based application for water quality performance management of local government in South Africa

P.F. de Souza[1], *A. Wensley, G.S. Mackintosh and M.H. Ramba*

[1] *Emanti Management, PO Box 1264, Stellenbosch, South Africa, 7599 (E-mail: philipds@emanti.co.za)*

Keywords: water quality performance management; web-based application; local government

Edited by Enrique Cabrera Jr and Miguel Ángel Pardo. ISBN: 9781843391913.
Published by IWA Publishing, London, UK.

1 INTRODUCTION

South Africa, like many developing countries, faces significant challenges in the sustainable provision of adequate and safe water services notwithstanding a good legislative framework. Despite considerable success in addressing water services backlogs, many municipalities continue to have inadequate water and effluent treatment and associated water quality management practises in place.

On-going surveys by the Department of Water Affairs and Forestry (DWAF) on water quality management have shown that in many instances drinking-water quality in non-metropolitan areas of South Africa is unacceptably poor. South Africa's Water Services Act of 1997 (DWAF: 1997) stipulates the standard of service provisioning, and specifies the responsibility of the municipality including operation and maintenance of infrastructure, monitoring and management of water quality, etc. In order to ensure an effective and sustainable water service, the above mentioned aspects must be addressed by municipalities.

As very few municipalities have satisfactory water quality monitoring programmes and even fewer utilise the available water quality monitoring data as intended, and in order to drive improvement, DWAF and other water sector partners have undertaken various initiatives to assist municipalities with operation and management of water services. In particular, it was evident that a need existed for a water quality data capture and information dissemination tool, which would both assist municipalities to meet their responsibilities, and meet DWAF's needs to monitor and regulate the operation of municipalities in a proactive cooperative governance fashion (Wensley et al: 2007). Consequently DWAF, together with the Institute of Municipal Engineering of Southern Africa (IMESA) have rolled out an internet-based electronic Water Quality Management System (eWQMS) to all 169 municipalities in South Africa (Stephens et al: 2007).

This paper highlights the various features and functions of the eWQMS, with focus on how outputs are being used for indicating performance at municipalities and driving progressive improvement.

2 ESSENTIAL ELEMENTS OF MUNICIPAL WATER QUALITY MANAGEMENT

Studies have shown that current water quality management related weaknesses in South Africa include (Mackintosh *et al*, 2004):

- Inadequate monitoring (i.e. water treatment plant and within networks)
- Lack of well structured maintenance (crisis management basis)
- Lack of management awareness and buy-in

- Lack of capacity to perform water quality management functions
- Lack of budget for water quality management at all levels
- Lack of structured programmes deal with water quality issues

In order for municipalities to improve water quality related performance, it is essential that municipalities have the ability to learn and understand, and are able to utilise information to formulate strategies and implement actions to address issues of concern. Experience has shown that if a municipality is to achieve optimum water quality performance, it will need to advance via a stepwise process. A five step Business Capability Model is presented as follows (adapted from Ingham, 2007):

- **Step 1: Unconscious incompetence** – the municipality is unaware of what they don't know (i.e. they do not know the requirements for effective water quality management)
- **Step 2: Conscious Incompetence** – the municipality is aware of what they don't know (i.e. the municipality is ignorant and they are aware of their ignorance). The municipality now knows the basics of water quality management but is unsure of how to practically implement a water quality management programme or respond to issues of concern.
- **Step 3: Conscious Competence** – the municipality is aware of how to do things properly (i.e. the municipality has the ability to do something but they have to concentrate on doing it properly). The municipality implements a basic water quality management programme and starts to practice identifying and resolving issues of concern.
- **Step 4: Unconscious Competence** – the municipality is unaware of how they do the things they know how to do (i.e. the municipality does things without even thinking about it). At this stage the municipality (and especially the staff comprising the water quality team) automatically responds to and resolves issues of concern. The municipality is also striving to continuously improve and would like to benchmark its performance vs. other municipalities.
- **Step 5: Mastery** – this would apply to a municipality that has taken a step further than unconscious competence and has optimized the total water management cycle within its area of jurisdiction. Municipalities at this stage are very proactive, identify issues, formulate strategies to close any gaps and timeously resolve issues.

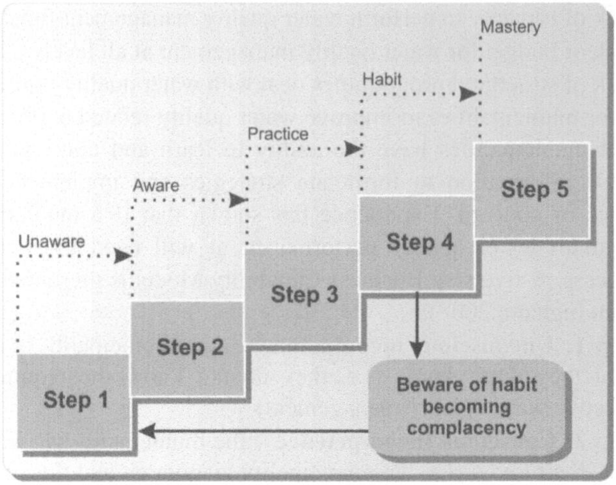

Figure 1. FIVE STEP BUSINESS CAPABILITY MODEL FOR ACHIEVING OPTIMUM WATER QUALITY PERFORMANCE (ADAPTED FROM INGHAM, 2007)

Although experience has shown that each municipality in South Africa can be considered unique, it is hypothesized that all municipalities will need to follow the above steps to master water quality management. Bearing the above in mind, the paper will focus on how the eWQMS has assisted municipalities (at each of the various stages) to achieve improve water quality performance.

3 THE ELECTRONIC WATER QUALITY MANAGEMENT SYSTEM (EWQMS)

Considering the five steps, it is clear that any IT system implemented to improve the current situation in South Africa should therefore:

- Ensure raised awareness (this should lead to increased budget allocation and improved water quality)
- Build on existing Good Practice (i.e. not be counter-productive, instead add value) (e.g. link to existing systems/LIMS)
- Bottom-up approach – i.e. the system must be useful to municipalities (e.g. provision of monthly reports, notification of issues of concern)
- The system must be easy to use, robust, reliable and low cost (preferably Free Open Source Software based to prevent payment of high software licenses by municipalities)
- Enable intervention in areas facing public health threats
- Provide strategic data to municipalities, DWAF, other role players, etc
- Satisfy municipal Governance Requirements

- Support DWAFs regulatory function and other role player requirements
- Undergo iterative enhancements via municipal and sector feedback

The above key aspects have been incorporated into the development of the eWQMS. The eWQMS is a novel Open Source Software based system which is accessible via the internet (www.wqms.co.za), and is a very useful means for allowing a range of participating parties (including municipalities and Provincial and National Government, etc) to guide the tracking, reviewing and improving of water quality. Importantly, the eWQMS has been developed in a "bottom up" approach with municipalities, IMESA, DWAF and the Water Research Commission. Features of the eWQMS include (a) *Management Dashboard* (highlights key issues including sample sites satisfying and/or failing drinking-water quality requirements), (b) *Overview* (summary of compliance), (c) *Data Analysis* (ability to dynamically generate tables and graphs), (d) *Reports* (archive of water quality management reports), (e) *Monthly Summary Reports* (automatically generated reports), (f) *Information* (water related information and references), (g) *Infrastructure* (capture details of water system infrastructure), (h) *Administration* (configure and manage system set-up) and (i) *Risk Profile* (municipalities can perform a self-assessment of the status of WQM Programmes, water supply system infrastructure, etc).

4 DEMONSTRATION OF DERIVED BENEFITS FROM USING THE INTERNET ACCESSIBLE EWQMS

Considering Figure 1 in Section 2, the following section will demonstrate the various steps of water quality management competence in municipalities, and include examples of how the eWQMS has assisted municipalities.

4.1 Step 1: Unaware

In order to improve water quality management, municipalities must be aware of the requirements for effective water quality management ("they don't know that they don't know"). This lack of awareness/knowledge was clearly evidenced during initial piloting and roll-out of the eWQMS, where by way of example, in 2004 in the Western Cape (a province in South Africa), a study of drinking-water quality management yielded the following key outputs (see figure below).

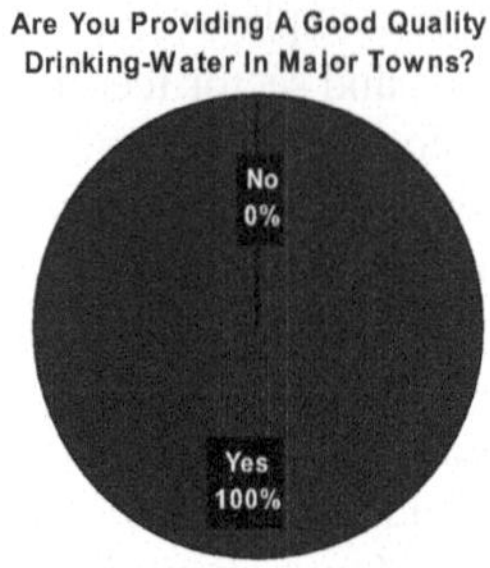

Drinking Water Quality Summary

Bacteriological :: Physical :: Chemical

Area	Total Coliforms		Faecal Coliforms	
	Samples	Compliance (%)	Samples	Compliance (%)
Western Cape (Masibambane)	***162***	***58***	***162***	***81***
Boland DM	30	47	30	90
Central Karoo DM	24	58	24	67
Eden DM	48	71	48	85
Overberg DM	30	53	30	73
West Coast DM	30	53	30	83

Figure 2. RESULTS OBTAINED FROM DRINKING WATER QUALITY MONITORING AND MANAGEMENT STUDY AND ASSOCIATED AUDIT OF DRINKING-WATER QUALITY

The above figures clearly highlight that although the municipalities stated that they were providing a drinking-water of good quality to consumers, the statement was in strong contrast to the actual results, which showed only 58% compliance with regards to total coliforms (a bacteriological indicator of treatment efficiency) and only 81% compliance with regards to faecal coliforms (a bacteriological indicator of health risk). South African drinking-water quality standards at the time of the study required 96% compliance with regards to both parameters for a drinking-water to qualify as "good". In addition, and of great concern was that a large proportion of municipalities (31%) were unaware of the drinking-water quality standards that needed to be maintained.

Similar findings were prevalent in other provinces of South Africa. To improve awareness, DWAF has implemented many water quality monitoring and management related initiatives (workshops, training material and sessions, conferences, etc). In conjunction, the eWQMS has supported improved awareness through discussions at forums, training at workshops, use of easy to understand colour coding highlighting issues of concern, indicating progress regarding data submission, the provision of water related information (e.g. guidelines for managing water system infrastructure, water quality parameters, their effects and how to rectify issues) and references (e.g. National Water Act, Water Services Act, National DWQM Framework, etc) (see figure below).

The approaches utilised (the eWQMS and other DWAF based initiatives) has lead to significant improvement in water quality monitoring and management awareness throughout South Africa.

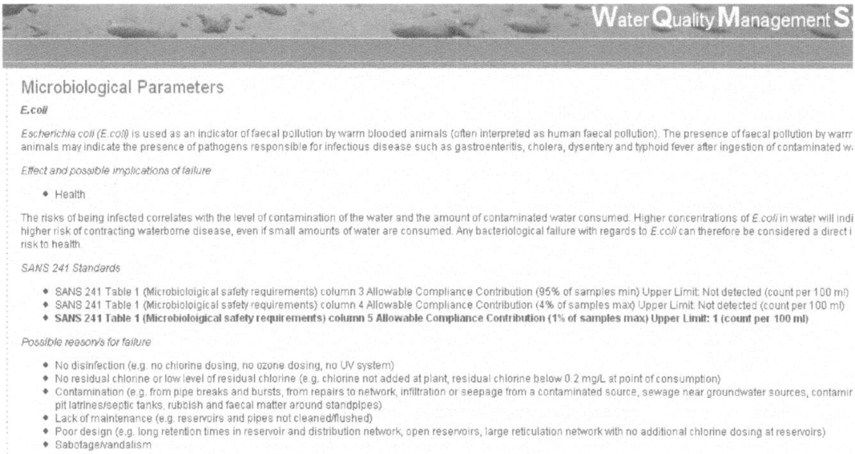

Figure 3. **INFORMATION PAGE EXPLAINING THE SIGNIFICANCE OF *E.COLI* AND ITS EFFECTS, THE APPLICABLE STANDARDS AND POSSIBLE REASONS FOR FAILURE**

4.2 Step 2: Aware

Even though municipalities become aware, they probably do not know what exactly is required from them, and thus the eWQMS has been configured to assist with this. By way of example, when data is loaded onto the eWQMS it is immediately interpreted vs. applicable water quality standards, with the eWQMS making use of summarised views (graphs and tables) of data which are colour coded to assist with interpretation and identifying issues of concern (for example, GREEN = Acceptable, RED = Poor) (see figure below). The eWQMS has therefore assisted the water sector to speak a common language ("I see you had two "REDs" last month – what did you do to rectify the issue?"). In addition, as municipalities are often poorly capacitated, and to assist municipalities with communication, a monthly summary report is automatically generated by the eWQMS. This report is also communicated to other sector stakeholders so that they are aware of issues of concern. These outputs make the municipality aware of the main water quality challenges they face. It does not, however, guarantee that they utilise the information effectively. In order to become more comfortable with water quality management and identify the underlying causes of issues, the municipality needs to practice (Step 3).

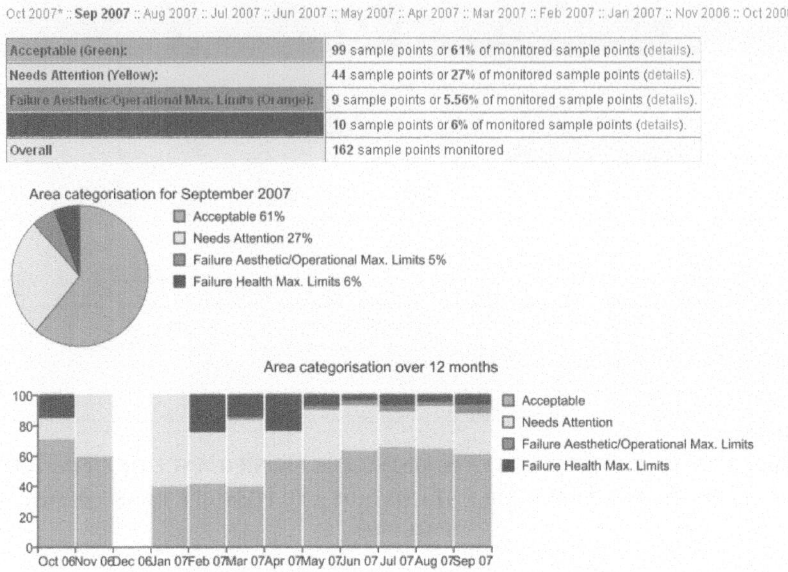

Figure 4. MANAGEMENT DASHBOARD FOR A MUNICIPALITY HIGHLIGHTING, FOR EXAMPLE, (I) SAMPLE POINTS THAT FAILED HEALTH RELATED MAXIMUM LIMITS AND (II) MONTHS LACKING DATA (E.G. DECEMBER 2006)

4.3 Step 3: Practice

Practice includes both effective monitoring and associated management, and utilising the various graphical and tabular outputs from the eWQMS to further investigate and address key issues of concern. At present, most municipalities do not use all the eWQMS features, with use largely limited to the automatically generated outputs (e.g. Management Dashboard, Compliance Overview, Reports). The use or lack of use of the more advanced features of the eWQMS (e.g. Analysis) provides an indication of the water quality management status of the municipality, with water quality management champion municipalities utilising a greater range of the available features. Importantly, technical staff utilise these outputs to track issue resolution and to convey to top management where an issue has been successfully addressed and resolved (see figure below).

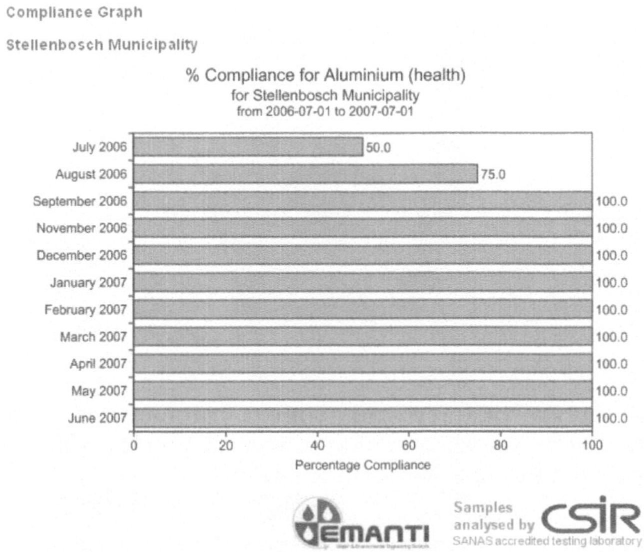

Figure 5. HIGH RESIDUAL ALUMINIUM VALUES IDENTIFIED THROUGH MONITORING WERE RESOLVED (THROUGH IMPROVED PH CONTROL) AND HAVE CONTINUED TO BE NON-PROBLEMATIC

4.4 Step 4: Habit

Through use of the eWQMS good habits can be grown. By way of example, the monthly summary report is automatically generated on the 15th of every month (e.g. October 2007 report was generated on the 15th November 2007). In order for the municipality to obtain an automatic report (which saves them time and effort), they need to load data onto the eWQMS before the end of the 14th. This incentive has been very effective, with some 87% of municipalities loading data before the cut-off date for the October 2007 report. In addition, once municipalities are more familiar with water quality management aspects, they increasingly want to meet stipulated minimum requirements and improve performance (i.e. they begin to become champions). In order to assist, an "audit" of the current municipal water quality management programme is automatically displayed on the eWQMS (see figure 6 overleaf). In addition, municipalities develop a need to benchmark their performance versus other municipalities, and therefore the eWQMS has been configured to contain benchmarking elements (see figure 7 below). This information is used to indicate to Municipal Council the status of the municipality (above/below national average), indicate how funding has been used to improve performance, etc.

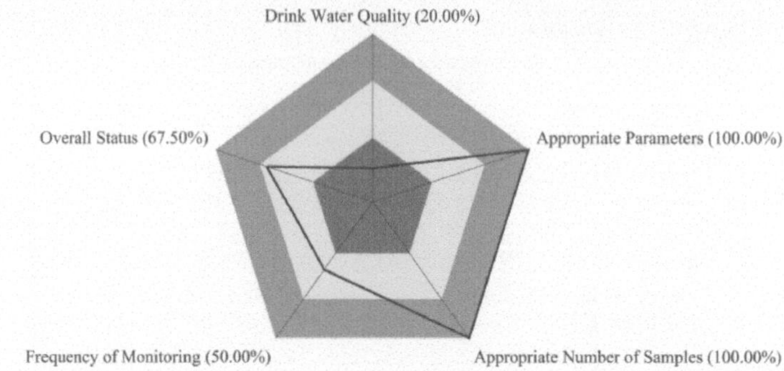

Figure 6. WATER QUALITY MONITORING PROGRAMME ASSESSMENT, HIGHLIGHTING KEY ISSUES AS DRINKING-WATER QUALITY AND FREQUENCY OF MONITORING (RED/YELLOW ON THE DIAGRAM)

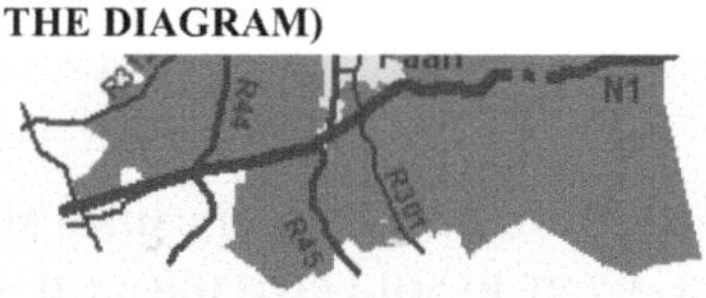

Drinking Water Quality Summary

Microbiological Safety :: Microbiological Operational :: Physical :: Chemical

Configure Parameters			Faecal Coliforms (health)		E.coli (health)	
Area	Population	Min Samples	SampleCount	Compliance %	SampleCount	Compliance %
South Africa	*48606643*	*58328*	*13265*	*98.0 (view)*	*30686*	*96.0 view*
Western Cape	*5083373*	*6100*	*1814*	*98.0 (view)*	*3746*	*95.0 view*
Drakenstein Municipality	*213914*	*257*	*181*	*100.0 view*	*525*	*100.0 view*
Gouda	0	0	4	100.0 view	12	100.0 view
Hermon	0	0	7	100.0 view	23	100.0 view
Paarl	0	0	108	100.0 view	321	100.0 view
Pearl Valley	0	0	4	100.0 view	12	100.0 view
Saron	0	0	10	100.0 view	26	100.0 view
Simondium	0	0	4	100.0 view	12	100.0 view
Wellington	0	0	44	100.0 view	119	100.0 view
Data Period	2006/11/01 to 2007/11/01					

Figure 7. OVERVIEW SCREEN INDICATING (I) PERFORMANCE OF MUNICIPALITY VS. PROVINCIAL AND NATIONAL PERFORMANCE AND (II) RECOMMENDED SANS 241 MINIMUM NUMBER OF SAMPLES THAT SHOULD BE COLLECTED BY THE MUNICIPALITY (BASED ON MUNICIPAL POPULATION)

A key concern related to Step 4 (Habit) is municipal complacency, where for example, a single individual is responsible for water quality management functions within the municipality. Where this person is ill or leaves the organisation, a gap immediately exists. This can literally turn a champion municipality into a struggling municipality overnight. Through storage of other water services related information on the eWQMS (e.g. not only water quality data but sample point locations, infrastructure types, laboratory information, etc), a municipality can maintain knowledge within the organisation, thus minimising disruptions as a result of staff migration. This information also assists with mentoring of junior staff.

4.5 Step 5: Mastery

Although the metropolitan municipalities and some of the better resourced local municipalities may argue that they have already "mastered" water quality management, investigations have revealed that this might not be the case. It is therefore argued that within the South African context, the vast majority of municipalities are at best at Step 4 (Habit). With increased competence in water quality management is the desire to perform strategic assessments to identify key gaps that need to be addressed to optimize water services. The eWQMS has started to address this need, and currently contains a number of risk based assessment tools, which can be used by municipalities to conduct self-assessments of infrastructure, water quality management programmes, etc. The outputs from these tools should be used to draft Municipal Action Plans, so that the municipality can master water quality management within its area.

5 CONCLUSIONS

The implementation of the web-based Water Quality based Performance Management System for Local Government in South Africa has:

- Created improved awareness of the requirements for effective water quality monitoring and management
- Driven progressive improvement in water quality
- Provided real-time reporting of water quality by municipalities
- Enabled early intervention in areas facing immediate public health threats
- Provided strategic data and information related to the quality of water services

Of importance to note, however, is that the use of eWQMS alone cannot solve issues, and a number of key factors are also required, including awareness and prioritisation (from Mayor to Technician), on-going communications between role players, operational test equipment and proficiency therewith, water quality data collection and assessment, and ongoing support and interaction.

Through use of the eWQMS, however, municipalities are made more aware of issues of concern which therefore assists with directing limited resources to areas of need. Thus by using the system significant improvements in water quality can be achieved despite little or no additional capacity. By simply having a structured water quality sampling programme and working "smarter" with better information, municipalities become more effective and thus make better use of the capacity it already has. Therefore, by using the eWQMS to its full potential, municipalities in South Africa will be empowered to master water quality management.

5 REFERENCES

De Souza PF, Ramba MH, Wensley A and Delport E (2006) Implementation Of An Internet Accessible Water Quality Management System For Ensuring The Quality Of Water Services In South Africa. Paper Presented at WISA Conference, Durban, South Africa, 2006.

Ingham G (2007) Motivate People. Dorling Kindersley Limited. ISBN-978-1-40531-751-1 pp 38 – 39.

Mackintosh GS, De Souza PF, Wensley A, and Delport E (2005) Multidisciplinary Considerations For Satisfying Water Quality Related Services Delivery Requirements - A Practical Tool For Municipal Engineers. Paper presented at IMESA Conference, Port Elizabeth, South Africa, 2005.

Mackintosh GS, de Souza PF, Wensley A, and Delport E (2004) Operationalising South Africa's Compulsory National Standards for Potable Water – Practical Considerations for Water Service Authorities. Paper presented at IMESA Conference, Mossel Bay, South Africa.

A Risk-based Strategic Assessment of Local Government's Capacity to Implement Sustainable Water Quality Management

A. Wensley[1], *G.S. Mackintosh*[2], *L. Hattingh*[3] *and T. Manxodidi*[4]

[1] *Department of Water Affairs and Forestry, Private Bag X313, Pretoria, South Africa, 0001 (E-mail: AllestairW@dwaf.gov.za)*
[2] *Emanti Management, PO Box 1264, Stellenbosch, South Africa, 7599 (E-mail: grantm@emanti.co.za)*
[3] *WRP Consulting Engineers, PO Box 14096, Hatfield, Pretoria, South Africa, 0028 (E-mail HattinghL@dwaf.gov.za)*
[4] *Emanti Management, PO Box 1264, Stellenbosch, South Africa, 7599 (E-mail: thabisam@emanti.co.za)*

 Performance Assessment of Urban Infrastructure Services
Edited by Enrique Cabrera Jr and Miguel Ángel Pardo. ISBN: 9781843391913.
Published by IWA Publishing, London, UK.

Keywords: Strategic Assessment; Sustainable Water Quality Management; Local Government Capacity

1 INTRODUCTION

South Africa, like many developing countries, faces significant challenges in the sustainable provision of adequate and safe water services not withstanding a good legislative framework (DWAF: 1997 & 2003). Despite considerable success in addressing water services backlogs throughout South Africa, many municipalities continue to have inadequate water and effluent treatment and associated water quality management practises in place. As the water services sector leader and national regulator, the Department of Water Affairs and Forestry (DWAF) has been developing and implementing a Water Quality Performance Assessment and Management System for Local Government (Stephens *et al*: 2007) in order to support municipalities to remedy the situation.

However, the challenges relating to the establishment of sustainable water quality management (WQM) at all South African municipalities are not insignificant. Each municipality has its own peculiar mix of circumstances and shortcomings. By simply providing a suitable WQM tool will in the majority of cases not resolve the current circumstances and shortcomings alone (no single, isolated initiative can be expected to ensure effective and sustainable WQM at a particular municipality).

One must however be able to first identify these circumstances and shortcomings and then assess the extent thereof before deploying any initiative to remedy the situation. As a result, a risk based Decision Support Tool which provides a strategic assessment of each municipality using performance indicators relating to the progressive attainment of the provision of safe drinking water and effective effluent treatment by municipalities have been developed and will be described in more detail below.

2 METHODOLOGY

The methodology utilised for the strategic assessment was based on a methodology initially developed during the initial implementation of the Water Quality Performance Assessment and Management System for Local Government at the Eastern Cape Province municipalities and refined in the subsequent implementation at the all other municipalities of the remaining eight provinces of South Africa.

To assist in ensuring that effective, sustainable WQM practises are established, each municipality is taken through a structured "self-assessment" interview process to capture the performance of the municipality as regards to six key areas of performance necessary to attain sustainable WQM. The Six "Legs of Sustainable WQM" comprise the following:

- Water Legislation, Policies and Regulations;
- Water Resources and Water System Infrastructure;
- Water Quality Monitoring, Laboratories and Logistics;
- Human Resources;
- Management; and
- Finances.

Before the start of the interview process the implementation team collected as much as possible information with regards to each particular municipality using various sources including an annual survey conducted by DWAF on Drinking Water Quality and Service Quality performance of municipalities as well as the considerable knowledge base of the regional office staff members of DWAF resulting in a well prepared interview team. Using the information, the following structured interview process was used:

- The first contact with the municipality was at an introductory senior level with either the Technical Manager (TM) and/or Water Quality Manager to initiate the strategic assessment in presence of DWAF regional office staff;
- This was followed up with structured "One-on-One" on-site interviews with the Technical Manager and/or Water Quality Manager of each municipality to critically assess the current status of all six the different performance areas using a gap analysis methodology;
- Initially, a spreadsheet was developed to capture the responses from the interview process. Subsequently a simple and easy to use web-based tool (which forms part of a web-based Water Quality Performance Management system – De Souza *et al*: 2007) have been used to capture the responses in real time online. This system is therefore "live" and as soon as any changed response is immediately reflected. Each municipality also have access to their own strategic assessment via the web; and
- Finally, it is important to note that a core team of only five people were used over a period of less than 18 months to perform all 169 strategic assessments which provided a high level of consistency in the outputs.

Via a risk based methodology, from the results of the interview process weighted scores are used to summarise the current status of the particular category and highlight where municipal efforts, and capacity building efforts, should be focused. These scorings are then graphically represented in a so-called "spider diagram" which readily communicates to all parties the current standing and risk profile of the municipality as regards WQM. This graphical presentation is also available on the web-based tool. An example of the "spider-diagram" plot is shown below in Figure 1.

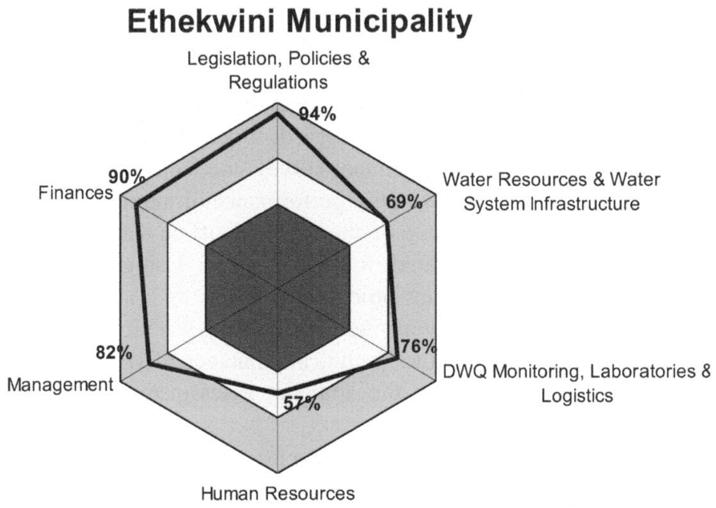

Category score	Characterisation	Comment
>70% to 100%	Acceptable	No or minor intervention required to ensure sustainability
>45% to 70%	Marginal	Interventions required to ensure sustainability
0% to 45%	Poor	External support and intervention required

Figure 1. **EXAMPLE OF A "SPIDER-DIAGRAM" HIGHLIGHTING THE MUNICIPAL RISK PROFILE**

3 RESULTS OF STRATEGIC ASSESSMENT

3.1 Local level

3.1.1 Water Legislation, Policies and Regulations

The majority of municipalities had an acceptable understanding of the requirements of the legislation governing water quality. The strategic assessments, however, indicated a lack of the implementation of the relevant regulations as well as the existence outdated by-laws requiring urgent revision.

3.1.2 Water Resources and Water Supply System

Water infrastructure is mostly maintained on a crisis management basis as provision of funding for preventative maintenance had been a major challenge. The municipal water sector is also going through a process of transfer of infrastructure and personnel from DWAF to municipalities in some areas as a result of the implementation of the current legislative framework.

3.1.3 Drinking Water Quality Monitoring, Laboratories and Logistics

The existing Drinking Water Quality Monitoring programmes varied from municipality to municipality and most opted for the use of external laboratories. In certain municipalities where acceptable drinking water quality programs were encountered it was noted that although systems were in place and perceived by the Municipality to be adequate, these programmes were not appropriate for proper evaluation and needed to be adjusted accordingly.

.3.1.4 Human Resources

Inadequate staff numbers and skills were reported amongst the majority of municipalities. This is compounded by restructuring and transfer of powers and functions. In addition, most of the IT support rested with senior management, with both middle to lower management and technical DWQM staff not having adequate support.

3.1.5Management

Most municipalities perceived themselves to have acceptable technical management, with communication between the departments and the municipal council (the political governing body) reported to be generally at an acceptable level. It was also noted that water quality awareness was increasing at council level although it is perceived to still be a problem which is highlighted below.

3.1.6 Finances

Appropriate and adequate budget for operation and maintenance aspects were lacking for a large number of municipalities. Budgeting constraints resulted in crisis maintenance, with preventative maintenance almost non-existent in most of the municipalities.

3.2 Provincial and national level

As described above, the risk based Decision Support Tool not only summarises the current status of key water quality management sustainability indicators for each of 169 municipalities in South Africa, the outputs are also used at a regional level and at a national level for guiding assistance to municipalities and prioritising areas of engagement with the water services community.

Three different perspectives are generally used, namely:

- A summary of the six performance areas for each province;
- A summary of the provincial perspectives for a particular performance area; and
- The use of a star-based ranking system using the characterisation results (see Figure 1) which is similar to the medal system used in major sport events.

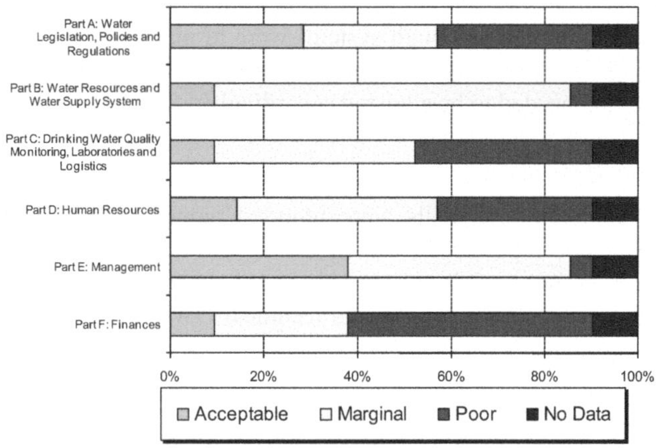

Figure 2. **A TYPICAL SUMMARY OF THE SIX KEY PERFORMANCE AREAS PER PROVINCE**

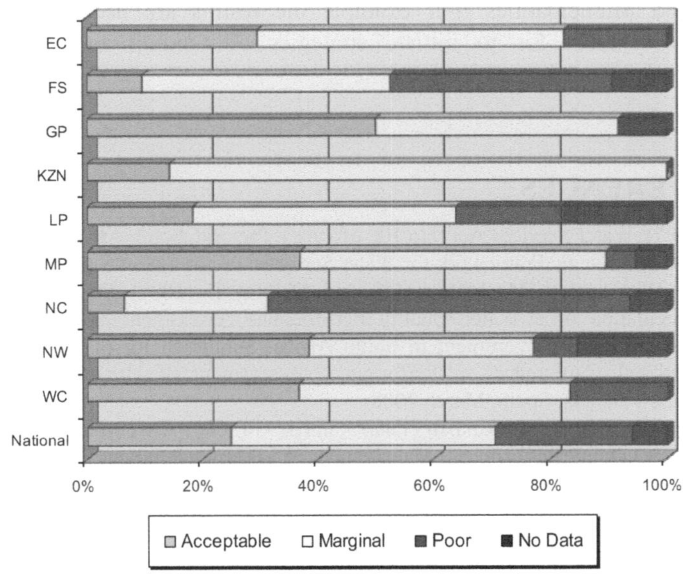

Figure 3. A TYPICAL SUMMARY OF THE PROVINCIAL PERSPECTIVES FOR DRINKING WATER QUALITY MONITORING, LABORATORIES AND LOGISTICS

4 CONCLUDING REMARKS

The national deployment of this risk-based Decision Support Tool amongst all municipalities of South Africa can be seen to be making a direct and significant contribution towards the ultimate goal of ensuring the provision of sustainable water services (in particular water quality management) to all South Africans. The tool has been shown to be a key tool in assisting:

- Municipalities to address key weaknesses and especially communication critical technical related issues to political decision-makers;
- Municipalities to communicate challenges easily to all stakeholders through the use of the "spider diagrams" and in particular the colour coding system of green, yellow and red (see Figure 1);
- Regional government to prioritise assistance to Municipalities. For example three of the nine regional governments (Free State, Western Cape and Northern Cape) has already used the outcomes of the strategic assessment in some or other form to assist municipalities; and

- National government to direct strategic efforts to Regional government and Municipalities. For example the results are used by national government for to initiate strategies of funding and support at 40 prioritised municipalities.

5 REFERENCES

DWAF (1997). Water Services Act (No 108 of 1997). Pretoria, South Africa.

DWAF (2003). Strategic Framework for Water Services. Pretoria, South Africa.

Stevens F, Wensley A, Mackintosh GS and Delport E (2007) Development and implementation of a Water Quality Performance Assessment and Management System for Local Government in South Africa. Submitted for publication at the International Conference on Performance Assessment of Urban Infrastructure Services, Valencia, Spain, March 2008.

De Souza P, Wensley A, Mackintosh G and Ramba, M (2007) A web-based application for Water Quality Performance Management of Local Government in South Africa. Submitted for publication at the International Conference on Performance Assessment of Urban Infrastructure Services, Valencia, Spain, March 2008.

Sewer asset management tool: dealing with experts' opinions

C. Werey[1], *F. Cherqui*[2], *M. Ibrahim*[2] *and P. Le Gauffre*[2]

[1] UMR Cemagref-ENGEES Gestion des Services Publics, 1 quai Koch, BP 61039, F-67070 Strasbourg Cedex, France, caty.werey@cemagref.fr, Tlf +33 38824 8253, Fax: +33 38824 8284
[2] Université de Lyon, F-69000, Lyon, France INSA-Lyon, F-69621 Villeurbanne, France;Université Lyon 1, F-69622, Villeurbanne, France ; Laboratoire Génie Civil & Ingénierie Environnementale ; pascal.le-gauffre@insa-lyon.fr, tel.: +33 47243 8858 ; fax: +33 47243 8521

Keywords: sewer network, CCTV inspection, dysfunction indicators, expert opinion

Edited by Enrique Cabrera Jr and Miguel Ángel Pardo. ISBN: 9781843391913.
Published by IWA Publishing, London, UK.

ABSTRACT

Asset management requires the development of performance indicators (PIs) and decision procedures. Within the French RERAU methodology each rehabilitation criterion is assigned a grade out of four possible ones. This grade results from an aggregation of complementary PIs that use information derived from various sources: visual inspection, O&M data, network monitoring, etc.

This paper focuses on the development of dysfunction indicators derived from visual inspection results (WP1 of the French INDIGAU program). Inspection reports provide sequences of observation (defect) codes. On this basis, three complementary procedures are proposed so as to assign a condition grade to the sewer segment: (a) expert rules identifying major defects (b) calculation of a single score and comparison to three thresholds and (c) rules based on the analysis of scores distribution along the segment.

Calibrating procedure b) means defining parameters used in the calculation of a single score and defining three (crisp or fuzzy) thresholds. The calibration also requires experts' judgments that will be used as references. A sample of 45 links has been studied by 8 experts, regarding 10 indicators. The results display a lot of conflicts between experts' opinions. Three types of situation are defined: 1) no conflict, a consensus can be identified; 2) one expert disagrees with a majority, a consensus may be defined; 3) major conflicts between answers.

1. INTRODUCTION

Asset management is an increasing concern for wastewater utilities and companies. Criteria are developed for supporting the definition of investigation and rehabilitation programs. Dysfunction indicators contribute to the calculation of criteria, using expert rules. Indicators based on visual inspections provide major information. However, difficulty remains in the translation of a visual inspection survey into dysfunction indicators (Rahman & Vanier, 2004). In the framework of the French RERAU program (Rehabilitation of urban sewer networks) a methodological approach has been developed (Le Gauffre *et al.*, 2004; Le Gauffre *et al.*, 2007). 10 dysfunction indicators are defined and assessed with visual inspections reports of sewer segments; these dysfunction indicators are assigned a grade $G \in \{1, 2, 3, 4\}$: (from the best to the worst).

The different steps of the dysfunction indicators' valuation are detailed in the 2nd section. The 3rd section presents the ongoing work within the French project "INDIGAU" (Performance Indicators for asset management of urban sewer networks"), dealing with experts' judgment of CCTV reports. The first results and conclusions are presented in the 4th section.

2. DYSFUNCTION INDICATORS VALUATION

Inspection reports provide sequences of observation codes, using a coding system (for example European standard EN 13508-2), that are quantified in order to obtain a score distribution on each sewer segment. The score of a section considers all defects, their gravity and their extent. Moreover, a single score is calculated for the sewer segment: the global density of defects relating to each dysfunction. Three complementary procedures are then executed so as to obtain a condition grade related to the sewer segment: (a) expert rules based on sequences of observation codes, (b) comparison of single score to threshold and (c) rules based on analysis of segment profiles (score distribution), as presented on Figure 1.

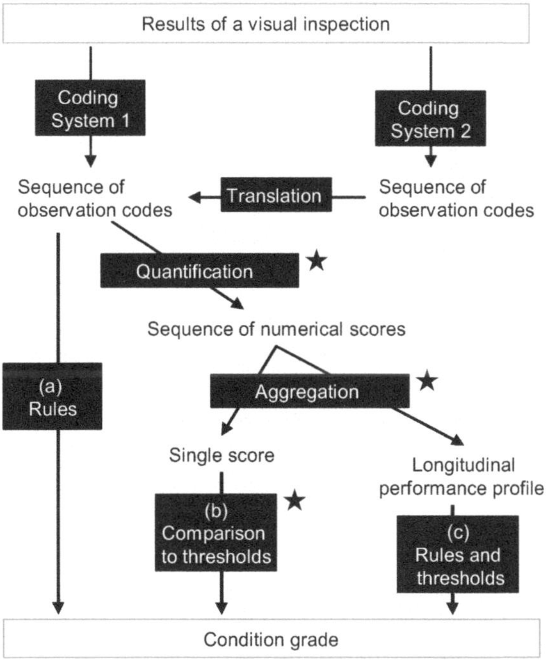

Figure 1. TRANSLATION OF VISUAL INSPECTION ENCODING INTO CONDITION GRADE

The translation of visual inspections into condition grades depends on the coding system that is used. The methodology proposed in the RERAU project needs observation codes using the European standard EN 13508-2. So if the existing data are coded either in a self-made codification (case for the application done in the department of Bas-Rhin (Dorchies, 2005; Werey *et al.*, 2006) or in a national coding (in France, AGHTM, 1999), it is necessary to translate them within the EN 13508-2 codification. We can notice that new material exists since a few time on the market making the inspection directly with this new framework.

When the defaults are available into EN 13508-2 standard, the calculation of the dysfunction scores can begin using tables provided within the RERAU project (Le Gauffre *et al.*, 2004; Le Gauffre *et al.*, 2007) where dysfunction indicators are defined concerning: infiltration *(INF)*, exfiltration *(EXF)*, decrease of hydraulic capacity *(HYD)*, sand silting *(SAN)*, blockage *(BLO)*, destabilisation of ground-pipe system *(SPD)*, ongoing corrosion *(COR)*, ongoing degradation from roots intrusion *(ROO)*, ongoing degradation from abrasion *(ABR)*, risk of collapse *(COL)*. Figure 2 and Figure 3 present the corresponding tables for the infiltration dysfunction.

Dysfunction	INFILTRATION
Indicator	INF4: Infiltration risk, estimated from visual inspection
Valuation scale	Segment
Valuation type	Observation-based estimation of a dysfunction
Unit or gravity levels	Grades : 1/2/3/4
Valuation	1 – coding Ci of observations Oi according to EN 13508-2 ; 2 – translating Ci into scores Ni according to the following table ; 3 – calculation of **density** D = N / LT, with N = Σ Ni, and LT: length of the segment (m) $Ni = \alpha^n \times P$ (or L_i), with n = 0, 1, 2 or 3 and α = 2, 3 or 4; 4 – comparison of D with thresholds S1, S2, S3 : level 1 if D ≤ S1 ; 2 if S1 < D ≤ S2 ; 3 if S2 < D ≤ S3 ; 4 if S3 < D.

Figure 2. DENSITY CALCULATION FOR INFILTRATION (LE GAUFFRE *ET AL.*, 2004)

Observation Oi	Code Ci	1	α	α^2	α^3	← Gravity Extent ↓
Deformation	BAA		BAA			P
Fissure	BAB	BAB B		BAB C		L
Break/collapse	BAC			BAC A	BAC B/C	P
Missing mortar	BAE		BAE			P
Defective connection	BAH			BAH B/C/D		P
...	...					

Figure 3. DEFECTS CONTRIBUTING TO INFILTRATION (LE GAUFFRE *ET AL.*, 2004)

3. ELICITING EXPERTS' OPINION

We present here the first results of the expert survey we made within the INDIGAU project. For each dysfunction indicator, experts' opinions are used for calibrating thresholds for the 4 different states presented in the previous part. Then we will use either a crisp approach (Ibrahim *et al.*, 2007) or a fuzzy approach (Le Gauffre *et al.*, 2008) for threshold determination; these methods offer the possibility to take into account simultaneously false positive and false negative errors between the expert valuation and the calculated scores. Figure 4 illustrates the assignment errors related to score value comparison to thresholds.

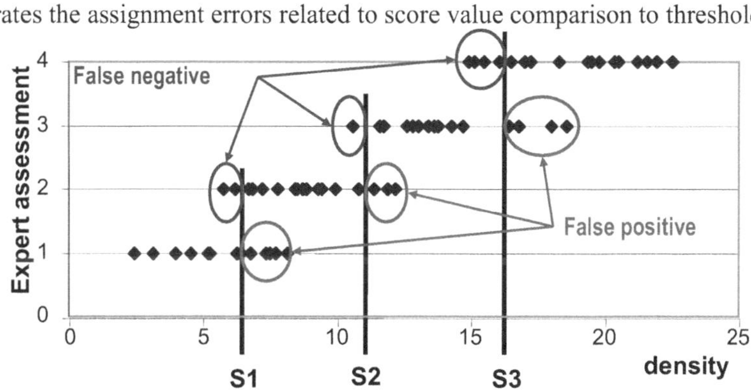

Figure 4. EXPERTS' OPINION VERSUS CONDITION GRADES OBTAINED BY COMPARING SCORES WITH THRESHOLDS (63 SEWER SEGMENTS); IDENTIFICATION OF ASSIGNMENT ERRORS.

In the present application we gathered data on 45 links within 3 different utilities and we asked each of the 8 experts involved in the project to valuate 22 links, that is to say to assign a grade to each dysfunction indicator for each link. This means that for each link we expected 3 or more answers. Thus we want to take into account in our model the gap between calculated scores and expert's opinions and also the fact that on a same link different experts can propose different results.

First of all, the four grades have been defined and discussed in accordance with experts' opinion. These grades are defined in the Table 1.

Table 1. DEFINITION OF EACH GRADE

Grade	Definition
G1	no or few noticed defects (regarding the considered dysfunction)
G2	situation with low gravity, link to be kept under watch
G3	situation with a certain gravity, intervention to be prioritized
G4	unacceptable situation in any context, action needed

Next, a sample of 45 links was built up: one third of the sample in the French coding system, the second third in EN 13508-2 system after recodification from French system, the last third available directly in the EN 13508-2 system.

Then, each expert had to answer in a spreadsheet; the generic response sheet is presented in Figure 5. Each answer with Grade 3 or 4 must be justified, following procedures (a) *major(s) defect(s) or combination of defects*, (b) *density* or (c) *concentration of defects* presented in Figure .

Link id: CG03 *Time spent:* 20 minutes
Expert id: FJ + MW

PI	*G*	*Judgment justification* Density	Concentration of defects	Major(s) defect(s)	Combination of defects	Comments
INF	**G3**	**X**				
EXF	**G2**					
HYD	**G2**					
SAN	**G1/G2**					
BLO	**G1**					
SPD	**G3**	**X**				
COR	**G1**					
ROO	**G3**			**BBA A 4.7 + BCA E A 23.7**		
ABR	**G2**					
COL	**G4**					**Risk of road collapse + BAP 18.9 + BAH B 24.5**

Figure 5. EXAMPLE OF EXPERT'S JUDGMENT OF A LINK

4. RESULTS OF THE SURVEY

We now describe the results of the expert valuation we have just collected on 34 links on which we have 3 or more answers. Conflict between experts occurs when the difference between 2 answers is 2 or more levels; in fact when difference between experts' opinions is 1 level, consensus may be obtained at the middle of the levels (*i.e.* consensus of D1 + D2 is D1/D2).

Table 2 presents the answers for link 12. It illustrates the situations observed for the 33 links.

Table 2. EXPERTS' JUDGMENT FOR LINK 12, GRADING OF EACH DYSFUNCTION INDICATOR. THREE CASES ARE IDENTIFIED: 1) A CONSENSUS IS OBTAINED (WHITE COLUMNS); 2) ONE EXPERT DISAGREES BUT CONSENSUS IS KNOWN (WHITE COLUMNS AND BLACK CELL); 3) CONSENSUS IS DIFFICULT (GREY COLUMNS).

Expert	*INF*	*EXF*	*HYD*	*SAN*	*BLO*	*SPD*	*COR*	*ROO*	*ABR*	*COL*
X1	G3	G3	**G4**	G4	**G4**	**G3**	G1	G1	G1	**G4**
X5	G3	G3	**G3**	G3	**G1/G2**	**G2**	G1	G1	G1	**G3/G4**
X6	G4	G4	**G1**	G4	**G3**	**G1**	G1	G1	G2	**G1**
X7	G3	G3	**G4**	G4	**G4**	**G2**	G1	G3	G2	**G1**
X8	G3	G3	**G1**	G2	**G1**	**G2**	G1	G1	G1	**G2**

Three cases occur. In case 1, a consensus is identified, there is no conflict; consensus can be "no dysfunction" as for *COR* and *ABR* in Table 2 or consensus can be "major dysfunction" as for *INF* and *EXF*. In case 2, consensus is also known because only one expert disagrees and this answer may be excluded. For example, in Table 2, *SAN* is assigned grade G3/G4 and *ROO* is assigned G1. In the last case (case 3), consensus is hardly known since there are major conflicts between all experts (*HYD*, *BLO* and *COL* in Table 2); further investigations are needed in order to conclude (ask other experts to assess this link, question again experts, characterize expert behaviour to see if expert is optimist or pessimist, etc.).

If we now consider all the 34 links, the number of links for each case is presented in Table 3.

TABLE 3. COMPARISON OF EXPERTS' JUDGMENTS FOR 34 LINKS; CASE 1: CONSENSUS, CASE 2: ONE OPINION DIFFERS, CASE 3: NO CONSENSUS

Cases	*Dysfunction indicators*									
	INF	*EXF*	*HYD*	*SAN*	*BLO*	*SPD*	*COR*	*ROO*	*ABR*	*COL*
1	23	19	20	18	20	19	31	33	27	17
2	6	10	5	5	9	7	3	1	6	10
3	5	5	9	11	5	8	0	0	1	7

For *COR* and *ROO*, there are no major conflicts, however there are few defects concerning these dysfunctions: grade G1 and G2 have been assigned by the experts. At the opposite, *SAN* presents many conflicts so as *HYD*, *SPD*, and *COL*. From Table 3, we can conclude that experts agree with each other in the majority of links; however there is a surprisingly high rate of conflict. Discrepancies between experts' judgement must be investigated and although further work is needed, several directions have been identified:

- expert's background: an expert may be lenient (few D3 or D4) because the asset stock of its utility is in poor condition and rehabilitation budget is slim…
- expert's skills: an expert may be mainly preoccupied with some aspect such as hydraulics (HYD, BLO, etc.) and not with some others such as water tightness (INF, EXF);
- human error: a major defect or a harmful combination of defect may have been omitted;
- interpretation differences: each procedure (a, b, c, Figure) may not have the same limits and meanings for each expert.

Taking into account these different points of view will be our challenge but we will first of all have to consolidate these results by presenting them to the experts to validate or not their positions.

When we look at the results, expert by expert, we see different behaviours as we mentioned it already before. Let us have a look to the answers given for INF presented on Figure 6. First of all, it is noticeable that results are well spread between grade 2 and grade 4; grade 1 results are considered less interesting because the aim of the INDIGAU project is to provide a support for the prioritization of rehabilitation projects. Expert 6 may be identified as more severe than other experts: he often gives the highest grade amongst other answers. On the contrary, expert 2 may be considered as less severe than others: he has never assigned grade 4 and often gives the lowest grade. Some other experts, such as expert 8, appear as "moderate" because they often give median answer. The study of other indicators shows that experts are not systematically optimist, lenient or median; it depends on the considered dysfunction.

5. CONCLUSION – OUTLOOK

We have reported an ongoing work that aims at calibrating dysfunction indicators based on the results of visual inspections. In a first step, three complementary procedures have been proposed for making the most of the observations, and rules have been defined for translating defect codes into deduct values that can be aggregated in a single score. This score may be compared to thresholds for assigning a grade to each link. Calibration requires experts' opinions used as references. We have shown that eliciting a reference condition grade from experts is not obvious. Further works will go on with the elicitation of reference values and calibration procedures will be implemented within crisp and fuzzy approaches.

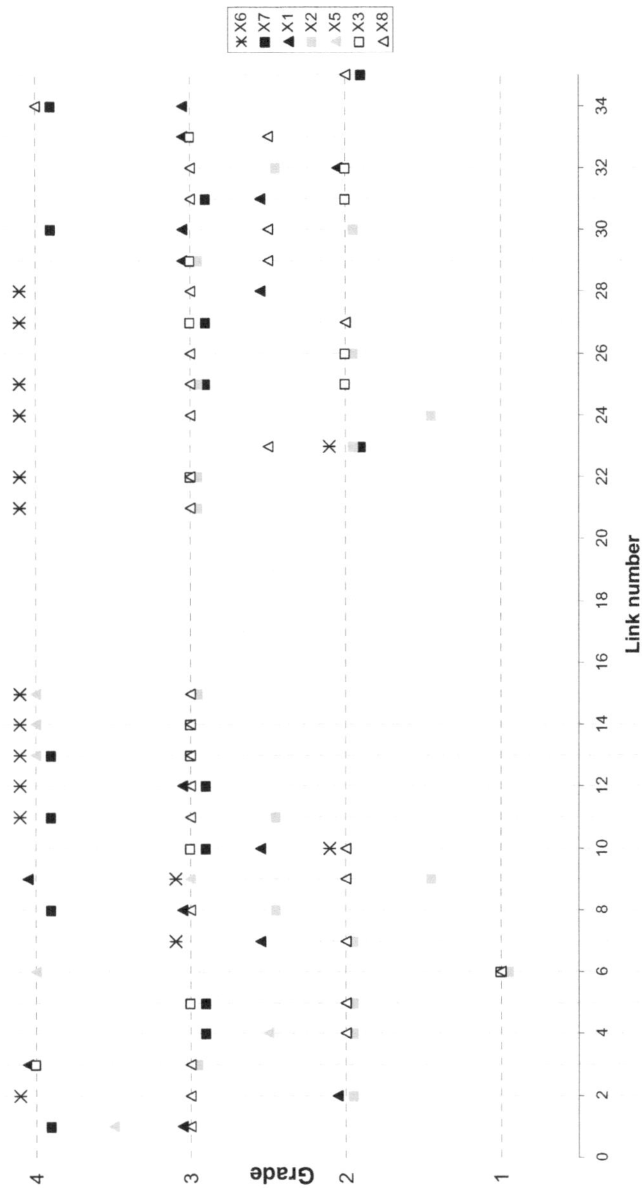

Figure 6. RESULTS OF EXPERTS' OPINION FOR *INF* INDICATOR AND FOR 30 LINKS *(PLEASE ROTATE THE FIGURE FOR BETTER UNDERSTANDING)*.

6. ACKNOWLEDGEMENTS

The research presented in this paper is carried out in the frame of the INDIGAU research project (www.indigau.fr) supported by the French National Agency for Research – ANR, within the PGCU-2006 program. The authors thank the experts of the utilities that are involved in the project: *Brest Métropole Océane, Communauté d'Agglomération Caen la mer, Communauté Urbaine de Strasbourg, Conseil Général du Bas-Rhin, Grand Lyon, Lille Métropole Communauté Urbaine, Nantes Métropole Communauté Urbaine.*

7. REFERENCES

AGHTM (1999). Les ouvrages d'assainissement non visitables : fiches pathognomoniques. Techniques Sciences et Méthodes, octobre 1999, 23-90

Dorchies, D. (2005). Etude de l'applicabilité de l'outil d'aide à la décision RERAU pour la réhabilitation des réseaux d'assainissement sur l'inventaire du département du Bas-Rhin, *Mémoire de fin d'études, ENGEES Strasbourg, UMR Gestion des Services Publics,* 62 p.

Ibrahim M., Cherqui F., Le Gauffre P., Werey C. (2007). Sewer asset management: from visual inspection survey to dysfunction indicators. *2nd Leading Edge Conf. on Strategic Asset Management, Lisbonne, Portugal, October 17-19, 10p..*

Le Gauffre P., Ibrahim M. and Cherqui F. (2008). Sewer asset management: fusion of performance indicators into decision criteria. Pi08 – Performance Assessment of Urban Infrastructure Services, Valencia, Spain, 12-14 March. In press.

Le Gauffre P., Joannis C., Breysse D., Gibello C. and Desmulliez J.J. (2004). Gestion patrimoniale des réseaux d'assainissement urbains. Guide méthodologique. Paris, *Lavoisier Tec&Doc*, 416 p., ISBN 27430-0748-6.

Le Gauffre P., Joannis C., Vasconcelos E., Breysse D., Gibello C., Desmulliez J.J. (2007). Performance Indicators and Multicriteria Decision Support for Sewer Asset Management. *Journal of Infrastructure Systems, ASCE*, **13**(2), June 2007, 105-114.

Rahman, S. and Vanier, D.J. (2004). An evaluation of condition assessment protocols for sewer management. Report B-5123.6. Ottawa (Canada): CNRC-NRC. Available online at: http://irc.nrc-cnrc.gc.ca/pubs/fulltext/b5123.6/b5123.6.pdf.

Werey C., Dorchies D., Mellac Beck I. (2006). Sewer asset management: assessing criteria for a multicriteria decision support on a county level data base. *Joint int. conf. on Computing and decision making in civil and building engineering, Montreal, June 14-16,* 886-895.

Managing utilities with performance indicators

Objectives and performance assessment of water supply services

P. Duarte[1]*, H. Alegre*[2] *and A. J. Monteiro*[3]

[1] *LNEC – National Laboratory of Civil Engineering, Urban Water Division, pduarte@lnec.pt, Av. Brasil 101, 1700-066 Lisboa, Portugal, Tlf: +351 218443842 Fax: +351 218443032*
[2] *LNEC – National Laboratory of Civil Engineering, Urban Water Division, halegre@lnec.pt, Av. Brasil 101, 1700-066 Lisboa, Portugal, Tlf: +351 218443625 Fax: +351 218443032*
[3] *IST – Instituto Superior Técnico, Technical Unversity of Lisbon, Civil Engineering and Architecture Department, ajm@civil.ist.utl.pt, Av. Rovisco Pais, 1049-001 Lisboa, Portugal, Tlf: +351 218418369 Fax: +351 218497650*

Keywords: assessment criteria; ISO 24500 standards; objectives; performance assessment; water supply

Edited by Enrique Cabrera Jr and Miguel Ángel Pardo. ISBN: 9781843391913.
Published by IWA Publishing, London, UK.

ABSTRACT

A highly performing water utility shall have clear objectives and targets that need to be monitored on a systematic way. Performance assessment is therefore an essential part of the management and decision-making processes in general and of infrastructure asset management in particular. Strategic, tactical and operational objectives should be considered, integrating a specific methodology for evaluating interventions in water supply networks, and performance measures should be used to assess their achievement, according to assessment criteria. The alignment of objectives and performance measures, recommended in the recently approved standards ISO 24500, is far from trivial.

This paper aims at being a step forward to the application of these standards, proposing a systematic approach for objective formulation at the three levels referred and corresponding performance measures, using not only indicators but also other measures, such as performance indices and levels. Examples are presented for two different strategic objectives. Whenever feasible, performance indicators from the IWA PI system are adopted. Some new performance measures are also proposed.

1 INTRODUCTION

During the last decade, performance assessment has been subject to much discussion and research. A lot of work has been carried out on performance assessment of water and wastewater utilities. Two manuals of best practice containing sets of performance indicators were published by the International Water Association: Alegre *et al.* (2000) for water supply services and Matos *et al.* (2003) for wastewater services. For water supply, IWA released recently a second edition (Alegre *et al.*, 2006), enriched with all the experience achieved during the three-year international field testing of the water supply indicators and also with opinions, experiences and examples shared by participating undertakings. Alegre (1992) introduced the technical performance assessment concept and Coelho (1997) proposed a methodology for using performance indices in water distribution systems. Cardoso (2007) proposed a technical performance assessment methodology for wastewater systems.

The elaboration of international standards on "Activities relating to drinking water and wastewater services", providing guidelines for the "assessment and for the improvement of the service to users" (ISO, 2007a) and "for the management and for the assessment of drinking water services" (ISO, 2007c) and "wastewater services" (ISO, 2007b), encourages undertakings to assess performance. Also initiatives like the Global Reporting Initiative (GRI, 2006), and the publication of Publicly Available Specification PAS-55 (IAM/BSI, 2004) are good examples of the importance of using performance assessment.

The COST Action C18 - Performance assessment of urban infrastructure services: the case of water supply, wastewater and solid waste (www.costc18.org, 2004-2008) provides an excellent state-of-the-art on performance indicators and proposes further research needs after identifying some knowledge gaps (Sjøvold *et al.*, 2007). As examples, this report suggests a better integration of performance assessment in the information structure and management of urban services and the formulation of formal approaches for infrastructure asset management, including relationship between performance at strategic, tactical and operational levels.

It is desirable that performance assessment-related concepts are universal and its measures are precisely defined, so that efficiency and effectiveness of different utilities can be easily understood and compared. Evaluating performance should not be seen as an independent process, but rather be part of decision processes, in line with Plan-Do-Check-Act approaches, and helping to choose between alternative solutions. Performance assessment allows for a clearer vision of alternatives and their advantages and disadvantages, not only in the immediate future but also during the expected life cycle of a system. The ISO 24500 standards establish general principles that fully fit these considerations. However, there is a need for more concrete guidance for their application by the utilities. The authors consider that the application of the standards may be more effective if the decision process is based on objectives split into the three levels referred.

The next sections of this paper focus on the use of performance assessment in the framework of the decision-making processes on investment prioritization, anchored on the principles of the ISO 24500 and taking in to account the strategic, tactical and operational levels of planning and intervention. A new framework for evaluating water distribution network investments using performance assessment analysis, subject of an on-going PhD thesis, has been proposed by Duarte *et al.* (2007) and is adequate for this analysis (see Figure 1).

The present work takes also into account a research program on water infrastructure asset management recently launched at LNEC (Alegre, 2007).

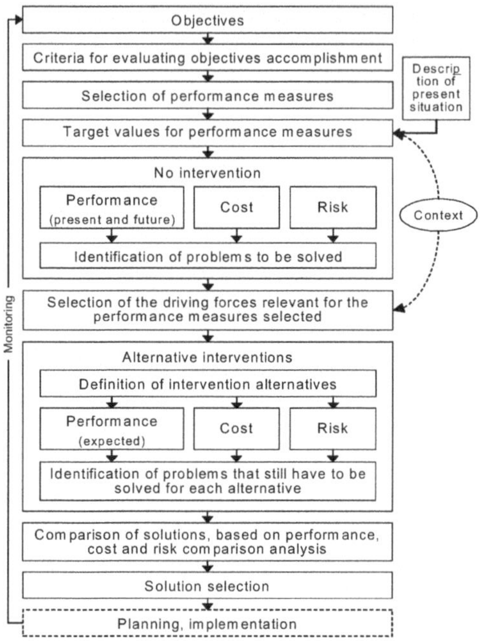

Figure 1. DECISION PROCESS FOR INFRASTRUCTURE ASSET MANAGEMENT (DUARTE *ET AL.*, 2007)

2 METHOD

Three levels of decision in infrastructure planning are considered: strategic, tactical and operational. At each level, objectives shall be formulated and assessment criteria will lead to performance measures, so that objectives achievement can be monitored without bias (Figure 2).

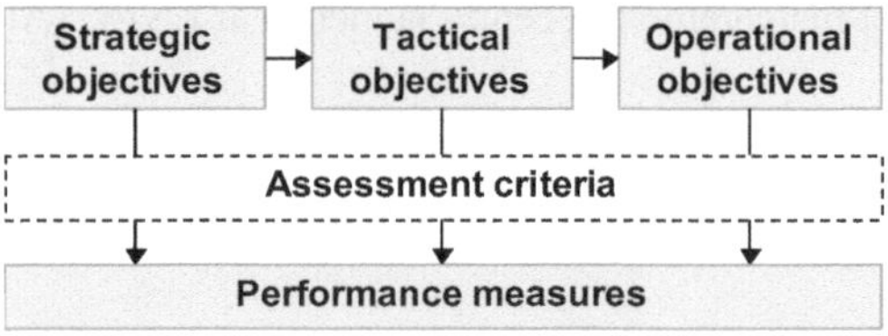

Figure 2. ASSOCIATING PERFORMANCE MEASURES TO OBJECTIVES IN STRATEGIC, TACTICAL AND OPERATIONAL LEVELS.

Taking as a basis the examples included in the ISO 24500 standards, several strategic objectives were considered. For each one, tactical and operational objectives were specified with assessment criteria and associated performance measures.

This task is part of an on-going PhD thesis work, aiming at producing a reference list to assist in decision processes on asset management, especially in the first three steps of the process in Figure 1.

3 RESULTS

Due to size limit, in this paper only two strategic objectives are presented: "protection of public health" (Table 2) and "provision of service under normal conditions" (Table 3), but this task was carried out for other strategic objectives, for example: "provision of service under emergency situations", "sustainability of the water undertaking", "promotion of sustainable development of the community" and "protection of the environment".

Some of the tactical and operational objectives were considered under different strategic objectives, so some lines are repeated in different tables.

Details on performance measure definitions may be provided under request.

Table 1. EXAMPLE TABLE FOR OBJECTIVES PRESENTATION.

Strategic objective		
<Strategic objective>		
<Assessment criterion>		
<Performance measure>	Type	Applicable to
...		

Tactical objectives	Operational objectives
<Tactical objectives>	**<Operational objectives>**
<Assessment criterion>	*<Assessment criterion>*
<Performance measure> .... Type Applicable to	**<Performance measure>** .... Type Applicable to
...	...

Type of performance measure:

PI.......Performance Indicator
TPI.....Technical Performance Index
PL......Performance Level

Applied to:

B........applicable to bulk water supply infrastructures
D........applicable to water distribution infrastructures
B/D ....applicable both to bulk water supply and to water distribution infrastructures

Table 2. ASSESSMENT CRITERIA, TACTICAL AND OPERATIONAL OBJECTIVES AND PERFORMANCE MEASURES FOR STRATEGIC OBJECTIVE "PROTECTION OF PUBLIC HEALTH".

Strategic objective

Protection of public health

Access to the service
Household and businesses supply coverage (%) .... PI B/D

Maintenance of an adequate quantity of drinking water in the distribution system that is greater than minimum requirements for cooking, cleaning and body hygiene
Water available for domestic demand (l/inhab./day) .... PL B/D

Quality of drinking water supplied
Quality of supplied water (%) .... PI B/D

Security of systems under normal situations
Security of systems .... PL B/D

Emergency plan for drinking water supply provision in case of contamination or catastrophe
Population covered by the emergency plan (%) .... PI B/D
(specific objective for this criterion: "Provision of service under emergency situations")

Pollution and waste generated by water supply activity
Public health risk from pollution and waste generated by the activity .... PL B/D

Tactical objectives

Promote total population coverage by the water supply system

Connections missing that may be constructed (personnel and budget allocation)
Connections that may be constructed in a period of time/ missing connections (%) .... PI D

Meet drinking water quality standards

Meet drinking water standards about aesthetic, microbiological, chemical and radiological parameters
Aesthetic tests compliance (%) .... PI B/D
Microbiological tests compliance (%) .... PI B/D
Physical-chemical tests compliance (%) . PI B/D
Radioactivity tests compliance (%) .... PI B/D

Operational objectives

Service connections constructed

New service connections constructed and connected
New service connections (no./year) .... PI D
Service connections missing
Water supply coverage after new service connections are built (%) .... PI D

Cleaning water supply system

Tank cleaning
Storage tank cleaning (-/year) .... PI B/D
Pipe cleaning
Network discharges (-/year) .... PI B/D
Cleaning and disinfection after repairs (%) .... PI B/D

Eliminate negative pressures in nodes

Nodes were pressures are lower than minimum requirements, from the simulation model
Nodes with low pressures (%) .... PI D

Quality control of supplied water

Aesthetic, microbiological, chemical and radiological analysis
Aesthetic tests carried out (%) .... PI B/D
Microbiological tests carried out(%) .... PI B/D
Physical-chemical tests carried out (%) . PI B/D
Radioactivity tests carried out (%) .... PI B/D

Strategic objective

Protection of public health (contd.)

Promote efficient use of water

Domestic use of water

Water consumed for domestic purposes (l/inhab./day) .... PI B/D

Promote customer information about importance and ways of saving water

Customers informed about water saving

Clients informed about efficient use of water (% customers/year) .... PI D

Maintain system integrity

Failures reduction

Mains failures (no./100 km/year). .... PI B/D

Service connection failures (no./1000 connections/year) .... PI D

Potential contamination points

Density of points where contamination may occur (no./km) .... PI B/D

Rehabilitate, renew or replace system components in case of malfunction

Rehabilitation of system components, including renewal and replacement

Mains rehabilitation (%/year) .... PI B/D

Mains renovation (%/year) .... PI B/D

Mains replacement (%/year) .... PI B/D

Service connection rehabilitation (%/year) .... PI D

Replaced valves (%/year) .... PI B/D

Assure security of system under normal conditions

Abstract protection

System input volume from intakes where protecting measure against contaminations where carried out (%) . PI B/D

Quality of water abstracted (Good/Poor) .... PL B

System surveillance and against contaminations on purpose

Plant video surveillance against intrusion (Total/Satisfactory/ Unsatisfactory) .... PL B/D

Improve plant protection

Access to plants

Plants with restricted access (% volume of water) .... PI B/D

Intrusion detection in plants

Plants with video surveillance and alarm system against intrusion of unknown people (% volume of water) ... PI B/D

Improve reservoir and groundwater protection

Personnel availability

Water resources and catchment management personnel (No./(10^6m^3/year)) .... PI B

Install and maintain online equipments for detecting contaminants in the water

Presence of undesired substances in the water

Unwanted substances concentration (mg/l) .... PI B

Reduction of pollution and waste generated by the water supply activity that may cause harm to public health

Pollutants emission

Amount of each pollutant released to environment (ton/(10^6m^3/year)) .... PI B/D

Waste production

Production of each type of residue (ton/(10^6m^3/year)) .... PI B/D

Control pollutants emission regularly

Pollutants identification and quantification

Surveys to identify and quantify pollutants (no./year) .... PI B/D

Reduction of pollutants production

Reduction of pollutant production (%/year) .... PI B/D

Control residues produced regularly

Residues identification and quantification

Surveys to identify and quantify residues (no./year) .... PI B/D

Reduction of residues production

Reduction of residues production (%/year) .... PI B/D

Table 3. ASSESSMENT CRITERIA, TACTICAL AND OPERATIONAL OBJECTIVES AND PERFORMANCE MEASURES FOR STRATEGIC OBJECTIVE "PROVISION OF SERVICE UNDER NORMAL CONDITIONS".

Strategic objective		
Provision of service under normal conditions		
Access to the service		
Household and businesses supply coverage (%)	PI	B/D
Amount of water in the system		
Available water / minimum requirements water ratio (-)	PI	B/D
Restrictions to water service		
Population experiencing restrictions to water service (%)	PI	B/D
Days with restrictions to water service (%)	PI	B/D
Adequate pressure in the system		
Percentage of nodes where insufficient or excessive pressures occur (%)	PI	D

Tactical objectives			Operational objectives		
Promote total population coverage by the water supply system			**Service connections constructed**		
Connections missing that may be constructed			*Service connections constructed and connected*		
Connections that may be constructed in a period of time/ missing connections (%)	PI	D	**New service connections** (no./year)	PI	D
			Service connections missing		
			Water supply coverage after new service connections are built (%)	PI	D
Rapid re-establishment of service in case of failure of components of the system			**Maintain spare pieces**		
			Spare pumps		
			Pumps with spare units/total pumps(%)	PI	B/D
			Spare valves		
			Valves with spare units/total valves(%)	PI	B/D
Promote distribution network redundancy			**Reduce branched links**		
Importance of pipes for the rest of the network in case of failure			*Importance of pipes*		
HCI (-)	TPI	D	**Total demand of the nodes in branched network/total network demand** (%)	PI	D
Isolating valves			**Improve the possibility of isolating parts of the network**		
Valve density (no./km)	PI	D	*Isolating valves*		
			Valve density (no./km)	PI	D
Maintain system integrity			**Rehabilitate, renew or replace system components in case of malfunction**		
Failures reduction			*Rehabilitation of system components*		
Mains failures (no./100 km/year)	PI	B/D	**Mains rehabilitation** (%/year)	PI	B/D
Service connection failures (no./1000 connections/year)	PI	D	**Mains renovation** (%/year)	PI	B/D
Potential contamination points			**Mains replacement** (%/year)	PI	B/D
Density of points where contamination may occur (no. /km)	PI	B/D	**Service connection rehabilit.** (%/year)	PI	D
			Replaced valves (%/year)	PI	B/D
			Pump refurbishment (%/year)	PI	B/D
			Pump replacement (%/year)	PI	B/D

Strategic objective

Provision of service under normal conditions (contd.)

Adequate pressure management in the system

Assuring positive pressures in the network

Nodes with negative pressures / total number of nodes (%) .... PI D

Assuring pressures under maximum values pre-established

Customers affected by excessive pressures/registered customers (%) .... PI D

Assuring pressures above minimum requirements

Pressure adequacy in distribution (%)... PI D

Pressure adequacy in bulk supply (%) .. PI B

Assuring pressure fluctuations that allow for a comfortable use of the service by customers and avoid failures in equipments

Service connections with pressure fluctuation related problems in a period of time (%)................................... PI D

Respect maximum pressures

Identify problems related to excessive pressures

Maximum pressures accomplishment in distribution (%).................................. IDT D

Correct excessive pressures in distribution

Nodes where excessive pressures were reduced / nodes with excessive pressures (%) .... ID D

Respect minimum pressures

Identify problems related to too low pressures

Minimum pressures accomplishment in distribution (%).................................. IDT D

Correct too low pressures in distribution

Nodes where too low pressures were reduced / nodes with too low pressures (%) .... ID D

Respect pressure fluctuations maximum

Identify problems related to excessive pressure fluctuations

Pressure fluctuation (%) .... IDT D

Correct excessive pressure fluctuations

Control legal access to water sources

Water abstraction legal access

Licensed abstracted water (%) .... PI B/D

Water availability management

Water resources availability (%)............ PI B/D

Personnel availability

Water resources and catchment personnel (no./(10^6 m^3/year)) .... PI B/D

4 CONCLUSIONS

The aim of this paper is the presentation of a set of objectives, assessment criteria and performance measures considering a decision process methodology for infrastructure asset management. Taking ISO 24500 standards examples as a basis, a more exhaustive task was developed: several strategic objectives for infrastructure planning were considered and completed with tactical and operational objectives, according to assessment criteria and performance measures are proposed for each objective. The full chain is synthetically presented in this paper for two of the strategic objectives.

The authors are confident that the result may have a significant leverage effect in the implementation of the ISO 24500 standards in practice, providing detailed guidance to water utilities for applying the referred standards, not only in a more effective way but also as part of decision processes.

5 REFERENCES

Alegre, H. (1992). *Tools for supporting the technical management of urban water supply systems*. PhD thesis, Technical University of Lisbon and National Laboratory of Civil Engineering, Portugal, ISBN 972-49-1608-1, in Portuguese (590 pages).

Alegre, H. (2007). *Water and wastewater infrastructure asset management*, Research Program, LNEC, Lisbon, Portugal (375 pp.) (in Portuguese).

Alegre, H.; Baptista, J.M.; Cabrera Jr., E., Cubillo, F.; Duarte, P.; Hirner, W.; Merkel, W.; Parena, R. (2006). *Performance indicators for water supply services*, second edition, Manual of Best Practice Series, IWA Publishing, London, ISBN: 1843390515 (305 pp.).

Alegre, H.; Hirner, W.; Baptista, J.M.; Parena, R. (2000). *Performance indicators for water supply services*, first edition, Manual of Best Practice Series, IWA Publishing, London, ISBN: 1900222272 (150 pp.).

Cardoso, M.A. (2007). *Performance assessment of urban drainage systems*. PhD thesis submitted to Technical University of Lisbon, Lisbon, Portugal (in Portuguese).

Coelho, S.T. (1997). *Performance in water distribution - a systems approach*. Water engineering and management, Research Studies Press Ltd., Somerset, England, ISBN 0 86380 219 2 (pp. 222)

Duarte P., Alegre H., Monteiro A.J. (2007). *A new framework for evaluating water distribution network investments using performance assessment analysis*. Paper submitted to IWA World Water Congress, Vienna, 7-12 September 2008.

GRI (2006). *Sustainable Reporting Guidelines*. Global Reporting Initiative, Amsterdam, The Netherlands (www.globalreporting.org).

IAM/BSI (2004). *PAS-55 – Asset management, Part 1: Specification for the optimized management of infrastructure assets*, Institute of Asset Management & British Standards Institution – Business Information (24 pp.)

ISO (2007 b). ISO/FDIS 24511:2007(E). *Activities relating to drinking water and wastewater services - Guidelines for the management and for the assessment of wastewater services*.

ISO (2007a). ISO/FDIS 24510:2007(E). *Activities relating to drinking water and wastewater services - Guidelines for the assessment and for the improvement of the service to users*.

ISO (2007c). ISO/FDIS 24512:2007(E). *Activities relating to drinking water and wastewater services - Guidelines for the management and for the assessment of drinking water services*.

Matos, R.; Cardoso, A.; Ashley, R.; Duarte, P.; Molinari, A.; Shulz, A. (2003). *Performance indicators for wastewaters services*, IWA Publishing., ISBN 19002229061, London, (192 pp.).

Sjøvold, F.; Alegre, H.; Guérin-Schneider, L.; Stahre, P.; Cabrera Jr., E.; Ancarani, A. A.; Vagliasindi, F.; Beja Neves, E.; Mallard, P. (2007). *Performance assessment of urban infrastructure services: state-of-the-art in the drinking water, wastewater and solid waste sectors*. COST C18 report (draft version dated October 2007).

Activity Based Costing and Benchmarking – the dual Approach of Ruhrverband

A. Bäumer[1]

[1] *abu@ruhrverband.de,Ruhrverband, Kronprinzenstr. 37, D-45128 Essen, Germany. Tlf: +49 / 201 / 178 1023, Fax: +49 / 201 / 178 1025*

Keywords: Activity based costing, benchmarking, performance indicators

Edited by Enrique Cabrera Jr and Miguel Ángel Pardo. ISBN: 9781843391913.
Published by IWA Publishing, London, UK.

ABSTRACT

This paper describes two approaches for cost controlling and improvement of Ruhrverband in Germany, focuses on the role of performance indicators and discusses some of the experiences that were obtained.

1 INTRODUCTION

The Ruhrverband (Ruhr River Association) is responsible for the planning, financing, operating and maintaining of all assets for water quantity and water quality management in the catchment area of the Ruhr. Around 800 different facilities (reservoirs, wwtps, stormwater tanks, pumping stations, gauging stations and impounding lakes) are operated in an area of 4,485 km^2 and by around 1,100 employees. The yearly turnover amounts to around 280 million €. The Ruhrverband is a corporate, self-governing and non-profit body under public law.

Of course the members - the communities and districts, all wastewater discharging industries and the waterworks along the Ruhr - are highly interested in reasonable fees and so the costs always have to be focussed. Today the operational controlling and cost management relies on two approaches.

2 INTERNAL ACTIVITY BASED COSTING

The traditional cost accounting is sorted in cost centres (e.g. departments, wwtps, reservoirs) and cost categories (e.g. for depreciation, interest, salaries, energy) using SAP R/3 for assistance.

In the past this traditional way of cost accounting left some questions open. In larger utilities like Ruhrverband (large both in terms of enterprise size and in terms of work area) there often is a high degree of bundling, dividing and specialising labour. Some departments and groups have a very broad field of work (e.g. groups of staff operating simultaneously several different facilities), others like the service departments have more specialised responsibilities (e.g. lawyers or central maintenance groups).

Sorting costs only to cost centres and cost categories left some questions open. The most products and processes are worked out in close cooperation by different departments and so by different cost centres. For a goal-orientated controlling it seemed to be reasonable to get more knowledge about the different products, to know which of these departments or groups is responsible for which part of the product costs. Further on it seemed reasonable to focus not only on the costs, but also on the amounts of work performed.

So today additionally all costs are allocated to so called external and internal products by Activity Based Costing (ABC). These products represent three main fields of activities:

"External Products" (total number 44) comprise all activities and processes which are directly generated for the members or customers of Ruhrverband. Examples are wastewater treatment, operation of reservoirs, stormwater treatment, operation of hydropower plants or timber industry activities.

"Construction Products" (total number 5) are directly generated for the planning or construction of facilities. Examples are the planning of plants or the financial controlling of construction sites.

"Service Products" (total number 63) comprise all activities and processes for the internal "customers" in the company. Examples are IT-Services, fleet management, human resources management or financial management.

Each product is exactly defined and described by a performance indicator dividing the costs by a parameter, which is reflecting the outputs of the activities (products). These performance indicators by that indicate a specific price for each product. The following tables present some examples.

Table 1. EXAMPLES FOR EXTERNAL PRODUCTS

No.	Product name	Performance indicator	Definition
E141	timber industry	€ / sold cubic meter wood	all services for the selling of wood from the water protection areas
E401	wastewater treatment	€ / wastewater unit	comprises all processes of wastewater treatment, sludge treatment and residual disposal in wwtps
E402	stormwater treatment	€ / m^3 tank volume	all services for stormwater treatment
E463	impounding lakes	€ / m^3 im-poundment vol.	comprises all services for the operation of the lakes
E 623	hydropower plants	€ / KW instal-led capacity	comprises all services for the hydropower plants in the dams of the reservoirs

Activity Based Costing covers all courses of business and all departments of Ruhrverband. Once a year all cost centres allocate their costs for personnel, depreciation, interest, energy, procurement etc. to the different products. These calculations are running by assistance of MS-Excel. The results are summarized in a detailed report. This report is published in the intranet of the company so that everyone is able to inform himself.

Table 2. EXAMPLES FOR SERVICE PRODUCTS

No.	Product name	Performance indicator	Definition
S131	IT-Service	€ / personal computer	comprises purchase, configuration etc. of all desktop personal computers
S310	Personnel administration, salary accounting	€ / employee	personnel and material costs of human resources department, costs for external service providers
S331	Office buildings	€ / m² office space	depreciation of buildings and maintenance costs ("rent excl. charges, heating etc.")
S320	Legal advice	€ / workday	personnel and workplace costs
S342	Facility management	€ / m² office space	Internal and external watchman and cleaning services
S530	Central Maintenance	€ / workday	all costs for maintenance works by the responsible department

The performance indicators can be analysed in different ways.

Activity Based Costing first gives an overview about the cost structures of all products. Figure 1 shows the proportion for the different cost categories for some products of the water quality sector of Ruhrverband. In average and considering the costs for the water levy and internal services the capital costs amount to 40 % of the costs for wastewater treatment, to 65 % of the costs for stormwater treatment and to around 30 % of the costs for the disposal of sludge and other residues. Here the costs for sludge transport and incineration are determining nearly half of the costs.

By building up time series the cost history of each product can be analysed over the years.

Typical service products worked out in the company can be compared to market prices. Also different plants in the company (e.g. wastewater treatment plants) can be compared with each other and by Benchmarking to external plants.

As Figure 2 presents the performance indicators of some larger plants of Ruhrverband are differing in a wide range. These differences are mainly resulting from

- the conditions of the site or the process,
- differing constructions costs
- different years of construction and
- sporadic expenses e.g. for sludge dewatering or maintenance.

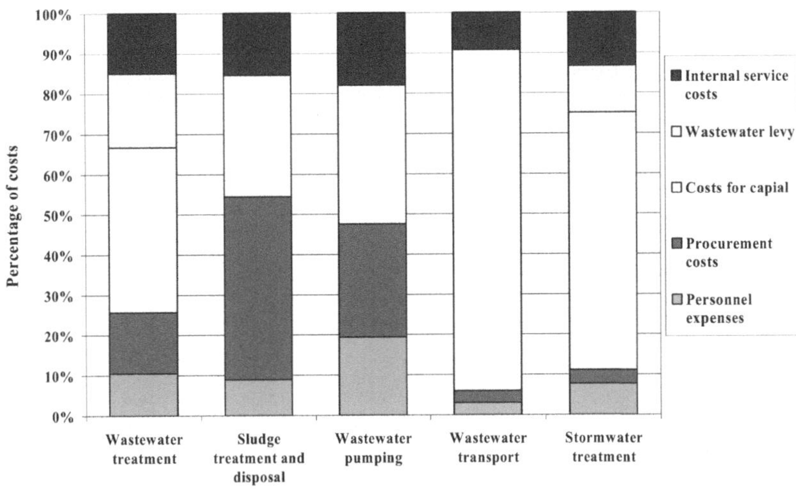

Figure 1. PROPORTION OF DIFFERENT COST CATEGORIES FOR SOME PRODUCTS OF THE WATER QUALITY SECTOR OF RUHRVERBAND

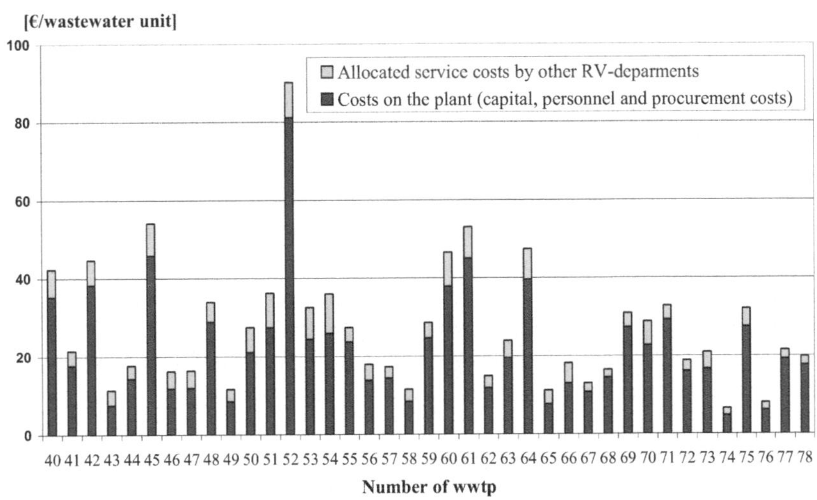

Figure 2. SPECIFIC COSTS OF GREATER WWTPS OF RUHRVERBAND

Further on the system displays which department recorded an effort to the different activities and by that is responsible for a part of the price of the product. Figure 3 presents this by example of a service department. The figure displays that this department is putting his main efforts into just six products. Alongside minor efforts arise from the work for another 8 products.

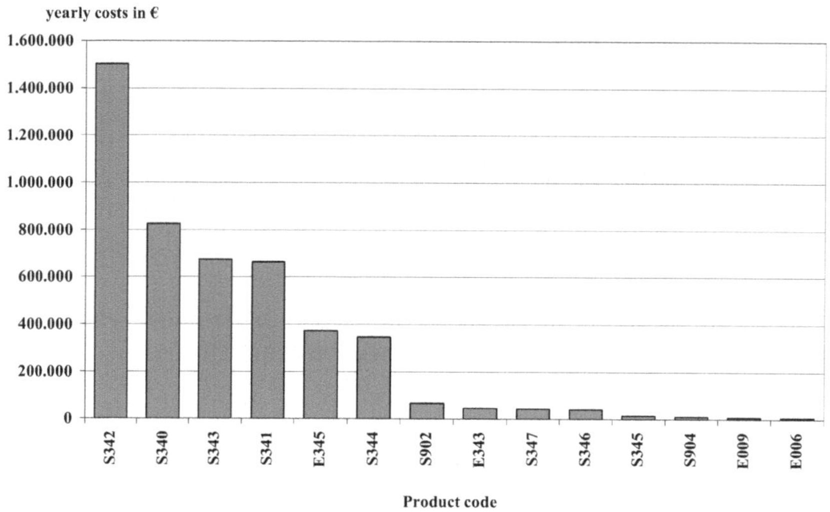

Figure 3. ANALYSING THE WORK OF A SERVICE DEPARTMENT – INTO WHICH PRODUCTS DID THEY INVEST THEY BUDGET?

From another point of view it can be identified by Activity Based Costing to which extent there is division of labour. Figure 4 shows by example of two service products (S904 - traineeship and B001 – service for construction works) which departments had been involved into these works. Both products are maintained by several departments. The service for constructions works relies mainly on the planning department (abbreviation code A-TB) and the four operating departments (abbreviation codes R-N, R-S, R-M and R-W) who are responsible for the supervision of the constructions sites.

Overall the combination of traditional cost accounting and Activity Based Costing is a multidimensional approach opening different perspectives for cost analyses to the cost centre managers.

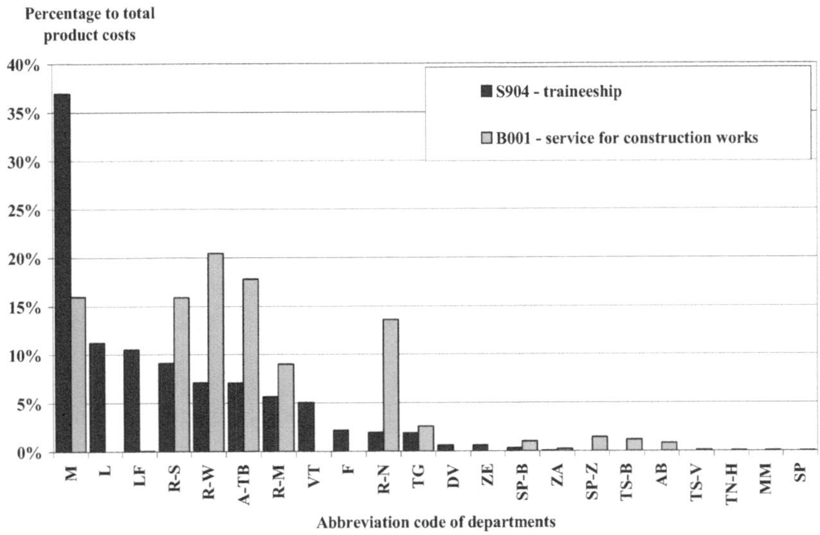

Figure 4. ANALYSING THE COSTS OF PRODUCTS – WHICH DEPARTMENTS TOOK HOW MUCH EFFORT?

3 EXTERNAL METRIC BENCHMARKING AND PROCESS BENCHMARKING

In the water sector of Germany benchmarking by initiative of some larger utilities and some state ministries and supported by the sector associations like DWA - German Association for Water, Wastewater and Waste since 10 years is in the focus of interest.

On the one hand there is a fast growing attendance in metric benchmarking approaches with today hundreds of participating utilities. From the authors experience metric benchmarking gives very aggregated, concentrated information and by that just allows rather superficial analyses of entire utilities. In practice it is complex to figure out the reasons for differing levels of performance indicators. When recognizing gaps between the utilities it is challenging to separate the impacts of the different general frameworks the utilities are working in (e.g. rural or urban region, amount and quality of wastewater, different effluent standards) from those of potential economical improvements. Nonetheless metric benchmarking reveals with some figures the general relations of fees and costs in different regions and cities and by that enables a utility to classify the economic situation.

Simultaneously a number of process benchmarking approaches for different fields of activity have been developed by consulting companies. The Ruhrverband is engaging not only his wwtps and reservoirs in benchmarking projects but also supporting functions like human resources management or the laboratories. In relation the costs of these departments are not that high, e.g. the costs of the human resources management division of Ruhrverband amount round about just to the costs of a wwtp with 35,000 population equivalents. Nonetheless it is an important signal to the workers and employees on the plants and facilities, that all functions and services in the company are focused by benchmarking.

Process benchmarking is much more detailed and concentrates on working out reasons for differing performance indicators. Some general observations and experiences are:

High performance indicators do not directly indicate potential savings. For wastewater treatment plants as the decisive factors for the complexity and differing costs

different local conditions,

characteristics in process engineering,

local prices for energy esp. for electricity,

way and costs of sludge and residue disposal have to be considered.

Because of these points a general „valid“ ranking of plants or a general Best Practice“ cannot be worked out. Improvements of external plants mostly are not directly applicable to the own plants.

Despite of these limiting factors process benchmarking is a very essential tool. It enables the persons, being responsible for the process and the costs, to analyse and assess the plants and to reflect themself in a intensive and structured share of experience with other plants and utilities.

4 EXPERIENCES WITH PERFORMANCE INDICATORS

In Germany performance indicators are likely defined in the form "costs per output" (e.g. amount of wastewater or population equivalents). By theory a performance indicator has to have a direct, at least a strong interrelation to a variable cost generating factor.

In the field of wastewater treatment several factors are determining the costs. Besides the points above-quoted in chapter 2

- the costs for capital are mainly a function of the size of the plant and also the performance that has to be achieved
- the costs for maintenance depend on the technical standard and the age of the plant
- the costs for operation depend on the loads of COD, nitrogen, phosphorus, suspended solids etc. which are eliminated in the plant.

The answer to the question, which of these parameters could be the most appropriate one depends on the individual plant. Still it is important to know, that focussing on just one parameter may lead to misinterpretations.

By means of a systematic analysis for eight feasible performance indicators it was described how much a ranking of 91 wwtps was depending on the choice of the parameter describing the output (Bäumer 2005). By way of example figure 5 draws a comparison between the performance indicators "€ per population equivalent (calculated by the load of COD)" and "€ per m^3 treated wastewater". For this comparison first the positions of the plants were calculated. Then these positions were correlated with each other. Figure 4.1 points out that the very most plants have differing positions. The systematic analysis of all correlations between all rankings had the result that around 30 % of the plants had no relevant changes in the ranking-positions, but nearly 30 % of the changes were relevant like that for the plant highlighted in the figure. This means 30 % of the plants are changing their position in a way that just by changing the parameter in the denominator a top position could shift to an average one, or an average position could shift to a substandard, or vice versa.

So the definition of the performance indicator has a significant effect on the results when comparing wwtps. This should be regarded in economic assessments, the eventuality of misinterpretation has to be kept in mind. Especially the publication of performance indicators without anonymisation ("naming and shaming") should be avoided until it is guaranteed that no company can be shamed by a unreliable "bad" assessment.

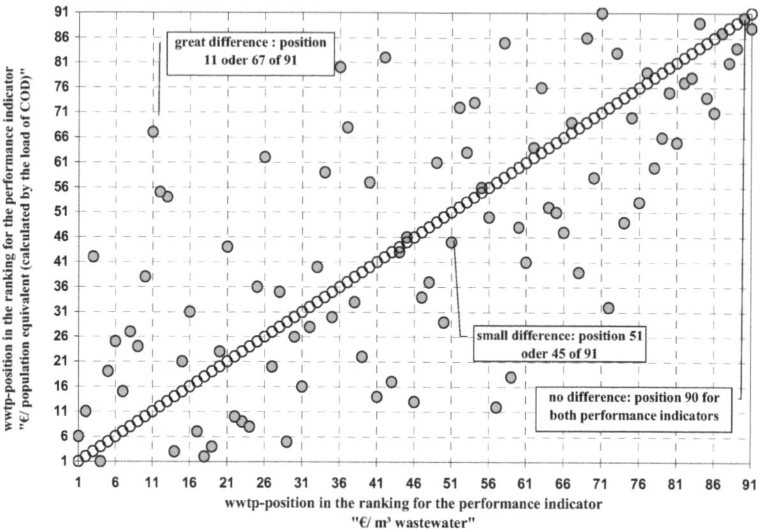

Figure 5. CHANGES IN THE RANKING OF 91 WWTPS DUE TO A CHANGE OF THE PARAMETER IN THE DENOMINATOR OF THE PERFORMANCE INDICATOR

To minimize the impacts of this context Ruhrverband uses the above quoted "wastewater unit" as the reference parameter in the denominator of the performance indicator for wastewater treatment (see table 2.1). The wastewater unit considers several parameters like wastewater flow, eliminated COD, nitrogen, phosphorus and the relation of COD to BOD_5 (for details see Evers P. et al., 1999, Bäumer et al., 2007) .

5 REFERENCES

Bäumer A., Evers P. and Grünebaum T. (2007): Kosten- und Leistungsrechnung für kaufmännische und technische Fragestellungen, conference papers of BEW-conference on 3/4 September 2007 in Essen

Bäumer, A. (2005): Methodically Remarks to the Definition and Scaling of Economic Performance Indicators for Wastewater Treatment Plants, GWF-Wasser Abwasser 06 / 2005 (in German)

Evers, P., Grünebaum, T., Wilde, J.: Leistungskennzahlen als Grundlage für vergleichende Wirtschaftlichkeitsbetrachtungen in der Siedlungswasserwirtschaft. GWF Wasser-Abwasser, 140. Jahrgang (1999), 04/1999

Management of corporate development based on performance indicators: Implementation of a balanced scorecard within the Emschergenossenschaft/ Lippeverband

H. Goebel[1], J. Stemplewski[2] and T. Fock [2]

[1]*heikegoebel@eglv.de Emschergenossenschaft / Lippeverband Kronprinzenstraße 24 45128 Essen Germany*
[2]*Emschergenossenschaft / Lippeverband Kronprinzenstraße 24 45128 Essen Germany*

 Performance Assessment of Urban Infrastructure Services
Edited by Enrique Cabrera Jr and Miguel Ángel Pardo. ISBN: 9781843391913.
Published by IWA Publishing, London, UK.

ABSTRACT

'Think private to stay public'. Acting according to this maxim dictum demands the implementation of an up-to-date management and controlling instrument. With this in mind the Emschergenossenschaft and Lippeverband (EG/LV) introduced an organisation-wide integrated management system based on a full-coverage process structure.

The integrated concept of the management system within the EG/LV combines a number of previously introduced and proven instruments for business management – as, for example, objectives, budgeting, performance indicators, benchmarking, reporting and action plans. To optimize the translation of corporate strategies into operative business a balanced scorecard (BSC) system - a strategic management instrument - has been implemented.

This lecture will introduce the BSC at EG/LV and consider how this instrument interacts with other instruments used within the EG/LV. As well as the conceptual design of the BSC a further focus of this lecture will be the implementation of this instrument within the EG/LV. Acceptance of management instruments on the part of management and employees is quintessential for successful implementation.

1 INTRODUCTION

The Emschergenossenschaft and Lippeverband (EG/LV) are publicly-owned regional water boards with statutory responsibility for water management within the Emscher and Lippe catchment area. This responsibility includes treating wastewater from 3.8 million inhabitants as well as from industry and commercial businesses equivalent to a further population of 3 million inhabitants in an area covering 4 200 km². In addition to the treatment and disposal of wastewater within the terms of an integrated river basin management, the palette of products and services offered by the EG/LV includes: maintenance of water bodies, environmentally responsible development of water bodies, flood protection and preservation of ground water levels. With a balance sheet total of approximately €2.5 billion and a turnover of around €300 million there are some 1500 employees involved with mastering these water management tasks.

With regard to our entrepreneurial mission the EG/LV have undergone a significant transformation over the last few years. In particular this included the rigorous introduction of management tools. In keeping with our legal obligations these guarantee technically appropriate and, above all, economically sound activities on behalf of our customers.

2 THE BALANCED SCORECARD INSTRUMENT

The balanced scorecard (BSC) is a performance measurement system based on performance indicators that allows both ex-post checks and leading strategic control of activities within an organisation. This instrument was initially designed for use in the private sector at the beginning of the nineties by Professor Robert Kaplan (Harvard Business School) and by David Norton (then consultant at KPMG); later both authors recommended this instrument for the public service and non-profit sector. By 1996 some 60% of the Fortune 1000 companies had implemented the BSC system. 46% of Germany's top companies were using the BSC system by the year 2000.

The BSC is based on both monetary and non-monetary performance metrics. It simultaneously takes into consideration both short-term and long-term targets. A BSC system monitors lagging target figures (in terms of a traditional 'ex-post' check) as well as leading indicators (performance drivers, 'ex-ante' check). It combines objectives and actions from the point of view of a cause and effect relationship, which, as a consequence, allows the organisation's management to document and examine the 'recipe for success' of the organisation. The BSC provides an overall view of the company's performance from the perspective of performance results (these in particular from the point of view of investors and customers) as well as from the perspective of performance realisation (meaning the processes and ability necessary for the creation of products and services as well as continual improvement of these through a process of learning). Thus, the BSC is divided into the following four decision-relevant perspectives, which set the field for performance indicators: finance, learning and growth, customer and internal business process. Hence, a company's (or organisation's) performance is depicted as a balance between the different perspectives on a clearly structured scorecard.

3 APPROACH TO IMPLEMENTING A BALANCED SCORECARD AT EG/LV

To gradually implement the organisation's vision and strategy, the EG/LV resolved in 2005 to introduce a BSC system as a controlling instrument. As an organisation-wide controlling instrument, the BSC is being implemented step-by-step.

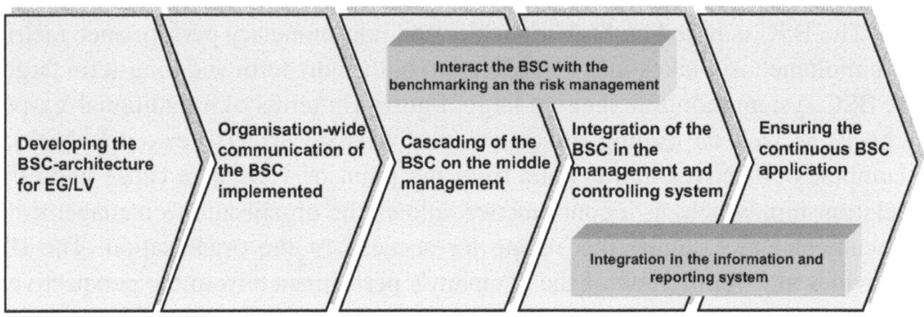

Figure 1. BSC-IMPLEMENTATION (STEMPLEWSKI, LANGE, SCHÄFER, 2005)

The first step saw the BSC being developed for the entire organisation and the top level of management. Particular importance was attached to ensuring that the structure and content of the instrument corresponded to EG/LV requirements as a non-profit organisation operating in the water management business. The organisation's vision and business objectives formed the basis for developing the performance indicator system. A 'strategic topic' from which the respective individual objective can be derived was formulated for each of the BSC perspectives. These individual objectives were assigned a manageable number of indicators relevant to decision-making processes while also taking into consideration potential risk/opportunity factors.

In 2005 the board of management adopted the BSC system together with objectives and performance indicators; it was also agreed to test the system intensively for a period of two years. Throughout this time there was an intensive exchange of views between those responsible for the BSC architecture and its users. The optimised BSC was adopted for the entire organisation and top management in autumn 2007. Every year beginning 2008 this BSC level will be assigned target values and measures to be undertaken. BSC roll-out for downstream decision-making entities will commence at the same time.

4 STRUCTURE OF THE BALANCED SCORECARD AT EG/LV

In view of the numerous environmental standards that water management companies must comply with, the EG/LV supplemented the four classic perspectives by integrating 'sustainability/environmental quality' as a fifth perspective.

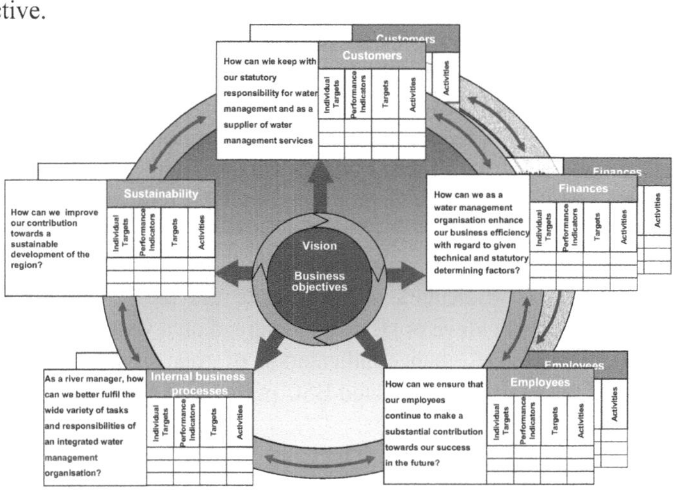

Figure 2. STRUCTURE OF THE BSC AT THE EMSCHERGENOSSENSCHAFT / LIPPEVERBAND

Financial targets are defined in the Financial Perspective. At the same time a decision is made as to which performance figures are to measure and monitor the financial situation. The strategic question at this point is: 'How can we as a water management organisation enhance our business efficiency with regard to given technical and statutory determining factors?' Targets consistent with fulfilling the requirements of EG/LV customers are defined in the Customer Perspective; these are defined in keeping with our statutory responsibility for water management and as a supplier of water management services. The 'Internal Business Processes Perspective' examines the process of creating products and services. This is where targets concerning the quality of the value added process are defined and rendered quantifiable through the use of operative performance indicators; these targets must be regarded as a prerequisite for achieving the success to which we aspire from the point of view of our customers. The key question is: "As a river manager, how can we better fulfil the wide variety of tasks and responsibilities of an integrated water management organisation?"

The EG/LV is aware of the fact that our employees have a huge contribution to make towards the success of the organisation. The key question posed in the Employee Perspective is concerned with how we can ensure that our employees continue to make a substantial contribution towards our success in the future. The basis for this success is to be found in improving motivation and commitment as well as boosting expertise and enthusiasm for innovation. The Sustainability Perspective pursues the strategic question of how we can improve our contribution towards a sustainable development of the region. In the Sustainability Perspective, targets are applied to the elements ecology, economy and social responsibility. These include improving quality, optimising consumption of resources as well as promoting social activities.

Each perspective contains three to five individual targets allied to the respective key question of the perspective. Individual targets are measured by means of performance indicators. To formulate the performance indicators, targets and activities to achieve objectives are agreed on a yearly basis.

Choosing the right performance indicators is of great significance. From the point of view of the EG/LV the selected BSC performance indicators must meet the following criteria:

- Top-Management relevance: Does the performance indicator represent an issue which is of importance to top-management?
- Measurability / objective orientation: Does the performance indicator provide information concerning the objectives to be achieved?
- Influence: Does the performance indicator represent an issue that can be influenced by the organisation?
- Comparability: Is the performance indicator formulated in a way that it is possible to compare performance indicators from other periods?
- Addressee orientation: Does the performance indicator give its addressee information relevant to his or her needs?
- Relation to the benchmarking.

Benchmarking between companies of this branch of business underlie the contents design of the BSC.

It supports aligning objectives with internal processes in such a manner that allows them to be continually optimised in accordance with process requirements. Consequently, benchmarking contributes towards improving efficiency with regard to security of supply, quality, customer service, sustainability and economic viability.

Aligning performance indicator objectives to best practice solutions helps to improve decisions when planning strategic activities. Balanced scorecard and benchmarking not only support providing internal stakeholders with information, but can also contribute towards providing external stakeholders (customers, the general public and politicians) with information.

5 INCORPORATING THE BALANCED SCORECARD IN THE CONTROL PROCESS AT EG/LV

The BSC is intended to implement the organisation's vision and strategy step-by-step by means of strategic targets, measurable goals in terms of performance indicators and objectives through to individual activities.

The combination of the BSC system with previously introduced instruments enables the EG/LV to take a holistic approach in the following steps:

- Translation of the business vision and strategy into a balanced system of objectives
- Communication of the strategy to the next levels in the hierarchy as well as further implementation and clarification
- Coordination of department-specific objectives and individual objectives with the strategy to ensure that all those involved understand and can identify with the long-term objectives
- Management of the business plan and with that an allocation of resources based on the strategy
- Reviewing the achievement of objectives, learning and adapting

All told, the BSC system increases the transparency of the process of creating products and services, which as a consequence enables the organisation's management to exercise effective control. To this end the BSC must not be an isolated instrument. Explicit agreement and transparent portrayal of the interaction of the BSC with other instruments and systems are a basic precondition for establishing a BSC system. The graphic below provides an overview of the instruments at the EG/LV with which the BSC interacts.

Figure 3. OVERVIEW OF THE INSTRUMENTS AT THE EMSCHERGENOSSENSCHAFT/LIPPEVERBAND

The BSC is integrated in the process 'Organisation Development and Controlling' at EG/LV. This is a management process within the integrated management system. Presently the EG/LV is in the process of implementing an organisation-wide integrated management system (IMS) to support management tasks. This will systemise future management tasks, which will be continually improved with the aid of organisational 'tools' such as BSC and benchmarking. With this in mind, the organisation's management responsibilities include:

- Planning targets and activities
- Determining and monitoring work processes
- Ensuring adherence to legal guidelines
- Guarantee training of personnel to meet workplace requirements
- Ensure that the organisation continues to develop.

Thus, the IMS will serve to implement both organisational responsibilities and support the continual development of the organisation.

The basis for the development of the organisation is the so-called continual improvement process which takes a similar approach to Deming's PDCA cycle.

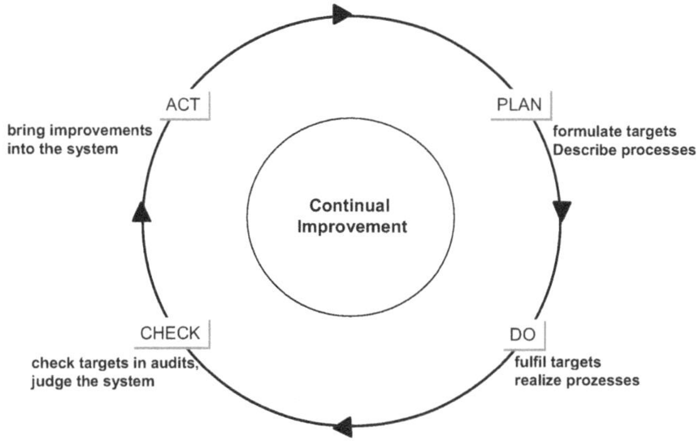

Figure 4: **PDCA CYCLE AS THE BASIS FOR THE DEVELOPMENT OF THE ORGANISATION**

To this end existing instruments such as the BSC, benchmarking, audits and risk management are aligned coincident with one another and further developed. Newly developed instruments (including organisation-wide process analyses, organisation-wide management programmes) are combined with established instruments. As an early warning system at EG/LV the IMS will also appropriately integrate aspects of strategy, determining and pursuing objectives as well as the necessary controlling so as to form a uniform controlling system. To this end, the BSC is an essential instrument. The BSC will continue to be improved in conjunction with the implementation of the organisation-wide IMS; this will ensure that measurement parameters for achieving objectives and guidelines designed to outline the approach to achieving objectives are co-ordinated to an optimum.

6 FUTURE PROSPECTS

Controlling instruments such as the BSC are themselves subjected to a process or continual improvement. Constantly examining and adapting the instruments to suit organisation requirements are a prerequisite for their effective use. The BSC for the entire organisation and top management level has been through all four steps of the continual improvement process between 2005 and 2007 (please refer to Figure 5). The present cascading of the BSC down to the divisions and departments at EG/LV will also complete this loop.

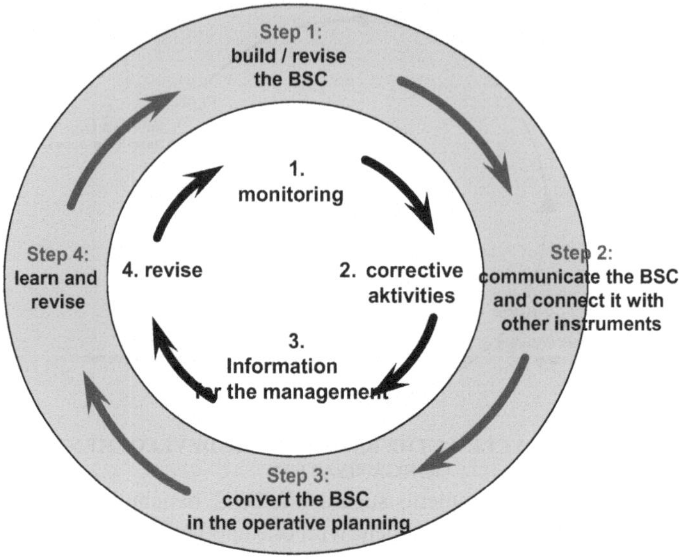

Figure. 5: CONTINUAL IMPROVEMENT PROCESS FOR MANAGEMENT INSTRUMENTS

In accordance with the principle 'learning from the best' it is intended that in a further step the BSC will be linked to benchmarking activities so that performance indicator objectives can be based on best class performers. A strategy map will be developed in order to be able to portray the most important cause and effect relationships in a graph. Discussions are presently being held about the BSC reporting system. In particular it is planned to produce yearly and quarterly reports in addition to a clearly structured scorecard containing the more important performance indicators.

REFERENCES

Jochen Stemplewski, Christoph Lange, Sigrid Schäfer „Balanced Scorecard als Führungsinstrument in der Wasserwirtschaft"; Korrespondenz Abwasser Nr. 8/2005

Managing utilities with performance indicators in Russia

S.B. Sivaev[1] *and A.Y. Rodionov*[2]

[1] *Director, Urban Economy Department, The Institute for Urban Economics, 20/1, Tverskaya str., Moscow, 125009, Russia. Tel/fax +7 495 787 45 20. E-mail*: sivaev@urbaneconomics.ru
[2] *Project Manager, Urban Economy Department, The Institute for Urban Economics, 20/1, Tverskaya str., Moscow, 125009, Russia. Tel/fax +7 495 787 45 20. E-mail*: rodionov@urbanconomics.ru

Keywords: Three- five Keywords to help identification, indicators, Russia, management

 Performance Assessment of Urban Infrastructure Services
Edited by Enrique Cabrera Jr and Miguel Ángel Pardo. ISBN: 9781843391913.
Published by IWA Publishing, London, UK.

ABSTRACT

The gradually increasing presence of private water companies in Russian municipal WSS market forces local authorities to defend their interests in contractual arrangements with private companies. One particular place in Russia where performance indicators are now being incorporated into the contract between the city and operator is Perm. Private operator of Perm water utility, “Novogor-Prikamye”, use performance indicators either for internal purposes (in order to identify weak points, increase efficiency and monitor company conditions) or for external purposes such as contract design and public relations. Implementation of the performance-based contract resulted in significant operational improvements for the city water supply.

1 GENERAL INFORMATION ABOUT THE WATER SUPPLY AND WASTEWATER SECTOR OF RUSSIA

At 2006 the water supply and sanitation (WSS) sector of the Russian Federation provided water supply and wastewater collection services to 78% and 62% of the population, respectively. Centralized water supply was organised in 100% of the cities and towns, 97% of urbanized localities and 38% of rural localities. Centralised wastewater collection was organised in 100% of the cities and towns, 79% of urbanised localities and 5% of rural localities.

The total average annual balance sheet value of production water supply and wastewater collection facilities reached 480.35 billion roubles (€ 13.96 billion), including 242.37 billion roubles (€ 7.05 billion) for the water supply sector and 237.98 billion roubles (€ 6.91 billion) for the wastewater sector.

Expenditures of water supply and wastewater collection enterprises related to the production and provision of services and resources to the users made 149.1 billion roubles (€ 4,33 billion), whereas their receipts over the same period made 144.1 billion roubles (€ 4,19 billion).

The bulk of communal resources and services produced in the Russian water supply and wastewater sector were provided by public and municipal enterprises. By the results of 2006, 87.3% of the total amount of water was provided to its users by municipal and state enterprises; the share of enterprises with other forms of property included 12.7% of the water provided to users, of which 9.7% was provided by private enterprises.

The Russian WSS sector is not devoid of some drawbacks; the main drawback is a stable deterioration of the condition of the enterprise fixed assets. For example, in 2006, 35% of the water supply networks and 29% of the wastewater collection networks used up their resources and needed replacement; in towns which are capitals of the Russian regions 41% of water supply networks needed replacement.

2. SOURCES OF INFORMATION ABOUT THE WATER SUPPLY AND WASTEWATER UTILITIES

Statistical observation over activities performed by the Russian WSS utilities is made by collecting information using a few forms of statistical reporting. These forms were developed during Soviet time and since then their content has not practically changed. They present information characterizing respective engineering infrastructure facilities (the length of networks, including networks requiring replacement, the power of pumping stations and treatment plants), the scope of provided services and their structure by the types of users, the number of accidents in networks, the number of basic staff, etc. Another one form of statistical reporting was introduced at the end of 1990s and contains information, partly characterizing economic outcomes of the water supply and wastewater enterprises by each type of services.

So, the current statistical reporting forms present information largely about the key outcomes of WSS utilities' operations, but they does not contain extremely important information necessary for a thorough understanding of the condition of the WSS sector, the quality of WSS services, and whether the volume, quality and charges for WSS services are reasonable.

3 INDICATORS AS A BASIS OF MONITORING OF THE WSS UTILITIES

The monitoring of the WSS utilities has become mandatory in the Russian Federation from 2006. According to the Federal Law No. 210-FZ of 30 December 2004 "On Principles of Regulation of Tariffs for Communal Organizations", "monitoring of implementation of production programme and investment programme by communal organizations shall be made by regulatory authorities so as to ensure electricity, heat and water supply, wastewater collection and treatment, disposal (burial) of solid communal waste and timely decision-making about the development of communal infrastructure systems". The law defines that "monitoring shall include collection and analysis of information on the fulfilment of indicators, specified by production and investment programmes, as well as analysis of information on the state and development of communal infrastructure systems". The law also specifies that "monitoring shall be made according to methods of the monitoring, containing a list of economic and other indicators used by regulatory authorities for analysis of information on the implementation of the production and investment programmes by communal organizations".

The gradually increasing presence of private water companies in municipal WSS market forces local authorities to defend their interests in contractual arrangements with private companies. Performance-based contracts are clear example of how performance indicators may be used by local authorities. Local authorities need to set clear objectives and targets in the contract with private operator and they must be able to monitor the efficiency of private operator. Performance indicators also play a public relations role by allowing both local authority and private operator to explain to the population the need and benefits for transfer of management to private company.

One particular place in Russia where performance indicators are now being incorporated into the contract between the city and operator is Perm. The contract essence between private operator of Perm water utility, "Novogor-Prikamye", and the local authority (proprietor of municipal WSS system) is expressed in the accurate, clear and quantitatively defined purposes of activity. Achievement of the purposes of activity is traced with use of system of performance indicators. In practice, the activity purposes are expressed in target values of performance indicators.

The selection of performance indicators was based on the following criteria:

- The indicator should lend itself to quantitative measurement at any point in time;
- The indicator should describe an outcome and not the process by which the outcome was reached;
- The indicator should yield the same outcome during a repeated measurement (for the same time period), independent of who is conducting the measurement;
- The indicator should have a clear definition that is understandable to anyone;
- The procedure for determining the value of each indicator should not require expenditures of materials or time that are incommensurate with that spent on other indicators.

All set of performance indicators can be divided into following basic groups:

- The indicators characterising a condition and potential of use of municipal system of water supply;
- The indicators characterising quality of activity;

- The indicators characterising efficiency of activity.

The result of comparison of target and actual values of performance indicators or testifies to successful activity of the company, or is the reason for penalty to the company.

Long term of contractual relations between the operator and the local authority has predetermined distribution of the purposes of activity on short-term, intermediate term and long-term.

Long-term purposes of activity defines the minimum size of the indicator which is necessary for observing during all period of the contract's validity. In particular, in the contract between "Novogor-Prikamye" and local authority the indicator of maintenance of round-the-clock water supply and water removal is used as the indicator of achievement of long-term purpose of activity.

The intermediate term purposes should be reached during period of the investment program's realisation. We will notice, that the part of the intermediate term purposes can become long-term on their achievement (the indicator of maintenance of 100 % of sewage treatment).

The short-term purposes are established for one year and allow to trace performance of specific targets, which can be solved within a year.

The use of performance indicators helps the local authority to answer the following questions:

- WHAT is happening in the municipal WSS sector?
- WHERE has positive/negative changes occurred in the volume, cost, and quality of WSS services?
- WHEN did these positive/negative changes happen?
- Did the cost for producing and delivering WSS services change in the period of analysis?
- What are the reasons for changes in the cost of producing and delivering WSS services?
- How well do changes in the cost of WSS services correspond to changes in service volume and quality?
- To what extent is budget financing of WSS projects reflected in the volume and quantity of WSS services?
- How did the population react to changes in the cost and quality of WSS services?
- Does the change in the cost of WSS services correspond to the population's ability to pay?

"Novogor-Prikamye" use performance indicators either for internal purposes (in order to identify weak points, increase efficiency and monitor company conditions) or for external purposes such as contract design and public relations.

Implementation of the performance-based contract resulted in significant operational improvements for the city water supply.

Implementation of the performance-based contract in Perm resulted in significant operational improvements for the city water supply.

3 REFERENCES

Sergey Ivanov, Sergey Sivaev and Ella Shalukhina (2007) *Municipal Water and Wastewater Services Reform in the Russian Federation*.

"Monitoring of the System of Water Supply and Wastewater in the Russian Federation and Tadjikistan". Prepared by the Institute for Urban Economics for The EU Water Initiative's EECCA Working Group. Paris, 2007

Performance indicators – Aquapor's three years activity

A. M. Monteiro[1] and T. Carvalho[2]

[1] *amonteiroåquaporservicos.pt*
[2] *tcarvalho@aquaporservicos.pt*

ABSTRACT

Águas de Portugal, which withholds 100% of AQUAPOR, detains a group of companies created in partnership with Portuguese municipalities for the management of public services of water supply systems, drainage and treatment of wastewater and solid waste and provides services to a total amount of about 70% of the Portuguese population.

Edited by Enrique Cabrera Jr and Miguel Ángel Pardo. ISBN: 9781843391913.
Published by IWA Publishing, London, UK.

AQUAPOR provides water distribution services in eleven concessions to a total population universe of 935 thousand inhabitants. It also offers services of wastewater systems management in the form of concessions and WWTP operational services to more than 1,57 millions inhabitants, being present in 40 municipalities.

AQUAPOR's role in water supply and wastewater drainage field will be addressed, in particular to what refers to the sector regulation institute and its evaluation system for measuring operator's service quality – effectiveness and efficiency - by using a relevant group of performance indicators.

It will be presented the implementation of performance indicators in AQUAPOR, how the variables and indicators' selection were chosen and the applied methodology for the collection and treatment of information (presented as annual operational reports).

All the work done to variables and indicators consolidation will be described, in particular its definitions, and the substitution of pre-defined variables and indicators by others with similar purposes (which have already being adopted by the regulation institute). In 2007, some of the company budgets were already produced based on the indicators matrix developed.

In parallel, software was developed which allows the information visualization and the possibility to create graphs and reports in an automatic way.

After the closure of 2004-2006 reports, the actual challenge is to consolidate the information and to use it to help companies to improve their performance and to define their goals based on the information obtained by the performance indicators.

1 GENERAL APPROACH

Public water supply in Portugal has more than 300 water suppliers, 19 bulk systems.

Bulk systems are companies created by "Águas de Portugal" (a company owned by the Portuguese government) in partnership with Portuguese municipalities for the management of public services of water supply systems, drainage and treatment of wastewater and solid waste, providing services to a total amount of about 70% of the Portuguese population

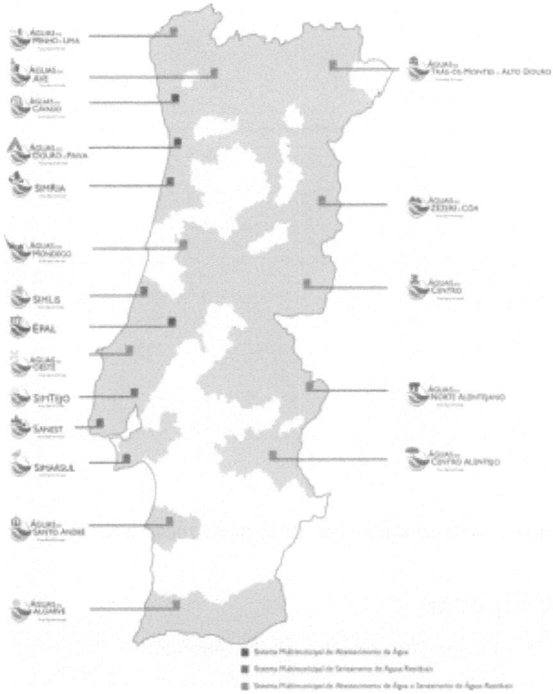

Figure 1. WATER SUPPLY AND WASTEWATER SANITATION (FROM WWW.ADO.PT)

1.1 The Holding AQUAPOR

AQUAPOR is a private society created in 1997 to participate in the Portuguese retail concessions market of wastewater and water supply, and it is 100% owned by Águas de Portugal.

Today, AQUAPOR provides water distribution services in eleven concessions to a total population universe of 935 thousand inhabitants, and also offers services of wastewater systems management in the form of concessions and WWTP operational services to more than 1,57 millions inhabitants, being present in 40 municipalities..

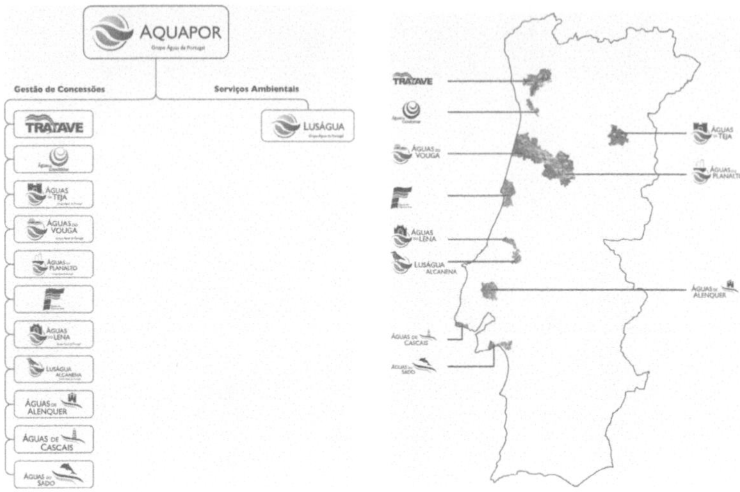

Figure 2 .AQUAPOR'S GROUP (FROM 2006 ANNUAL REPORT)

1.2 Main Figures

Concerning the retail market, AQUAPOR managed 33% of the private market and about 4% of the portuguese population at the end of 2006

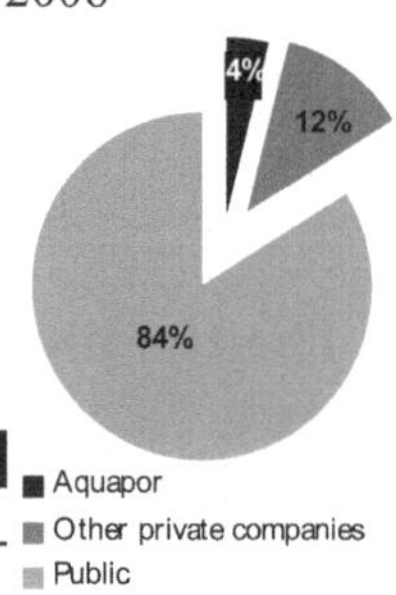

	AQUAPOR
Turnover (thousand Euros)	97,2
Number of employees (No)	1.039

Figure 3. PORTUGUESE WATER MARKET AT THE END OF 2006

Performance indicators – Aquapor's three years activity

	Water supply	Wastewater collection and treatment
Distribution/ Collection (Thousands m3)	55	92
Population (hundred inhabitants)	936	1.571
Costumers (hundred)	361	263
Treatment Plants (No)	10	29
Reservoirs (No)	377	--------
Pumping Stations (No)	140	144
Network (km)	5.802	2.096

Figure 4 .AQUAPOR'S KEY DATA (FROM 2006 COMPANIES ANNUAL REPORTS)

1.3 The role of the Regulation Institute

In Portugal, water supply, waste water collection and treatment services are regarded as a public benefit, essential to the citizens' welfare, public health, businesses and environmental protection. For that, these services need to be governed by a set of principles that include universal access, continuity and quality of service, efficiency and fair pricing.

The aim of regulation is, therefore, to protect the users' interest by promoting the quality of the services provided by the operators, and ensuring a fair balance in the charged costs. This must be achieved taking in account the need to safeguard the economic viability and legitimate interests of the operators in order to ensure the medium- and long-term sustainability of the service and to contribute to environmental sustainability.

The Institute for the Regulation of Water and Solid Waste (IRAR, established by Decree-Law 230/97 of 30 August) is the regulatory body in Portugal. Its Statutes are set out in Decree-Law 362/98 of 18 November, as amended by Decree-Law 151/2002 of 23 May. This institute only regulates the services established by concession contracts.

The regulation model comprises tools that self-monitor the operators' performance. The results thereof are validated by the regulating body with the help of assessment tools and further compared to the results of similar operators based at different geographic areas (benchmarking).

2 PERFORMANCE INDICATORS – THE START

This point deals with the implementation of performance indicators in AQUAPOR, the purposes presented for the variables and the selection of indicators, the methodology applied for the collection of information and its treatment in annual operational reports.

The need for the use of indicators came from the Group Administration which wanted indicators to compare the performance between different companies. Several points of discussion arose:

Which indicators should be taken into account?

The companies are different, they are not comparable!

The companies don't give us adequate and comparable data!

A work group including members of the Administration, Controller and Technical Areas was created with a specific goal,

Find a group of variables and indicator that respect the following aspects:

1. A small number of figures!
2. All the important information!
3. Easy data to obtain from companies!
4. Comparable indicators!

Afterwards, research work was conducted on these topics to learn about what has been made worldwide. Two relevant studies were taken in account . Nevertheless, those studies couldn't be directly transposed to our work aims since it includes more than two hundred variables. Moreover, many of the variables were not applicable or relevant and the data difficult to obtain.

I - A flowchart illustrating the process followed to choose the variables is shown:

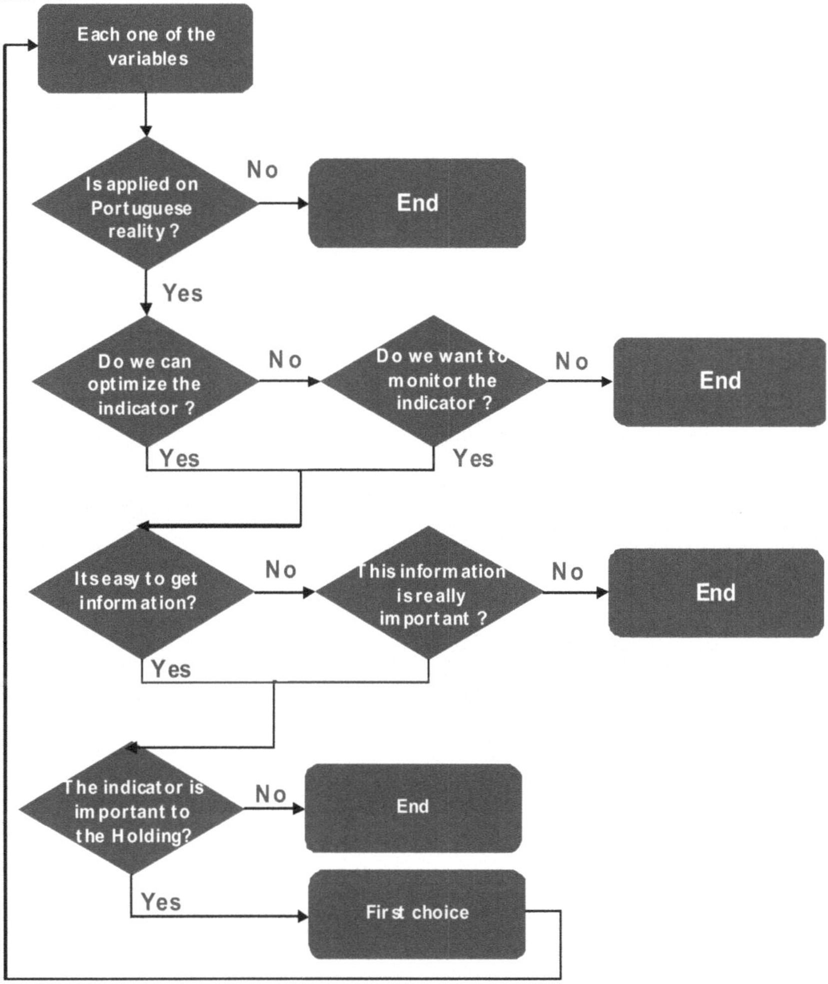

II - Criticise the "first choice" variables and indicators based on the following flowchart:

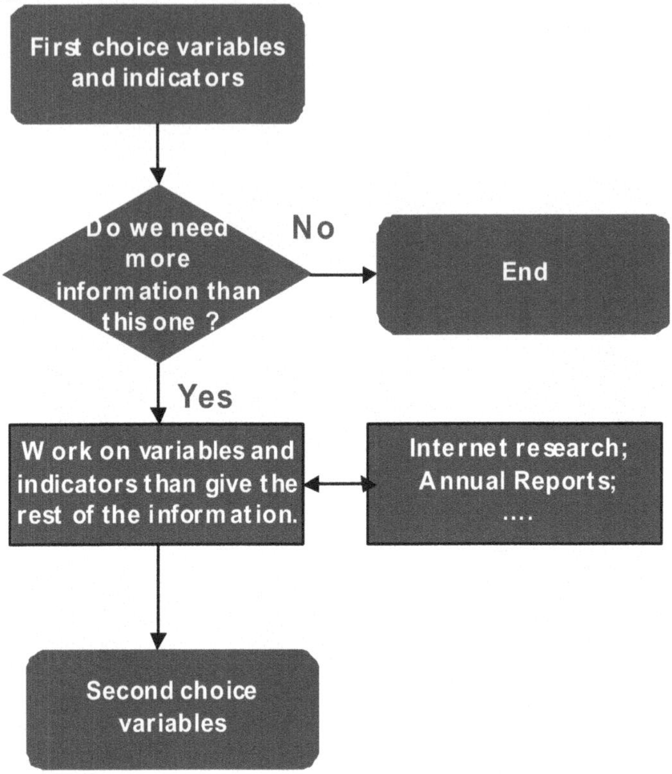

III – Discuss the definition of each variable and indicator and make final adjustments:

1. Easy-to-get data!
2. Data related with the holding needs!
3. Simple and comparable data between companies!

IV – Final questions were addressed:

1. When and how to collect the data, and at which frequency (monthly, etc…)?
2. What kind of reports; which frequency?

It was decided to start at the beginning of 2004 to collect data monthly by mail (checking the data and mail back validating the data). At first the companies send the information too late or did not did not respond. The Administration discussed with the Directors about the best way to get positive feed-back from the companies. It was decided to help the companies by means of appointed visits which explained the meaning of the variables, the best way to collect them and the improvements they need to make in their procedures/ infrastructures to collect them.

The data collected monthly was organized in an annual report at the end of the economic year. Copies were distributed by all the Administrators and Directors of the companies which were encouraged to comment on the report and give suggestions

3 PERFORMANCE INDICATORS – THE DEVELOPMENT

Based on the experience of the first year, some of the variables were changed. Some were replaced and other changed in order to fine-tune their definitions.

The year 2004 was also the first year for the IRAR variables and indicators. For that, the same variables were equally changed (similar to those of the Regulation Institute, IRAR), particularly on what concerns to their definition and the substitution of pre-defined variables and indicators by others with similar purposes that have already being adopted by the IRAR.

In the second year a more complete report was made which compared years 2004 and 2005.

Simultaneously efforts were made to develop software which is now routinely used to visualize information and which gives the possibility to create graphs and reports in an automatic manner. All users (board, directors and other key users) can monthly evaluate and monitor the current company performance with projections. It is a kind of "Tableau de Bord" with operational, commercial, human resources and finance information that is essential in order to support corrective and appropriate decisions.

In 2007 some of the budgets were already produced based on the indicators matrix developed

Code	Commercial - Water supply (Variables)
VCA1a	Export water (Euros)
VCA1b	Export water (m3)
VCA2a	Billed water (Euros)
VCA2b	Billed water (m3)
VCA3	Water clients (nº)
VCA4	Produced water (m3)
VCA5a	Imported water (Euros)
VCA5b	Imported water (m3)
VCA6	Replaced meters (No)
VCA7	New costumer' meters (No)
VCA8	Water amount pre-contracted (m3)
VCA9	Water costumer' cut (No)
Code	**Commercial - Water supply (Indicators)**
ICA1	Annual system input (m3)
ICA2	Total sold water (Euros)
ICA3	Total sold water (m3)
ICA4	Billed average price (Euros/m3)
ICA5	Exported average price (Euros/m3)
ICA6	Imported average price (Euros/m3)
ICA7	Monthly average billed (Euros) per costumer
ICA8	Monthly average billed (m3) per costumer
ICA9	Amount of water contract that was sold (%)
ICA10	Cuts' number / 1000 costumers
Code	**Operational - Water supply (Variables)**
VOA1	Number of hours that the supply was interrupted
VOA2	Number of interruptions of supply
VOA3	Number of network failures
VOA4	Number of connection failures
VOA5	Works - Water supply network replacement (km)
VOA6	Works - Water supply network extension (km)
Code	**Operational - Water supply (Indicators)**
IOA1	Non-revenue water (m3)
IOA2	Non-revenue water (%)
IOA3	Number of meters replaced/ Number of costumers

Figure 5. ACTUAL VARIABLES AND INDICATORS MATRIX FOR WATER SUPPLY

Code	Commercial - Wastewater (Variables)
VCR1a	Billed imported wastewater (Euros)
VCR1b	Billed imported wastewater (m3)
VCR1c	Collected imported wastewater (m3)
VCR1ai	Billed imported wastewater - Industrial costumers (Euros)
VCR1am	Billed imported wastewater - Domestic costumers (Euros)
VCR1bi	Billed imported wastewater - Industrial costumers (m3)
VCR1bm	Billed imported wastewater - Domestic costumers (m3)
VCR1ci	Collected imported wastewater - Industrial costumers (m3)
VCR1cm	Collected imported wastewater - Domestic costumers (m3)
VCR2a	Billed exported wastewater (Euros)
VCR2b	Billed exported wastewater (m3)
VCR2c	Delivered exported wastewater (m3)
VCR3a	Billed wastewater (Euros)
VCR3b	Billed wastewater (m3)
VCR3c	Collected wastewater (m3)
VCR4	Treated wastewater (m3)
VCR5	Wastewater costumers (No)
VCR5i	Wastewater costumers - Industrial costumers (No)
VCR5m	Wastewater costumers - Domestic costumers (No)
VCR6	Wastewater amount pre-contracted (m3)
Code	**Commercial - Wastewater (Indicators)**
ICR1	Total sold wastewater (Euros)
ICR2	Total sold wastewater (m3)
ICR3	Imported average price (Euros/m3)
ICR3i	Imported average price - Industrial costumers (Euros/m3)
ICR3m	Imported average price - Domestic costumers (Euros/m3)
ICR3	Exported average price (Euros/m3)
ICR5	Billed average price (Euros/m3)
ICR6	Monthly average billed (Euros) per costumer
ICR7	Monthly average billed (m3) per costumer
ICR8	Amount of wastewater contract that was sold (%)
Code	**Operational - Wastewater (Variables)**
VOR1	Number of network blockages
VOR2	Works - Wastewater network replacement (km)
VOR3	Works - Wastewater network extension (km)
Code	**Operational - Wastewater (Indicators)**
IOR1	% Treated wastewater
Code	**Quality (Variables)**
VQA1	Number of treated water tests carried out
VQA2	Compliance of treated water tests (No)
VQR1	Number of treated wastewater tests carried out
VQR2	Compliance of treated wastewater tests (No)
VQA3	Written complaints - Water supply
VQA4	Written responses - Water supply
VQR3	Written complaints - Wastewater
VQR4	Written responses - Wastewater
Code	**Quality (Indicators)**
IQA1	Quality of supplied water (%)
IQR1	Quality of the treated wastewater (%)
IQA2	Response to complaints (%) - Water supply
IQR2	Response to complaints (%) - Wastewater

Figure 6. ACTUAL VARIABLES AND INDICATORS FOR WASTEWATER AND QUALITY

2 PERFORMANCE INDICATORS – THE FUTURE

It is intended that all future budgets may be made in accordance to the established Matrix. Nevertheless, this will not be easy since historically the budgetary information is almost financial.

The actual challenge is the consolidation of information, critical analysis and the definition of companies goals based on information obtained by performance indicators

Performance measures to support infrastructure asset management

Proactive crisis management of urban infrastructure - report from C18s twin sister COST C19

J. Røstum[1]

[1]*SINTEF, Klæbuvegen 153, N-7465 Trondheim, Norway.*
E-mail: jon.rostum@sintef.no

Keywords: COST, urban infrastructure, risk, crisis management

ABSTRACT

Modern developed societies are heavily dependent on urban infrastructures like transport, railway, gas, electricity, and water supply. The breakdown of one of

Edited by Enrique Cabrera Jr and Miguel Ángel Pardo. ISBN: 9781843391913.
Published by IWA Publishing, London, UK.

such critical infrastructures may cause serious consequences for the safety of the citizens. Managing the safety and security of these infrastructures taking into account prevention and preparedness, crisis interventions and restoring normal conditions using optimised and risk based approaches are important. COST action C19 "Proactive Crisis Management of Urban Infrastructure" (2004-2008) deals with these issues. There final report from the action will be available from the webpage of the action March 2008.

1 INTRODUCTION

This paper aims at providing an overview from the work in COST ((European COoperation in the Field of Scientific and Technical Research) Action C19: "Proactive Crisis Management of Urban Infrastructure". The action was approved and officially started the 19th of March 2004 and will continue until the final conference/meeting in March 2008. COST Actions brings together researchers and experts in different countries working on specific topics. It finances networking of nationally funded activities in supporting meetings, conferences, short term scientific meetings etc. COST does not fund research itself.

The main objective of the Action is to define current knowledge gaps and identify possible measures to improve the multi-disciplinary research on urban infrastructure vulnerability and handling of crisis situations. Additional objective of the Action is to present a state-of-the-art on current know-how demonstrating its application in direct relation to crisis situations.

Modern developed societies are heavily dependent on urban infrastructures like transport, railway, gas, electricity, and water supply. The breakdown of one of such critical infrastructures may cause serious consequences for the safety of the citizens. Managing the safety and security of these infrastructures taking into account prevention and preparedness, crisis interventions and restoring normal conditions using optimised and risk based approaches are important. Urban infrastructure is highly vulnerable to earthquakes, political conflicts, terrorism, droughts, floods, and other natural and societal disasters. Failures of some of these structures, such as water supply and other pipeline systems, roads and bridges, cable communications and energy supply may have major impacts in terms of human lives and economic losses.

Even if an increased research effort within a single sector, e.g. transportation, may push forward the safety knowledge, overall safety research would gain more benefit from mutual co-operation involving other sectors as well. The scientific challenge is therefore more connected to the interaction of different sectors, and an increased co-ordination of the effort in safety research is preferred in the future.

The members of the actions represent 15 countries in Europe (Czech Republic, Finland, Germany, Italy, Netherlands, Norway, Portugal, United Kingdom, France, Switzerland, Serbia Montenegro, Slovenia, Cyprus, Iceland and Romania.) and with expertise covering different infrastructures (road and railway transportation, water and wastewater, electricity supply, gas).

1.1 Introduction to urban infrastructures, critical infrastructures and lifelines

COST C19 focuses on "urban infrastructures". Both the terms critical infrastructure and lifelines might be applied on these and within C19 both terms are used. The concept of critical infrastructure is well known in many countries, but the definitions vary slightly from country to country. There has also been a trend in adding more sectors within the definitions, from initially focusing on infrastructure in the public sector, such as roads, water and wastewater networks, power grid systems to covering other societal functions like financing, public health care facilities, banking and finance systems, information technology systems, telecommunications systems etc.

The concept of a "lifelines" might be interpreted as a subset of critical infrastructure and applies on larger, geographically distributed principal systems: electric power, gas, water supply, telecommunications and transportation. These lifelines are vital to a modern society, but are exposed to hazards which can have a critical influence on their operation. There are also interdependencies between the different lifelines. For instance, a disruption in the electricity supply can also impact traffic services and water distribution.

Table 1. CRITICAL INFRASTRUCTURES - SOME DEFINITIONS

Definition	Source
Critical infrastructures consist of those physical and information technology facilities, networks, services and assets which, if disrupted or destroyed, would have a serious impact on the health, safety, security	EU
Systems and assets, whether physical or virtual, so vital to the United States that the incapacity or destruction of such systems and assets would have a debilitating impact on security, national economic security, national public health or safety, or any combination of those matters.	US - Patriot act
«an infrastructure that is so vital that its incapacitation or destruction would have a debilitating impact on defence or economic and the national critical infrastructures»	US-PCCIP
National Critical Infrastructure refers to those physical and information technology facilities, networks, services and assets, that if disrupted or destroyed, would have a serious impact on the health, safety, security or economic well-being of Canadians or the effective functioning of governments in Canada.	Canada
Critical infrastructure is those facilities, services and information systems which are so vital to nations that their incapacity or destruction would have a debilitating impact on national security, national economy, public health and safety and the effective functioning of the government.	NATO
Organisations or institutions which are of essential importance to society, the failure or disruption of which will cause long-term bottlenecks in the supply chain or have other dramatic consequences for large sections of the population	Switzer-land
Critical infrastructures (CI) include organizations and systems with tremendous importance for the society that, if disrupted, would have sustained influence on supply chains and public safety and could lead to further dramatic consequences.	Germany
Infrastructures are deemed critical if they constitute an essential, indispensable facility for society, and if their disruption would rapidly bring about a state of emergency or could have adverse societal effects in the longer term.	Neder-lands
Those assets, services and systems that support the economic, political and social life of the UK whose importance is such that any entire or partial loss or compromise could: cause large scale loss of life, have a serious impact on the national economy, have other grave social consequences for the community or be of immediate concern to the national government	UK
Critical infrastructure is those assets and systems which are essential to uphold the critical functions of the society which again meets the basic needs of society and the feeling of security in the population. Critical infrastructure includes the following infrastructures: electric power, electronic communication, water and wastewater, transport, oil & gas and satellite-based infrastructure. Critical society functions includes banking, food, health and social security services, police, rescue and emergency service, crisis management, government, court of justice, defence, environmental monitoring and waste management.	Norway

2 WORK AND ACTIVITIES IN COST C19

The work in the action is organized in 3 working groups (WG), where two of them are closely linked and mostly are working and discussing together.

- WG A: Risk analysis and risk management with focus on theories, methods and tools for risk assessment.
- WG B/C: Planning of handling of acute crisis

The work in C19 covers most of the phases in the risk-/crisis management wheel as illustrated in Figure 1, from *prevention, preparedness, warning and alarm*, *response*, *limiting the extent of loss* and *recovery*.

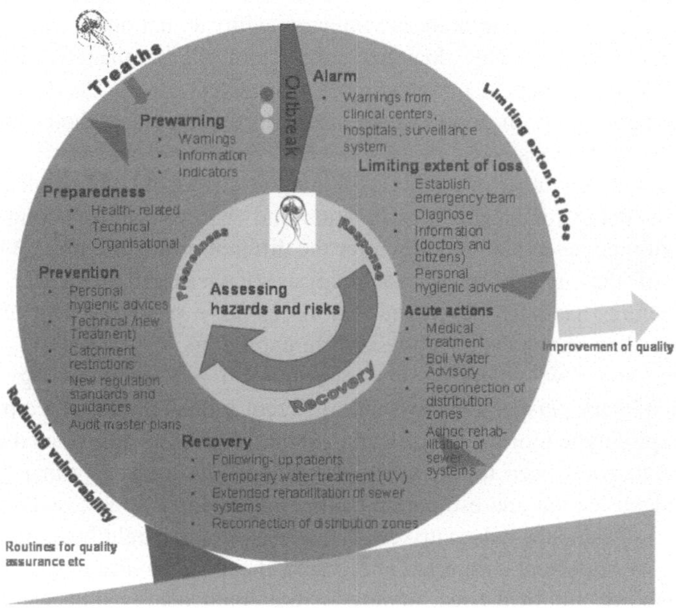

Figure 1. RISK WHEEL - ILLUSTRATION FOR CASE WATERBORNE OUTBREAK.

3 RESULTS FROM THE ACTION

The results from COST action C19 consist of a wide variety of outputs.

One important outcome of the action is the networking among different disciplines/infrastructures. Even though each sector/discipline is unique and has its own challenges, learning from each other has been valuable.

An important part of the work in the actions has been the organisation of a series of conferences/seminars. These conferences have addressed exchange of knowledge between urban infrastructure sectors, i.e. energy, traffic water. The conferences also have had a local/national aspect where case studies and field trips have been important and valuable for the knowledge/dissemination process. Series of conferences/seminars with a national/local focus on problems/challenges. The documentation from all seminars/conferences is available at the homepage of the action, www.COSTC19.eu. The documentation includes description on good cases illustrating the use of methods and lessons learned from previous crisis.

At present, a number of methods for analysing risks of technological systems that are suitable for hazard identification and risk assessment along with the criteria of their selection are known for the different infrastructures. Within C19 a state of the art report on risk analysis (RA) methodology for different infrastructures has been developed and discussed in plenary. Great differences exist regarding method and culture in different sectors.

There are few attempts on developing a RA methodology with a interdisciplinary perspective/crosswise different infrastructures. As a part of the action a prototype tool for identification and estimation of risk related to critical infrastructure has been tested within the action. The method primarily focuses on the identification and estimation of the risk. The risk is assessed and plotted directly into standard risk matrixes. The consequences might be expressed with respect to different aspects: life and health, environment, economy, manageability, political trust, supply failure (both quality and availability of delivery). The tool can be used for overall analysis at national, region or local level. Prototype of the tool will be available at the C19 webpage after the end of the action.

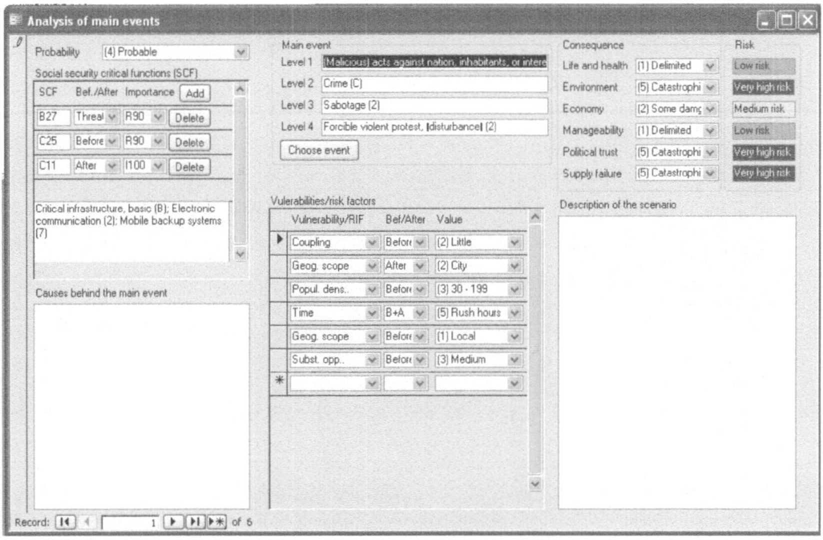

Figure 2 . PROTOTYPE TOOL FOR RISK ANALYSIS OF URBAN INFRASTRUCTURES.

Further research needs has been identified with respect to development of methods and tools for analysing risk and vulnerabilities between different sectors/infrastructures and the interrelationship between the urban infrastructures. The different stakeholders involved and their roles and influences on the processes should also be further analysed and understood.

STSM (Short Term Scientific Missions) has been carried out within the scope of the action. The STSM is a good tool for involving young researchers in the work in the action.

The C19 has and will also in the future form the basis for different EU proposals/initiatives within the EU 6FP and 7FP.

Final report is underway and will be available at the time of the PI08-conference. The report is divided into two main parts. The first session deals with generic methods for RA illustrated by cases where some of the methods have been applied on different infrastructures. The second part of the report deals with acute crisis and the handling thereof.

4 REFERENCES

Final report from the action available from the C19 webpage: www.COSTC19.eu. (by March 2008).

Sewer asset management: fusion of performance indicators into decision criteria

P. Le Gauffre[1], *M. Ibrahim*[1] *and F. Cherqui*[1]

[1] *Université de Lyon, F-69000, Lyon, France;*
INSA-Lyon, F-69621 Villeurbanne, France;
Université Lyon 1, F-69622, Villeurbanne, France;
Laboratoire Génie Civil et Ingénierie Environnementale;
pascal.le-gauffre@insa-lyon.fr, tel.: +33 4 7243 8858 ; fax: +33 4 7243 8521

Keywords: Aggregation, Fuzzy sets, Rehabilitation, Visual inspection

Edited by Enrique Cabrera Jr and Miguel Ángel Pardo. ISBN: 9781843391913.
Published by IWA Publishing, London, UK.

ABSTRACT

Within the RERAU methodology, each rehabilitation criterion is assigned a grade out of 4 possible ones. This grade results from successive aggregations of performance indicators. Issues related to base data (uncertain or imprecise data) and to fusion (conflicting or reinforcing indicators) lead to envisage a fuzzy method for processing the proposed set of performance indicators. Five issues are discussed in this paper: representing an indicator with fuzzy linguistic variables; representing each aggregation table with "if... then..." rules; combining two premises of a rule; fuzzy representation of the implication; aggregating the results of several rules.

1 INTRODUCTION

1.1 The RERAU methodology

A French R&D program (RERAU, 1999-2004) was dedicated to the definition of a set of criteria for prioritising investigation and rehabilitation measures for sewer segments. Each criterion was defined as a combination of complementary performance indicators (Le Gauffre *et al.,* 2004). These performance indicators use information that is obtained from complementary sources: visual inspections of sewer segments, network monitoring, data issued from network and treatment plant operation, data related to the vulnerability of the build environment and of the receiving waters.

The process that was used for defining indicators and decision criteria is explained in (Le Gauffre *et al,* 2007) with 3 main steps: 1) Setting up typologies for defects, dysfunctions and impacts; 2) Deriving indicators for defects, dysfunctions and impacts; 3) Combining dysfunction indicators and impact indicators into decision criteria.

Two types of dysfunctions are distinguished. Infiltration (INF), exfiltration (EXF), sand silting (SAN), blockage (BLO), destabilization of ground-pipe system (SPD), etc. are source dysfunctions occurring at the pipe scale. Flooding (FLO) and excessive spillage (CSO) occur on some specific sewer segments, but indeed refer to primary dysfunctions (like silting or blockage) which may be located some distance downstream, or in some cases upstream.

Each proposed criterion assesses a contribution of a particular dysfunction of a sewer segment to a particular impact. Each of the 8 defined impacts is linked to some of the 10 source dysfunctions (Figure 1). R/POL/BLO, R/OCP/INF, R/TRA/COL and R/SLC/SPD are four examples of criteria that may be used to defined rehabilitation needs and priorities:

- R/POL/BLO: sewer segment contributing to pollution of surface water due to spillages induced by repeated blockages;
- R/OCP/INF: sewer segment contributing to infiltration inducing treatment plant operation surplus costs;
- R/TRA/COL: risk of traffic disruption due to collapse;
- R/SLC/SPD: shortened lifetime cost due to destabilization of the ground-pipe system.

Dysfunctions: INF: infiltration, EXF: exfiltration, HYD: decrease in hydraulic capacity, SAN: sand silting, BLO: blockage, SPD: destabilization of the ground-pipe system, COR: degradation due to corrosion, ROO: degradation due to roots intrusion, ABR: degradation from abrasion, COL: risk of collapse.

IMPACTS \ DYSFUNCTIONS	INF	EXF	HYD	SAN	BLO	SPD	COR	ROO	ABR	COL
POL	X		X	X	X					
POG		X	X	X	X					
NUH			X	X	X					
TRA				X	X					X
DAB		X								X
OCS	X			X	X					
OCP	X									
SLC						X	X	X	X	

Impacts: POL: pollution of surface waters, POG: pollution of ground and groundwater, NUH: nuisances of a hydraulic nature (flooding, etc.), TRA: disruption of surface activities (traffic, etc.), DAB: damages to the built environment, OCS: networks operation surplus costs, OCP: treatment plant operating surplus costs, SLC: cost associated to shortened lifetime.

Figure 1. DECISION CRITERIA DEFINED BY LINKING IMPACTS AND SOURCE DYSFUNCTIONS

1.2 The INDIGAU research program

A new R&D program (INDIGAU, 2007-2010) is dedicated to calibration and experiments of models that were defined within the RERAU program.

Another goal of the INDIGAU program is to address issues linked to the characteristics of available data and knowledge: uncertain, imprecise and incomplete data. This research task refers to the aggregation or "fusion" of complementary data provided by several sources. In the case of sewer asset management, the various sources are results of visual inspection, on-site measurements, hydraulic modelling, registered failures, etc.

Within the RERAU methodology each criterion is assigned a grade out of 4 possible ones (from G1 - good condition ... to G4 - very bad condition). This grade results from successive aggregations of indicators. Figure 2 provides three examples of aggregation tables: PI1 and PI2 are two indicators combined with a "mean" operator; the result (PI3) and PI4 are combined with a "Max" operator; finally criterion C is evaluated by combining PI5 and PI6 with a "min" operator.

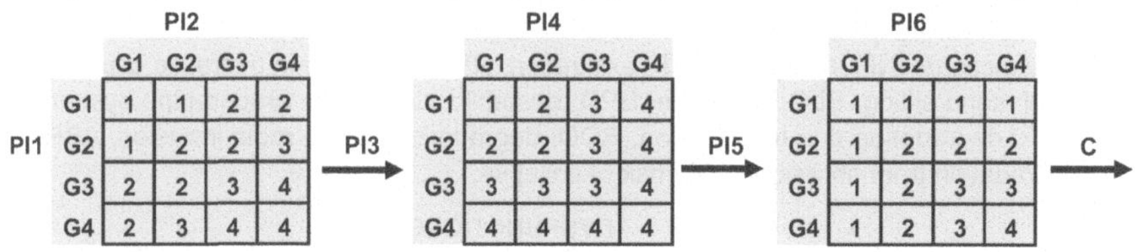

Figure 2. EACH CRITERION RESULTS FROM SUCCESSIVE AGGREGATION OPERATIONS

2. ISSUES INDUCED BY BASE DATA

2.1 A typology of base performance indicators

We define several types of base performance indicators together with the associated sources of uncertainty.

Type A: indicators directly derived from inspection results; for instance infiltration may be observed during a visual inspection; for these PIs sources of uncertainty are: non registered dysfunction, wrong code, etc.

Type B: indicators derived from a synthesis of the results of a visual inspection; for instance "watertightness deficiency" (risk of infiltration) can be assessed by considering the number and the seriousness of observed defects; for these PIs sources of uncertainty are: non registered defects, wrong codes, cut-off values, etc. (see 2.2).

Type C: indicators directly derived from O&M data, i.e. registered events (blockages) or actions (cleansing operations); sources of uncertainty are: non registered events, cut-off values, etc.

Type D: indicators derived from O&M data (flooding or spillage events) associated with a hydraulic calculation procedure for defining sewer segments that may have caused these events; for these PIs sources of uncertainty are: non registered events, cut-off values, hydraulic calculation procedure, etc.

Type E: indicators derived from monitoring data (for instance infiltration flows); sources of uncertainty are: uncertainty on measurements, cut-off values, etc.

Type F: indicators related to external factors of defects and dysfunctions (type of soil, water table level, etc.); sources of uncertainty are: poor knowledge of the environment, etc.

Type G: indicators representing urban vulnerabilities (possible consequences of dysfunctions); sources of uncertainty are: poor knowledge of the urban environment of the network, etc.

Type H: indicators derived from O&M data for assessing consequences on O&M costs; sources of uncertainty are: available cost data.

2.2. PIs derived from a synthesis of the results of visual inspections

Results of a visual inspection of a particular segment, regarding a given dysfunction (let say: infiltration), can be synthesised by calculating a score *S* aggregating single scores associated to observed defects (cracks, missing wall, etc.). Translating this score *S* into a grade out of 4 possible ones (G1, G2, G3 or G4) requires to compare score *S* with 3 cut-off thresholds. This procedure leads to 3 problems. Problems 1 & 2 refer to the calibration process while problem 3 relates to the use of the cut-off thresholds.

Problem 1. Score *S* is not an exact expression of the expertise: if we compare scores with expert assessment of a sample of sewer segments we can see that it is not possible to deduct the condition grade from the score without any errors (Figure 3).

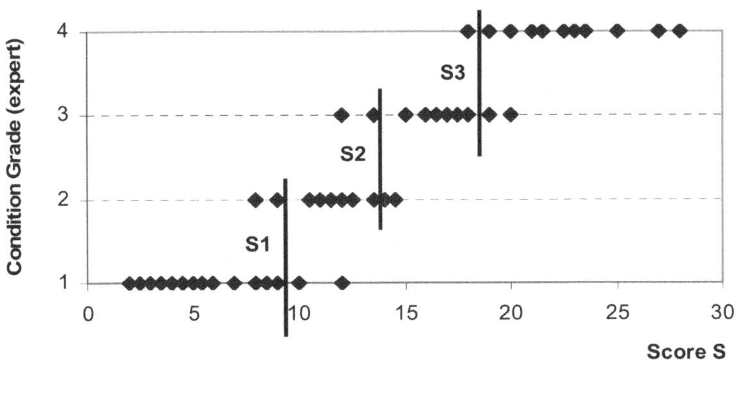

Figure 3. CUT-OFF THRESHOLDS (S1, S2, S3) ARE USED TO ASSIGN A GRADE AFTER CALCULATION OF A SCORE S.

Problem 2. The chosen cut-off thresholds do not rely on objective knowledge but will be derived from a calibration procedure using expert judgments as references (Ibrahim *et al.* 2007). However, for a given segment, the judgments expressed by several experts may be very different from each others (Werey *et al* 2008). That poses a problem for defining a crisp reference.

Problem 3. A score that is close to a cut-off threshold may underestimate the right score (for instance if some defects have not been registered, Dirksen *et al.* 2007) or it may overestimate the right score (for instance if a code has been used instead of the right one). These errors may lead to two types of misclassification: the grade is underestimated (e.g. G3 instead of G4) or overestimated (G4 instead of G3).

In (Ibrahim *et al.* 2007) we proposed a calibration criterion for fixing cut-off thresholds: *MC misclassification cost* integrates simultaneously underestimation and overestimation errors. Thus imprecision of assessment is taken into account: the thresholds are chosen in order to be the most cost-efficient. However, this procedure has a meaning exclusively at the scale of the asset stock; it induces punctual errors (at the scale of some sewer segments) that may be important. Fuzzy sets theory provides a way of addressing these issues. Grades G1, G2, G3 and G4 may be seen as four fuzzy sets. With this approach a sewer segment will be assigned a fuzzy description, for instance: 0.7 for grade G3 and 0.3 for grade G4, on a given indicator (see 3.2.2.)

3. FUSION OF PERFORMANCE INDICATORS

3.1. Fuzzy rules for assessing rehabilitation criteria

Issues related to base data (2.1) and to fusion (see 3.2) lead to envisage a fuzzy method for processing the proposed set of performance indicators. Five issues are discussed in this section (Mauris *et al.* 1996): representing an indicator with fuzzy linguistic variables; representing each aggregation table with "if… then…" rules; combining two premises of a rule; fuzzy representation of the implication; aggregating the results of several rules.

3.1.1. Representing an indicator with fuzzy linguistic variables

The four terms G1, G2, G3 and G4 may be viewed as four fuzzy subsets allowing a numeric/symbolic conversion. The fuzzy subset theory led to the development of the concept of fuzzy meanings and fuzzy description (Mauris *et al.* 1996). Fuzzy meanings are the representation of fuzzy subsets corresponding to linguistic terms: $\mu_{Li}(x)$ denotes the membership function associated to the linguistic term L_i (Figure 4).

The fuzzy description is a simple way of describing a measurement with words: $\mu_x(L_i)$ denotes the membership value associated to the linguistic term L_i, for a given numeric value (e.g.: $x = 12$ in Figure 4).

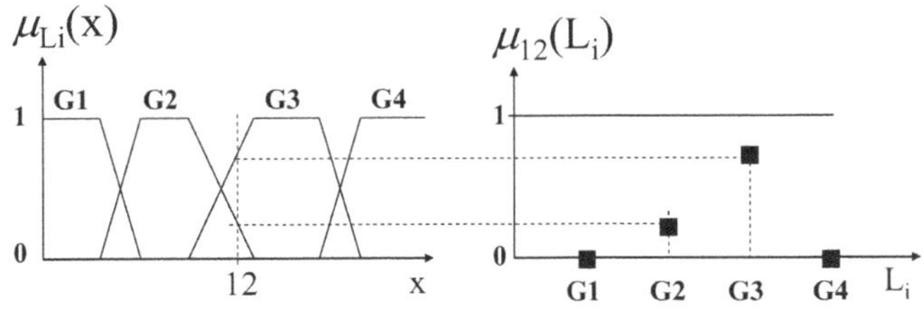

a) FUZZY MEANINGS b) FUZZY DESCRIPTION

Figure 4. LINK BETWEEN FUZZY MEANINGS AND FUZZY DESCRIPTION

3.1.2. Representing each aggregation with "if... then..." rules

Each aggregation table can be expressed as a set of if/then rules. We denote:

- $R_{i,j,k}$ the rule: "if X is L_i and Y is L_j then Z is L_k",
  with $L_i, L_j, L_k \in \{G1, G2, G3, G4\}$.
- $\mu_\Gamma(i, j, k)$ the degree of validity (or weight) of the rule $R_{i,j,k}$ for operator Γ (for instance: min, max, mean, etc.)

	G1	G2	G3	G4
G1	1.0 ; 0.0 ; 0.0 ; 0.0	0.5 ; 0.5 ; 0.0 ; 0.0	0.0 ; 1.0 ; 0.0 ; 0.0	0.0 ; 0.5 ; 0.5 ; 0.0
G2	0.5 ; 0.5 ; 0.0 ; 0.0	0.0 ; 1.0 ; 0.0 ; 0.0	0.0 ; 0.5 ; 0.5 ; 0.0	0.0 ; 0.0 ; 1.0 ; 0.0
G3	0.0 ; 1.0 ; 0.0 ; 0.0	0.0 ; 0.5 ; 0.5 ; 0.0	0.0 ; 0.0 ; 1.0 ; 0.0	0.0 ; 0.0 ; 0.5 ; 0.5
G4	0.0 ; 0.5 ; 0.5 ; 0.0	0.0 ; 0.0 ; 1.0 ; 0.0	0.0 ; 0.0 ; 0.5 ; 0.5	0.0 ; 0.0 ; 0.0 ; 1.0

Figure 5. EXAMPLE OF AN AGGREGATION TABLE WITH FUZZY CONCLUSIONS (EACH CELL PROVIDES 4 VALUES: $\mu_\Gamma(i, j, k)$ FOR K = 1 ... 4)

3.1.3. Combining two premises of a rule

Each aggregation table uses 2 indicators for defining a high level indicator. Each associated rule has 2 premises connected by an "and" operator (X is L_i and Y is L_j).

We denote $\mu(L_i, L_j)$ the membership value associated to the pair (L_i, L_j). If we make the assumption that the two indicators are independent, $\mu(L_i, L_j)$ is decomposable (Mauris *et al.* 1996) and can be derived from $\mu_x(L_i)$ and $\mu_y(L_j)$ by using an operator of intersection, denoted $\wedge_1$ (1).

$$\mu(L_i, L_j) = \mu_x(L_i) \wedge_1 \mu_y(L_j) \qquad (1)$$

Several operators of intersection may be used: *min*, *product*, etc.

3.1.4. Fuzzy implication

As proposed by Zadeh (1975), the "if… then…" meta-implication could be viewed as a fuzzy relation. We denote $\mu(L_i, L_j, L_k)$ the membership value associated to "Z is L_k" which is inferred from "X is L_i" and "Y is L_j" by using the rule $R_{i,j,k}$. $\mu(L_i, L_j, L_k)$ can be derived from $\mu(L_i, L_j)$ and from $\mu_\Gamma(i, j, k)$ (weight of the rule $R_{i,j,k}$) by means of an intersection operator, denoted $\wedge_2$ (e.g.: *min* operator or *product* operator):

$$\mu(L_i, L_j, L_k) = (\mu(L_i, L_j)) \wedge_2 \mu_\Gamma(i, j, k) \qquad (2)$$

Using (1) for $\mu(L_i, L_j)$, we obtain:

$$\mu(L_i, L_j, L_k) = (\mu(L_i) \wedge_1 \mu(L_j)) \wedge_2 \mu_\Gamma(i, j, k) \qquad (3)$$

3.1.5. Aggregating the results of several rules

The conclusion "Z is L_k", concerning a high level indicator, may be inferred from several rules $R_{i,j,k}$. We denote $\mu(L_k)$ the membership value associated to the linguistic term L_k. This value is calculated by considering all the rules inferring this conclusion, by means of a union operator denoted $\perp$:

$$\mu(L_k) = \perp_{i=1\ldots4,\ j=1\ldots4} \mu(L_i, L_j, L_k) \qquad (4)$$

3.1.6. Systems of operators that will be tested

Table 1 resents three possible systems of operators that will be tested.

Table 1 THREE POSSIBLE SYSTEMS OF OPERATORS

Systems	Combination of premises $\mu \wedge_1 \mu'$	Implication $\mu \wedge_2 \mu'$	Aggregation of rules $\mu \perp \mu'$
Min-Max	$\min(\mu, \mu')$	$\min(\mu, \mu')$	$\max(\mu, \mu')$
Min-Sum	$\min(\mu, \mu')$	$\min(\mu, \mu')$	$\min(1, \mu + \mu')$
Prod-Sum	$\mu \times \mu'$	$\mu \times \mu'$	$\min(1, \mu + \mu')$

3.2 Issues related to the fusion of data: types of reasoning

Grabisch *et al.* (1998) or Bloch and Hunter (2001) have discussed the issue of aggregating complementary information or assessments. Possible aggregation operators may be classified into several categories: conjunctive, disjunctive, compensative, and non-compensative. Aggregation tables used in the RERAU PI processing system refer to these categories.

For some of these fusion operations a fuzzy representation appears much more convenient than crisp operators. The case where two different sources provide different conclusions (in evaluating the same assumption) is a good example of a need for propagating uncertainty within the PI system: "*at the combination level, the choice of an operator is crucial. Conjunctive combination operators are discontinuous in the presence of conflicts and may provide no interesting results at all. Averaging is not realistic since such operators may provide answers that are given by none of the sources, or even rejected individually by each source. If some of the sources are assumed to be reliable, disjunctive operators can be used, which will retain all answers from all sources at the price of increased imprecision*" (Bloch and Hunter, 2001).

In Table 2 Case 1 is an example where two sources provide conflicting conclusions: $G4$ or $G3$ according to source S1, but $G2$ according to source S2. Combining these results (PI3) leads to 3 possible conclusions (from $G2$ to $G4$!). On the contrary, Case 2 refers to a situation where the two sources provide conclusions that are reinforcing each other, towards $G3$. In this type of situation the final conclusion will be a higher membership value associated to the common conclusion ($G3$).

TABLE 2. TWO EXAMPLES WHERE TWO SOURCES S1 AND S2 ARE COMBINED TO ASSESS A RESULTING PI (PI3)

	$\mu_{Si}(G1)$	$\mu_{Si}(G2)$	$\mu_{Si}(G3)$	$\mu_{Si}(G4)$
Case 1: conflict				
PI1 (source S1)			0.4	0.6
PI2 (source S2)		1		
PI3 = PI1 ⊗ PI2		*0.5*	*0.2*	*0.3*
Case 2: reinforcement				
PI1 (source S1)			0.7	0.3
PI2 (source S2)		0.3	0.7	
PI3 = PI1 ⊗ PI2		*0.05*	*0.90*	*0.05*

5. CONCLUSIONS

Different kinds of base performance indicators have been inventoried, depending on the means to acquire information, and sources of uncertainty have been identified. PIs resulting from CCTV reports may require fuzzy modelling in order to control and identify misclassification error associated to sewer segment condition grading. This paper provides a way of combining PIs in taking account the imperfection of available data (imprecise or uncertain data) and in modelling various types of reasoning such as the combination of conflicting or reinforcing indicators. Fuzzy assessment of sewer segments allows managing with imperfect information and knowledge (contradictory expert assessments, etc.); fuzzy modelling offers a way to control misclassification errors both at the scale of the asset stock and at the scale of the sewer segment. Next steps will be to calibrate fuzzy cut-off thresholds from experts' judgment for each base PI and fuzzy rules for each composite PI or rehabilitation criterion.

6. ACKNOWLEDGEMENTS

Research presented in this paper is carried out in the frame of the INDIGAU research project (www.indigau.fr) supported by the French National Agency for Research – ANR, within the PGCU-2006 program.

7. REFERENCES

Bloch, I. and Hunter, A. (2001) Fusion: general concepts and characteristics. *Int. J. Intelligent Systems*, 16(10), 1107-1134.

Dirksen, J., Goldina, A., Veldhuis ten J.A.E., and Clemens F.H.L.R. (2007) *2nd Leading Edge Conf. on Strategic Asset Management*, Lisbonne, Portugal, October 17-19.

Grabisch, M., Orlovski, S.A., Yager, R.R. (1998). "Fuzzy aggregation of numerical preferences". In : R. Slowinski (ed), *The Handbook of Fuzzy Sets Series*, *Vol. 4: Fuzzy Sets in Decision Analysis, Operations Research and Statistics,* Kluwer Academic, 1998, 31-68.

Ibrahim, M., Cherqui, F., Le Gauffre, P. and Werey, C. (2007) Sewer asset management: from visual inspection survey to dysfunction indicators. LESAM 2007 (2nd Leading Edge Conf. on Strategic Asset Management) IWA and LNEC, Lisbon, Portugal, 17-19 October.

Le Gauffre, P., Joannis, C., Breysse, D., Gibello, C. and Desmulliez J.J. (2004) *Gestion patrimoniale des réseaux d'assainissement urbains. Guide méthodologique*. Paris, Lavoisier Tec&Doc, 416 p., ISBN 27430-0748-6.

Le Gauffre, P., Joannis, C., Vasconcelos, E., Breysse, D., Gibello, C., Desmulliez, J.J. (2007) Performance Indicators and Multicriteria Decision Support for Sewer Asset Management. *Journal of Infrastructure Systems*, ASCE, **13**(2), June 2007, 105-114.

Mauris, G., Benoit, E., Foulloy, L. (1996). "The aggregation of complementary information via fuzzy sensors", *Measurement,* 17(4), 235-249.

Werey, C., Cherqui, F., Ibrahim, M., Le Gauffre, P. (2008). Sewer asset management tool: dealing with experts' opinions. Pi08 – Performance Assessment of Urban Infrastructure Services, Valencia, Spain, 12-14 March. In press.

Zadeh, L.A. (1975) The concept of linguistic variable and its application to approximate reasoning. *Information Sciences*. Part 1: 8(3) 199-249, part 2: 3(1), 43-80.

Technical Performance Indicators for Water Mains, the First Steps in Serbia on IWA Best Practise Application

D. Radivojević[1], D. Milićević[2], and M. Ivetić[3]

[1] *Faculty of Civil Engineering and Architecture, A. Medvedeva 14, 18 000 Niš, Serbia, dragan.radivojevic@gaf.ni.ac.yu, Tlf +381 18 588-200, Fax: +381 18 588-202*
[2] *Faculty of Civil Engineering and Architecture, A. Medvedeva 14, 18 000 Niš, Serbia, drgara@gaf.ni.ac.yu, Tlf +381 18 588-200, Fax: +381 18 588-202*
[3] *Faculty of Civil Engineering, B. Kralja Aleksandra 73, 11 000 Belgrade, Serbia, marko.ivetic@hikom.grf.bg.ac.yu, Tlf/Fax: +381 11 3370 206*

Keywords: Water Mains, Performance Indicators, IWA Best Practice

Edited by Enrique Cabrera Jr and Miguel Ángel Pardo. ISBN: 9781843391913.
Published by IWA Publishing, London, UK.

ABSTRACT

Active water loss management has became one of the primary interest of the water utilities in the world. Traditional indicator of water loss, widely adopted in the world, was % of the system input value, as it always sounded well to have water loss less than some prescribed value (for instance 10%). This approach has been proven to be misleading in many cases. This paper presents the basic principles of the methodology that compares actual with unavoidable losses from the water main and the results of the first step in attempt to approach Serbian water mains to the new standards.

1 INTRODUCTION

The level of water losses, both real and apparent, is one of the most important efficiency issues for water utilities across the world. Commonly used practice amongst water supply utilities to express water losses as a percentage of the system input volume has been proven to be unsuitable as a technical performance indicator as it does not reflect many of the influencing factors. For instance, systems with higher system input volumes will automatically have an (apparently) lower level of water losses if expressed in percentages.

Even though one would assume that proper performance indicators are used for benchmarking, international performance comparison, or target setting for internationally funded projects, unfortunately this is widely not the case - utility managers, consultants and the International Lending Institutions continue to use a very inappropriate indicator when talking about water losses.

2 IWA BEST PRACTICE WATER BALANCE

IWA Task Forces produced an international 'best practice' standard approach for Water Balance calculations (Figure 1), as the essential first step in practical management of water losses (Alegre at al, 2000, Lambert A & Hirner, 2000).

The IWA 'best practice' Water Balance is rapidly gaining international acceptance, and has already been adopted or promoted (with minor variations) by:

- DVGW (Germany), WSA and Queensland Environmental Protection Agency (Australia), Malta Water Services Corporation, South African Water Research Commission, New Zealand Water and Waste Association, American Water Works Association, and the Canadian Federation of Municipalities and National Research Centre,

- Utilities and/or consultants working in Austria, Brazil, Cyprus, Ghana, Jordan, Kazakhstan, Malaysia, Oman, Palestine, Saudi Arabia, Uzbekistan, and the many other countries.

System Input Volume (corrected for known errors)	Authorized Consumption	Billed Authorized Consumption	Billed Metered Consumption	Revenue Water
			Billed Unmetered Consumption	
		Unbilled Authorized Consumption	Unbilled Metered Consumption	Non-Revenue Water (NRW)
			Unbilled Unmetered Consumption	
	Water Losses	Apparent Losses	Unauthorised Consumption	
			Metering Inaccuracies	
		Real Losses	Leakage on Transmission and/or Distribution Mains	
			Leakage and Overflows at Utility's Storage Tanks	
			Leakage on Service Connections up to point of Customer metering	

Figure 1. IWA 'BEST PRACTISE' WATER BALANCE

3 TRADITIONAL PI FOR REAL LOSSES

The following 4 traditional performance indicators (PI) are used

- Water Losses and Real Losses as a % of system input volume
- Water Losses and Real Losses per property per day
- Water Losses and Real Losses per km of mains per day
- Water Losses and Real Losses per service connection per day

Water Losses, as a percentage of system input, is easily calculated and frequently quoted and is certainly the most common indicator. Various definitions for water losses are used, in developing countries the concept of Non-Revenue Water is most generally used. Thus the indicator is not meaningful for various reasons, mainly because of the sometimes enormous levels of unauthorized consumption ('illegal connections'). The IWA best practice manual suggests its use only as a financial performance indicator and states clearly it is 'unsuitable for assessing the efficiency of management of distribution systems'. Real Losses, as a % of system input, also suffer from deficiencies, mainly the level of (and changes in) consumption and variations in supply time (intermittent supply). For instance, a system with 12 hours supply per day may easily have only 20% real losses. In case of an uninterrupted supply situation all bursts would leak for 24 hours thus twice more water could be lost.

Real losses per property have to be rejected as the property has very little to do with leakage. Frequently an apartment block with 50 apartments is counted as 50 properties even though it has only one service connection which may leak.

This leaves the question of which of the remaining two indicators is more appropriate. Leakage component analyses in water distribution systems across the world have shown that the greatest proportion of annual real losses occur on services connections, including the connecting point to the main. This applies to all systems with a connection density of more than around 20 connections per km main. Only very rural supply systems normally have a lower connection density.

The Task Force therefore recommended (Lambert A & Hirner, 2000) that the basic traditional PI with the greatest range of applicability for real losses, to be referred to as the 'Technical Indicator Real Losses' (TIRL) is:

$$\text{TIRL} = \text{Real Loss Volume/Service Connection/Day w.s.p.} \qquad (3.1)$$

TIRL is the best of these traditional indicators - but should always be calculated as 'w.s.p.' - when the system is pressurized, to allow comparisons between systems with different levels of supply. However, this indicator still does not take operating pressure into account, which is a major disadvantage. The Task Force recommended further interpretation of the calculated TIRL value for an individual system by comparing it with a calculated value for Unavoidable Annual Real Losses (UARL), using a methodology which takes account of the local factors of density of connections, location of customer meters on service connections, and average operating pressure.

4 THE CONCEPT OF UNAVOIDABLE ANNUAL REAL LOSSES

Leakage management practitioners recognize that it is impossible to eliminate real losses from a large distribution system. There must therefore be some value of 'Unavoidable Annual Real Losses' (UARL) which could be achieved at the current operating pressures if there were no financial or economic constraints. If the UARL volume for any system can be assessed, taking into account key local factors, then the ratio of TIRL to UARL offers the possibility of an improved PI for real losses. In the most basic form, UARL in liters/day is

$$\text{UARL} = (18 \times \text{Lm} + 0.80 \times \text{Nc} + 25 \times \text{Lp}) \times \text{P} \;\; \textit{w.s.p.} \qquad (4.1)$$

where Lm is mains length in km, Nc is number of service connections, Lp is the total length in km of underground pipe between the edge of the street and customer meters, and P is average operating pressure in meters. This equation, based on component analysis of Real Losses for well-managed systems with good infrastructure, has proved to be robust in diverse international situations (Lambert A. and McKenzie R 2002), and is the most reliable predictor yet of 'how low could you go' with real losses for systems with more than 5,000 service connections, connection density (Nc/Lm) more than 20 per km, and average pressure more than 25 meters.

5 ILI – INFRASTRUCTURE LEAKAGE INDEX

The ratio of TIRL to UARL becomes a non-dimensional Infrastructure Leakage Index (ILI), which allows overall infrastructure management performance to be assessed independently of the influence of current operating pressure. The ILI is a measure of how well a distribution network is managed (maintained, repaired, rehabilitated) for the control of real losses, at the current operating pressure. It is the ratio of Current Annual volume of Real Losses (CARL) to UARL.

$$ILI = (18 \text{ x } Lm + 0.80 \text{ x } Nc + 25 \text{ x } Lp) \text{ x } P \quad \textit{w.s.p.} \qquad (5.1)$$

Being a ratio, the ILI has no units and thus it facilitates comparisons between countries that use different measurement units. As an excellent system can be rated one with the ILI value close to 1, while system with high ILI value, for instance 10, can be rated as a poorly maintained one. Part of the problem is that people in many countries are simply not aware of the ILI - and the other part of the problem is the limited understanding and acceptance of the ILI. Many practitioners prefer not to use the ILI for one or other of the following reasons:

- the accuracy of the UARL formula is questionable;
- data required to calculate UARL are not available;
- the ILI is not basically accepted in the industry;
- the ILI is not needed - the classical performance indicators (like real losses per km mains per day) are considered as sufficient.
- 10% water losses always sounds acceptable - while the ILI in many cases highlights that the true leakage performance is far from satisfactory
- warnings from the Water Losses Task Force that the ILI must not be used for systems with less than 25 m average pressure or less than 5,000 connections.

In spite of that, ILI technical performance indicator can be used for benchmarking water supply systems in different areas and countries. The following benchmarking criteria has been proposed (Liemberger M, 2002):

Table 1. BENCHMARKING CRITERIA FOR DEVELOPED AND DEVELOPING COUNTRIES

Technic. Perfor. Categ.		ILI	Litres/Connection/Day when the system is pressurized at average pressure of:				
			10 m	20 m	30 m	40 m	50 m
Developed Countries	A	1-2		< 50	< 75	< 100	< 125
	B	2-4		50-100	75-150	100-200	125-250
	C	4-8		100-200	150-300	200-400	250-500
	D	>8		>200	>300	>400	>500
Developing Countries	A	1-4	< 50	< 100	< 150	< 200	< 250
	B	4-8	50-100	100-200	150-300	200-400	250-500
	C	8-16	100-200	200-400	300-600	400-800	500-1000
	D	>16	>200	>400	>600	>800	>1000

As it can be seen from the Table 1, different ILI ranges have been provided for developing and developed countries. The proposal attempts to classify the leakage levels within the Water Utilities into four categories based on the ILI value as follows:

- Category A: Further loss reduction may be uneconomic unless there are shortages; careful analysis needed to identify cost effective improvement
- Category B: Potential for improvements; consider pressure management; improve active leakage control practices and network maintenance
- Category C: Poor leakage record; tolerable only if water is plentiful and cheap; even then, analyze level and nature of leakage and intensify leakage reduction efforts
- Category D: Horrendously inefficient use of resources; leakage reduction programs imperative and high priority

As an illustration, ILI values are presented for some water systems in England and Wales (the initiators of ILI methodology) and in Italy Fig 3 (Lambert A. and McKenzie R 2002, Fantozzi M, Bazzurro N, Mazzola M 2005). Since the vast majority of water utilities in the developing world will have ILI values exceeding the upper limit of the Table 1, reducing real losses to below 16 will be the starting point. As soon as utilities start to introduce active leakage control, carry out flow and pressure measurements, and improve overall data quality the bandwidth of the ILI will dramatically be reduced. Often leakage reduction will also lead to an improved supply situation and pressure increases that will make the calculation of the UARL formula more accurate.

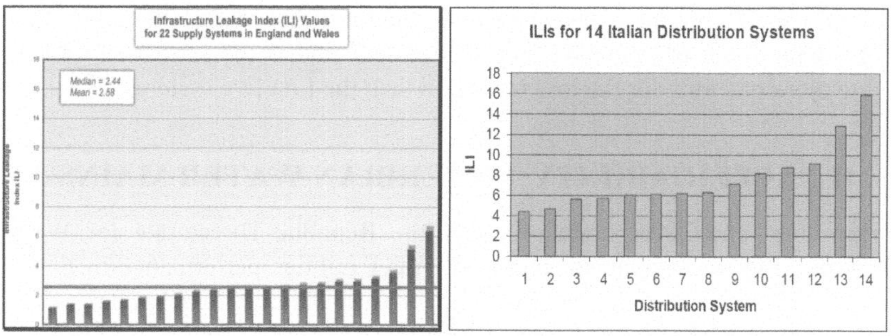

Figure 2. ILI FOR DATASET OF WATER MAINS FROM ENGLAND&WALES AND ITALY

6 COMPARING ILI VALUES TO PERCENTAGES

If real loss expressed in % of system input volume, it sounds to one as a real potential to decrease (or even) eliminate water loss. The meaning of TIRL is to estimate unavoidable real loss level, considering very good infrastructure conditions (mentioned above) and taking into account specific conditions of any water main, and ILI should indicate ratio between volumes of the current and unavoidable losses.

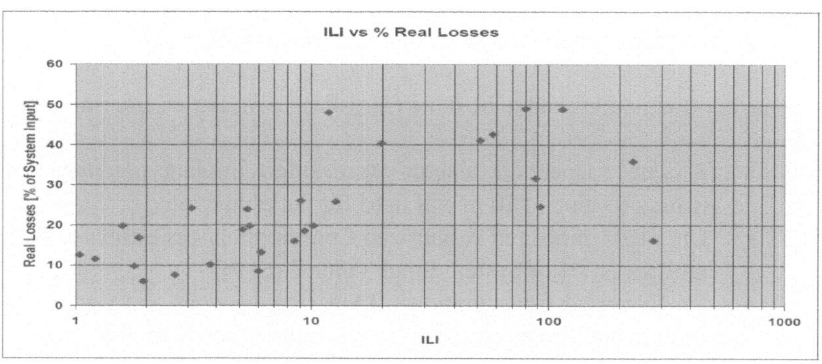

Figure 3. ILI VS % REAL LOSSES

Figure 4 shows the leakage management performance of 30 utilities from international data set (Water Loss Group, 2005) using the ILI and the respective losses expressed as percentage of total system input. It is obvious that there is no correlation, for example 50% real losses mean in one case an ILI of around 12 and in another case 114! As can be seen, real loss level of 10% is not necessarily an indication for good real loss management. Two systems with ILI value very close to 1 have more than 10% real losses, the third best ranked 20%, while the system with about 6% real loss has ILI value 2 and is the sixth best ranked, having more potential for the loss reduction than the first five systems.

7 ILI APPLICABILITY TO SERBIAN WATER MAINS

During 2004 and 2005, upon order by the Republic Directorate for Water Resources, the assessment on municipal water utilities performance in Serbia was done. Following collected data from 98% of the questioned water mains, water supply systems in Serbia are in the most of cases composed of rather old pipes (over 40 years), made of steel, asbestos cement, while the newer parts of the mains are constructed of plastic, polyethylene and ductile iron pipes. The similar set of problems was experienced like when introducing the ILI in the developing world:

- Not always reliable information on the true network length,
- Unreliable (or no any) system input volume metering,
- Accurate number of service connections is not known - number of customers is used instead,
- Neither pressure data nor pressure loggers available. Estimated average pressure usually too high ("wishful thinking"!),
- High level of apparent losses (difficult to estimate) and therefore unreliable and inaccurate volume of real losses.

Some approximations had to be done (Jaroslav Černi, Beoinženjering,2005):

- Service connection lengths were often missing parameter and common value of 10 m was used for the estimation,
- Unbilled Unmetered Authorized Consumption was estimated for all utilities as 1% of System Input Volume (SIV),
- Unauthorized Consumption – Illegal connections and usage water from hydrants was estimated for all utilities as 1% of SIV,
- Metering Inaccuracies due to low bulk meter sensitivity on minimal night consumption were estimated for all utilities as 1% of SIV,

- Number of days when system is pressurized was adopted as 365 days, while the real data are different; intermittent supply decreases UARL and increases ILI value.

Using collected data, traditional and IWA best practices performance indicators were estimated for 36 water utilities. UARL average value, like as ILI value, calculated from the collected data with an error margin of 12%, does not fit to the IWA Task Force recommended 95% confidence interval. Average Real Loss, expressed in percentage is 36%, while there were estimated some extreme values of almost 70%! The average ILI value is 11 (Figure 5).

Even though there is no correlation between these two indictors (Fig. 5 and 6), the both of them imply on bad technical condition of water mains in Serbia and high potential for improvement. The ILI values seem much better than in many other developing countries (Sri Lanka, Vietnam, ...), comparing them with the highly developed systems (Great Britain, Australia, Germany, USA, ...) will make one conclude that Serbian water mains are currently in poor condition, with high potential for significant real loss reduction.

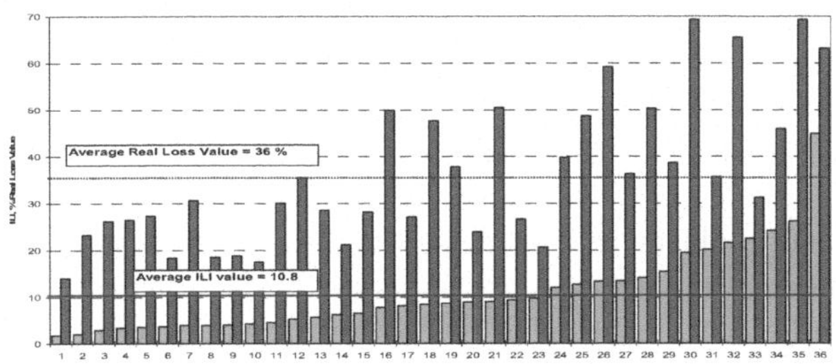

Figure 4. % REAL WATER LOSSES AND ILI IN DATASET OF SERBIAN MAINS

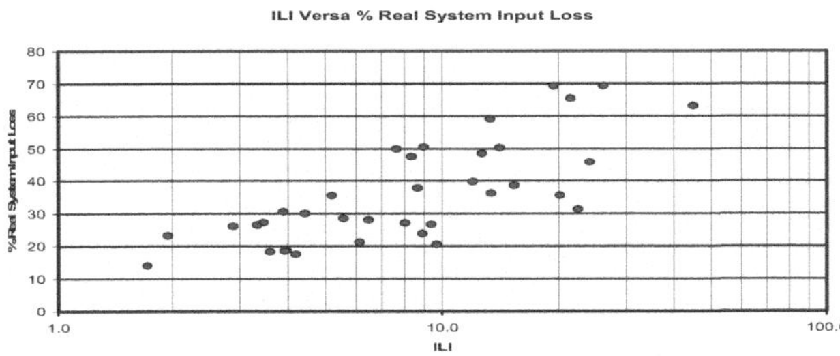

Figure 5. ILI VERSA % REAL SYSTEM LOSS IN SERBIAN WATER MAINS

In spite of problems in the initial steps of IWA best practice and UARL, TIRL and ILI performance indictors application in Serbia, this methodology can help to establish a new merit point of view and make water utilities to discover new facts on their functionality and potential to increase own efficiency, as water resources are not always so plentiful in Serbia (in fact, many small water mains face serious problems both with insufficient water resources and high level of leakages from water mains!).

New steps on the national level must be done to prescribe comprehensive systematic approach to this problem and define national strategy for water loss management. Further analysis should follow in order to recommend adoption IWA performance indicators in original or modified shape.

8 CONCLUDING REMARKS

The ILI has, in recent years, proved to be a very useful performance indicator when benchmarking leakage in water distribution systems. Although various limits on the use of the ILI have been proposed by it's original developer to safeguard the soundness of the results, it can still provide a useful indication of high leakage even when used outside the normally accepted limits. ILI is certainly also a most suitable indicator for water utilities in developing countries in spite of lack of accurate data, and it is now understood that "true" ILI of low pressure systems will always be higher than the calculated figures. This suggests that the leakage problem in developing countries is even more serious than previously anticipated. IWA best practice water balance and methodology used to estimate performance indictors in Serbian water utilities, can be rated as very useful tool for assessing and pointing on the real problems in Serbian water mains.

9 REFERENCES

Alegre et al, (2000) Losses from Water Supply Systems: Standard Terminology and Performance Measures, *IWA Manual of Best Practice*

Lambert A & Hirner, (2000) Losses from Water Supply Systems: Standard Terminology and Performance Measures, *W.H. IWSA Blue Pages*

Lambert A. and McKenzie R (2002) Practical Experience in using the Infrastructure Leakage Index, *Proceedings of IWA Conference 'Leakage Management: A Practical Approach' in Lemesos,* Cyprus

Liemberger M, (2002) Do You Know How Misleading the Use of Wrong Performance Indicators can be, *Proceedings of IWA Conference 'Leakage Management: A Practical Approach' in Lemesos*, Cyprus

Fantozzi M, Bazzurro N, Mazzola M (2005) Progress on Implementing the IWA approach in Italy: Case Studies, Dissemination and training activities of the Italian Water Loss User Group and Links with European Funded Projects, Halifax

Water Loss Group (2005) IWA Task Force: Best Practice Performance Indicators for Non Revenue Water and Water Loss Components: A Practical Approach, Halifax

Institute for Water Resources Jaroslav Černi, Beoinženjering, (2005) Vodovodi u Srbiji (*Water Mains in Serbia - Assessment*), Republic of Serbia, Ministry of Agriculture, Forestry and Water Resources

Model for Improvement of the IP Water Infrastructure in the Systems Featured by Annual Inequality in Water Production

O. Doklestić[1]

[1] *1M.Eng. Civil Engineering, "Vodacom" ltd., address, II dalmatinske 11, 85230 Tivat, Montenegro*
e-mail: olivera.doklestic@vodacom.cg.yu, tel 00382 88 672 779

Keywords: Montenegro Coastal region, infrastructure network, assessment of services

ABSTRACT

Consideration about the water supply of the Coastal Region of Montenegro implies two different periods in a with regard to water producing and water consumption. Summer water reaches nearly 500 l/s. This paper presents algorithm for situation modelling how to aimed at improving IP in the next ten years

1 RESEARCH SUBJECT

The research area encompasses the coastal region of Montenegro, including both the coast and inland area composed of six municipalities located along 280 km long coastal line. There are three municipalities in the Boka Kotorska Bay: Herceg Novi, Kotor and Tivat, and also three on the coast of the Adriatic sea: Budva, Bar and Ulcinj. There are approximately 160,000 of permanent inhabitants in the region. The main economic industry is tourism, growing in the past several years after undergoing stagnation in the 90ties. It is estimated that the number of tourists in the peak season reaches nearly 250,000. Apart from tourist economy, as the primary economic activity, the region developed also shipyard repairing industry, port economy, while agriculture is significantly less developed.

The subject of this research are independent, local water supply systems in this region, located within administrative (municipal) borders. Some of these systems are entirely independent, from their spring to consumption, and belong to the area of one municipality. Other systems are connected to two municipalities, or their waters have international status. Hydro-technical systems in the region were intensively developed in the period from early 60ties to late 80ties, while 90ties, due to economic stagnation, were featured by the lack of investments, uncontrolled and sporadic development of the system. Due to the economic weakness of municipalities and public community utilities, funds from collected water were barely sufficient to cover for rehabilitation of acute problems on pipeline and facilities. Over the several last years, more precisely from 2004, the coastal region of Montenegro, although still not unique, has taken a new organizational and management direction with regard to all decisions, in order to improve the state in its infrastructural systems. The basis for that "offensive" in access to hydro-technical infrastructure is a notion that water supply (and waste water channelling) is a direct, cause and effect indicator of the tourist economy development, i.e., urban development in the region.

2 MAIN FEATURES OF THE WATER SUPPLY SYSTEMS

This region is characterized by 80 – 90% of the total number of inhabitants being connected to the main water supply system. Water production is almost the same as water consumption in one part of the year (during summer period). During the summer tourist season, water consumption increases for two to three times, while capacities of the water resources are being significantly reduced. In addition, indicators of the state in the system have different features during tourist season in comparison with the other part of the year. In terms of terrain configuration, urbanization and new developments, and extension of the water supply system, the coast of Montenegro is featured by:

- Non-uniform system complexity,
- Different age and quality of facilities,
- Time and spatial inequality in production,
- Time and spatial inequality in consumption and distribution
- Different degree of collected water,
- Different IP of companies and systems.

Following basic data characterize the system in the coastal region:

- Water is produced from 34 production facilities (water intakes, springs, wells)
- Yearly water production varies from 1350 to 2000 l/s.
- The main part of water resources consist of springs 62%, of which 92% of the total water production in the winter period and 33% in the summer period; 33% are well fields with 52 wells; Herceg Novi's system is supplied with the water from the hydro-system on the Trebješnica River, i.e. from the Bilećko lake (Bosnia and Herzegovina, and is distributed through the territory of the Republic of Croatia).
- Water losses, pursuant to the conducted analysis (Master plan „Beller Consult“, 2005), are unequal as well, and it is considered that they range from 46 – 88%; out of this percentage non technical losses make nearly 25%.
- The expected summer water deficiency is 500 l/s.
- There are water restrictions during summer period, either only in parts of the system or in the entire system.

3 RESEARCH TASK AND THE AIM OF THE PAPER

The research aims to, through the analysis of the system's technical operation:

1. create data model with algorithm for determination of water supply system performance indicators, in accordance with the available input and expected accurateness;
2. present methodology for improvement of the water supply system performance indicators in terms of defining development strategy.

4 STARTING ASSUMPTIONS

Out of the numerous assumptions that can be used for the input matrix of model system, we will point to following general characteristics of all individual systems:

- The research subject are systems featured by seasonal inequality in water production and consumption.
- There are two types of flow in the water supply networks: under pressure and quasi-stationary.
- Water losses are functionally connected with working pressure in the system, but they also reflect the method of connection in knots, i.e., legality of connections is brought into a relationship with the so called administrative water losses, and its correctness with technical losses.
- Water quality, at some points in the system depends on qualitative features of raw water, type and the method of raw water treatment, concentration of chlorine and the type and age of pipes.
- Water consumption is functionally related to the working pressure, but also to the water quantity in the system. Moreover, it is related to restrictions, as well as to economic category, the water price.
- The number and type of water leakages depends on responsiveness and operation of repair teams, i.e., of the quantity and quality of staff.

By the model of numeric simulation of quazistationary flow in the water supply system, in realistic current time it is possible to perform space-time interpolation.

5 ALGORITHM

It is possible to reach the aim by establishing the so called goal function, i.e., by determining algorithm, which is a combination of mathematic operands and described variables. The algorithm consists of following steps:

5.1 The first step

Defining the model of the current state as real, movable model that has its data base and structure, and is indeed a data driven engineering model, being constantly filled with the new data, i.e., the base is updated, the structure is the same, but different variables are shown.

In accordance with the degree of organization, i.e., method of operation, i.e. degree of operation, current systems have their functional relations.

a) Assumptions and limitations:

- Water supply systems will surely become independent in the next 10 years
- The unique regional water supply, for the entire coastal area of Montenegro, will not become operational for the next 10 years.
- The method for managing water supply companies will pass through some transformations.
- Continuous trend of the increase of significance of the primary criteria for general operation and business "3E" (effectiveness, efficiency and economy),
- Falling behind regarding development plans, particularly tourism in terms of increase of accommodation capacities and raising quality level for five years.

b) The subject of the paper:

- The model system is based on the data and information of a particular water supply system and utilities (Herceg Novi) at the coast of Montenegro, featured by unequal yearly water production and consumption.
- General model for the entire coast is given, based on the individual one, which general characteristics are similar to the experimental water supply system,
- The general model is checked based on the target function on the other concrete systems and companies

c) Model forming

Individual water supply system model (company that manages it) has its structure, that can be made based on methodology and guidelines of IWA system of IP (Alegre H., Baptista J., M., Cabrera E., et al. 2006.), and which is composed of following indicators:

- performance indicators (PI);
- variables (V);
- connection between information and system data base (CI)

The following tables shows performance indicators and variables, with seasonal variability and ranking by importance.

Table 2. SEASONAL VARIABILITY AND RANK PERFORMANCE INDICATORS

	Performance indicators PI:	Seasonal variability	Rank
1.	Water resources (WR)	+	1
2.	Personnel (Pe)	-	6
3.	Physical (Ph)	+	5
4.	Operational (Op)	+	4
5.	Quality of services (QS)	-	3
6.	Economical and financial (Fi)	+	2

Table 3. SEASONAL VARIABILITY AND RANK PERFORMANCE VARIABLES

	Performance indicators PI:	seasonal variability	rank
1.	Water resources (WR)	+	1
2.	Personnel (Pe)	-	6
3.	Physical (Ph)	+	5
4.	Operational (Op)	+	4
5.	Quality of services (QS)	-	3
6.	Economical and financial (Fi)	+	2

Indicators and variables can be unchangeable entire year through, or they can be changeable, depending on the season. Water resources, at least with regard to the capacity of the local springs, reach a minimum during summer, the number of employees is unchangeable, but they are changing and adjusting to conditions of the intervention in the network or type and the method of maintaining network and facilities. Therefore, physical and operation indicators are different during two year period, and they depend on the water consumption. It is not a common practice of a water utility to increase the number of employees during the summer period regardless of significantly higher collection rate, so that the financial effect of utilities' performance is better than out of season. In addition, the efficiency and system operation is significantly better during summer than during winter. In technical sense, water leakages in the system are significantly less, and interventions in the network are more prompt and efficient. Service variable is a descriptive variable. It presents a combination of several indicators, from water delivery, method of payment, interventions in the network, administrative-technical measures on connections, water meters, as well as solving different legal, technical and economic issues. All this make a combination of service quality for consumers.

Thus, the claim that the system can be viewed through two different annual stadiums is completely founded. Rank, in the last column for PI and V, determines priorities for setting model system, of an ideal water supply system, and in accordance with them the prognosis model system as well.

The third group of system indicators presents a connection between information and system data base marked with CI, and these are following:

- service data base
- physical values
  - water resources
  - closed reservoirs
  - treatment
  - gate and distribution reservoirs/mainten.
  - pumping stations
  - gate and distribution network
  - connection maintenance
- consumption and peak indicators
- demography and economics
- environmental protection

Each of these groups of indicators can be subdivided to a great number of sub groups, and each sub group can be further subdivided to its lower rank groups. Out of each group of indicators it should select relevant representatives that will truly characterize the model of the current state, and in accordance with the target function it can serve for further system modeling to the so called „abstract model of prognosis“ for future operation of the system. Such ultimate model, „for the future“ should meet the target function in the assumed values.

Model 1, abbreviated MO1, is the model of the current state where it is disposed with well known data from certain "zero" period of procession, with as much as possible detailed data base of measured and particular variables.

When selecting parameters, group PI and information it should select from the great number of indicators those that are the most representative members of their groups and give the best results.

d) Forming of the model MO1

Model MO1 uses the data base of the experimental water supply system and its governing company. The model water supply system is Herceg Novi that uses the so called external water (inflow out of Montenegro) which flow ranges from 350 to 400 l/s and 40 to 250 l/s from local sources. The total quantity of water that enter the system (or possible quantities) are 650 l/s, in the winter period (out of tourist season), as maximal quantities, and 380 l/s in summer period, as minimal ones. The system operates through three height zones with the last level of water in reservoir on 250.00 mnm height. 80 % of inhabitants are concentrated in the narrower zone, of 100 meters of width along the coast, at a height of 150 mnm. In morphological sense the municipality area is of inequal configuration, from the low area of riviera with the most dense agglomeration of housing area to 50.00 mnm height, and with central city centre, located on a hill at a height of 150.00 mnm. Urban development increasingly reaches the inland area, of even over 400.00 mnm height.

Basic characteristics of the model water supply system are:

- The total volume of the water reservoir is 11,500 m3
- It is assumed that the real specific volume is 145 l/kor
- The real total volume is 27% of Qmax, dn
- the system has a total of 22 automatic pump stations
- there are 21 permanent flow meters
- the total lenghth of the distribution network is over 200 km
- the number of connections is 11,299
- 90% of inhabitants are connected to city water supply network.

Becoming involved in the monitoring of IP list the experimental ViK does permanent measurements:

- Of the raw water inflow
- Quantity of water that enters the system

Regular calculation is being done in order to build relations of system effectiveness:

- Quantity of invoiced water
- Quantity of unaccounted for water
- Record the number of connections (of home and legal subjects)
- Record replacement of water meters
- Record the degree of intervention on pipe network
- etc.

The basic characteristics of the system are given in the table 5. In the table 6 the lenghths of the main pipelines in the system are given, with diameters from 80 to 600 mm, with balance sheets from 2004 to 2007. Replacement of pipelines as a tendency for improvement of the state in distribution network can be seen in tables.

5.2 The step two– definition of the aim

The aim is composed of the framework model system given through border values. The aim has been formed in accordance with Water Supply Master Plan (Beller Consult, 2005.) which individual parameters are necessary to be reached in different time periods. Such an aim will, at the same time, check the validity of this study and its applicability to systems containing unequal annual water production and consumption.

Aim 1 – year 2015

For Herceg Novi norms of consumption in MO2' are presented in table 3.

Table 4. THE NORMS OF CONSUMPTION IN MO2'

Parameter	unit	2005	2010	2015	2031
Population growth	%		1.12		
Max. consumption during summer for residential population	l/č/d	266	245	220	180
Hotels 4* i 5*	l/č/d	500	450	400	385
Hotels 1* - 3*	l/č/d	350	300	250	225
Other Tourists (apartment, vacation dwelling, camping)	l/č/d	250	210	185	170

- „Beller Consult" study is based on the assumption that there will be a five year delay in planning and development.

- Maximal number of consumers (permanent inhabitants and tourists) is 72,000

- General, technical water losses: 35% (in 2010), 20% (in 2028).

The consumption norms for the coast of Montenegro in MO2 are presented on the table:

Table 5. THE NORMS OF CONSUMPTION IN MO

Parameter	unit	2005	2010	2015	2031
Population growth	%		1.12		
Max. consumption during summer for residential population	l/cd	271	237	220	180
Hotels 4* i 5*	l/cd	500	450	400	385
Hotels 1* - 3*	l/cd	350	300	250	225
Other Tourists (apartment, vacation dwelling, camping)	l/cd	250	210	185	170

5.3 The step three - selection of functions

It is possible to reach MO2' from MO1 by establishment of some functional variables. That functional variables are mathematic functions and descriptions of the state.

6 REFERENCES

Beller, Energoprojekt Water supply Master plan Study for Coastal Region of Montenegro and Cetinje, 2006.

Alegre, W., Baptista, Cabrera E., at al. Performance Indicators for water Supply Services, Published by IWA Publishing, Alliance House, 12 Caxton Street, London SW1H, UK, 2006.

COST C18 Activites / European Cooperation in the field of Scientific and technical research/ Draft memorandum of Understanding, Performance assessment of Urban Infrastructure services: The case if water supply, wastewater and solid waste, Brussels, 2003.

COST Action C 18 : Performance assessment of Urban Infrastructure services: The case if water supply, wastewater and solid waste WG 1, Seminar : The use of performance assessment tools for regulation of water supply, wastewater and solid waste services in Portugal, National context for the water supply, wastewater and solid waste services, Catania, 2005.

Benchmarking

Metric benchmarking as a driver for improved performance

The Right Choice of Denominators and Grouping Factors in Water Supply Metric Benchmarking

H. Theuretzbacher-Fritz[1], *R. Neunteufel*[2], *J. Kölbl*[1], *R. Perfler*[2] *and E. Mayr*[2]

[1] *Graz University of Technology, Institute of Urban Water Management and Water Landscape Engineering. theuretzbacher@sww.tugraz.at, Stremayrgasse 10/I, A-8010 Graz, Austria, Tlf +43 316 873-8884 Fax: +43 316 873-8376*
[2] *University of Natural Resources and Applied Life Sciences Vienna, Institute of Sanitary Engineering and Water Pollution Control. roman.neunteufel@boku.ac.at, Muthgasse 18, A-1190 Vienna, Austria, Tlf +43 1 36006-5805 Fax: +43 1 3689949*

 Performance Assessment of Urban Infrastructure Services
Edited by Enrique Cabrera Jr and Miguel Ángel Pardo. ISBN: 9781843391913.
Published by IWA Publishing, London, UK.

Keywords: comparability, metric benchmarking, performance indicators, water supply

ABSTRACT

Deriving comparable PI results is a core objective of any metric benchmarking project. Based on experiences of benchmarking activities in the Austrian sector of water supply services, examples are given for the interference of structural set-ups of the benchmarked utilities with the chosen PI set. At heart, it depends on the choice of the denominators of the PI formulas. It depends if the assessed parameters are related e.g. to cubic metres of water or kilometres of mains length or numbers of service connections. The Austrian way of coping with these challenges is elaborated in this paper.

1 INTRODUCTION

The linchpin of successful metric benchmarking activities is the question how comparable the PI results are. Besides homogeneous data collection, the required comparability depends on the fact whether the comparisons are carried out only within "peer groups" (of similarly structured water supply services) or not. Moreover, it is of importance how the structures of the utilities interfere with the chosen performance indicators. Especially the choice of the denominators in the PI formulas influences the PI results significantly.

Upon the experiences of the water supply benchmarking activities of OVGW, the Austrian Association for Gas and Water, these methodological issues of a high-quality metric benchmarking shall be elaborated within this article. The OVGW project "Stage B" (Theuretzbacher-Fritz et al. 2006) was carried out among 72 water supply services (WSS) representing more than 50 % of the supplied population in Austria. Extensive efforts were spent on analysing explanatory factors and incorporating them as grouping parameters. Thus, the sound empirical basis can be utilised for detailed methodological verifications.

1.1 Cubic metres, kilometres or number of service connections?

Core performance indicators of the IWA PI system (Alegre et al. 2006) are often referred to the central variables system input & consumption (m^3), mains length (km) and number of service connections. E.g. this applies to PIs of water losses (IWA code WR1, Op23 to Op29, Fi46) and of personnel deployment (Pe1, Pe2).

On the other hand financial indicators like total costs (Fi4), running costs (Fi5) and capital costs (Fi6) are only referred to cubic metres and not to km as well. As it could be verified within the Austrian project that the mains length is the primary cost driver, the charts of the financial PIs Fi4 to Fi6 (costs per cubic metre) were grouped by the context information network delivery rate (m^3 per km and year; IWA code CI76).

Nevertheless, the PI results for each participant depend on the relationship between the denominators in the PI formulas and the participant's structural circumstances of water distribution. More graphically, Figure 1 characterises the interdependencies between the central numbers cubic metre, kilometre and number of service connections.

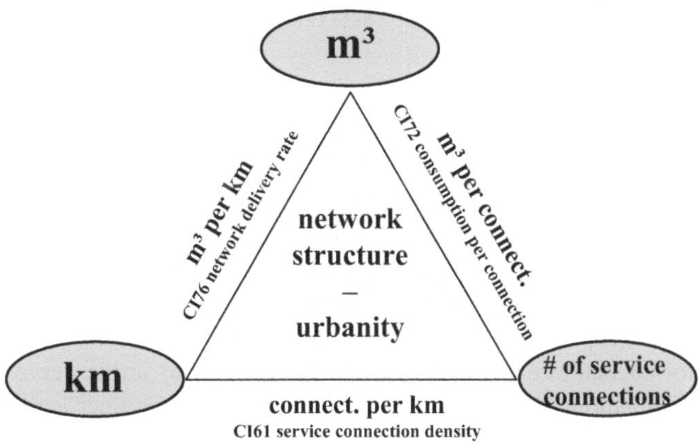

Figure 1. NETWORK STRUCTURES AND THEIR CHARACTERISATION IN FIGURES.

How this triangle is combined in real supply network structures, is defined by the context information network delivery rate (IWA code CI76), consumption per service connection (CI72) and service connection density (CI61). Especially when comparing PI figures for benchmarking reasons, it must be considered that the combination of these context information varies significantly among the participants.

Figure 2 shows the different distribution structures of the Austrian benchmarking participants. Visualising their network delivery rates in ascending order, their service connection densities do not follow any specific pattern, e.g. increasing connection densities with increasing network delivery rates. Hence, one has to be very careful in choosing the right PIs and grouping factors and, *vice versa*, each participant must consider his structural position within his peer group when interpreting the PI results.

For example, certain different charts would have come out for the PI unit total costs (EUR per m³, IWA code Fi4), if the same PI values had been grouped by the service connection density instead of the network delivery rate.

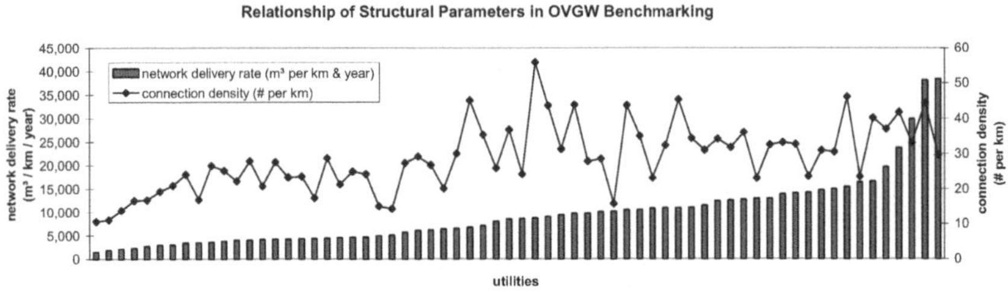

Figure 2. **NETWORK DELIVERY RATE AND SERVICE CONNECTION DENSITY WITHIN OVGW BENCHMARKING.**

Within the Austrian benchmarking project, a diligent procedure was chosen to derive a combined network structure parameter called "urbanity". Data from network delivery rate, consumption per service connection and service connection density were catenated to gain the urbanity categories "rural", "urban" and "metropolitan". Furthermore some revisions by the project team were done for special situations, e.g. when a typical rural utility serves with a disproportionate amount of bulk supply for an industrial customer leading to an untypical high network delivery rate.

The explanatory factor urbanity proved to work well as a grouping factor for both technical figures (e.g. water losses, failure rates) and business figures (costs, personnel).

2 COMPARING THE BEHAVIOUR OF DIFFERENT PI'S

2.1 The example of personnel PIs

Personnel PIs were chosen as an example for this elaboration because they are both sensitively discussed as well as very differently derived and visualised. The first precondition for comparable results of personnel PIs is to homogeneously collect the data for the numerator of the PI formulas: at heart, the correct confine of the amount of employees by fulltime equivalents requires the consideration if functions are outsourced and to what extent.

Water utilities in Austria are either embedded in larger organisations, e.g. in city works and municipal administrations, or have to fulfil their tasks "stand-alone" like water associations and water co-operatives (Theuretzbacher-Fritz et al. 2007). Therefore, besides outsourcing of tasks to external organisations, the fulfilment of tasks by overhead service units ("in-house outsourcing") has to be taken into account (Figure 3).

Once the personnel data are homogeneously defined and ratios of in-house and external outsourcing quantified, different personnel PIs can be derived using different denominators, e.g. employees per connection (IWA code Pe1), employees per water produced (Pe2) or transmission, storage and distribution personnel per 100 km (Pe12).

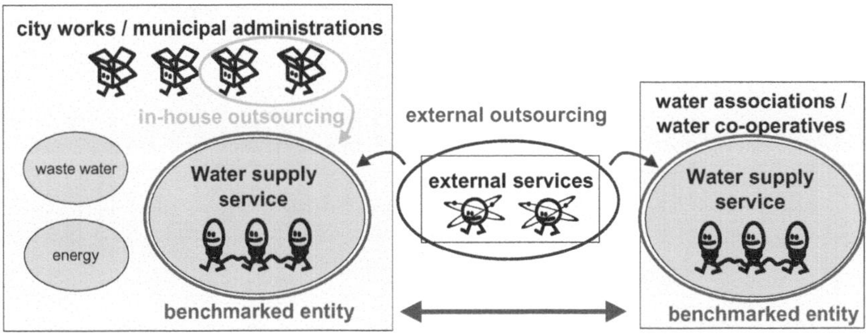

Figure 3. SCHEME OF CONFINING PERSONNEL DATA OF DIFFERENT LEGAL FORMS OF WSS

Figure 4 shows the results of two personnel PIs, both for total personnel and both double-grouped by urbanity and outsourcing total (in-house & external). Some of the groups had to be masked out because data confidentiality could not be granted due to too small group sizes. The group sizes are visualised by the small numbers in the boxes which lie at the level of the median values. The boxes define the range between the 25 % to 75 % quantile.

Furthermore, the PI values of two participants belonging to the group of urban WSS with high outsourcing ratio are shown (WSS "A" and WSS "B"). Whereas the value for the personnel PI referred to the service connections is pretty high for WSS A, the PI with cubic metres as denominator yields a low value for the same utility. For WSS B the same occurs just the other way round! So what – which utility is more efficient now?

Looking behind the scenes, both water supply services lie at the same level of network delivery rate (WSS A … 10,500 m^3/km/y, WSS B … 9,900 m^3/km/y). But they differ significantly in the service connection density (WSS A …

16 conn./km, WSS B … 44 conn./km) and in the consumption per service connection (WSS A … 645 m³/conn., WSS B … 223 m³/conn.). So, each of the two utilities *must* perform the way they do within these results. This example leads us to the assumption that structural frame conditions would dominate the comparisons. Therefore two aspects must be mentioned not to throw out the baby with the bath water: the two water supply services chosen for the example do have extreme structural figures; most of the utilities lie closer to each other in their general set-ups than these two networks. Secondly, this rigorous example shall unveil the necessity of deriving and interpreting results based on more than just one PI, based on more denominators!

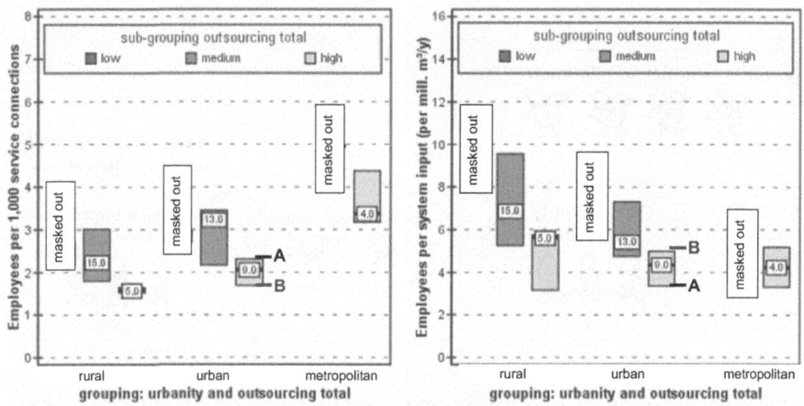

Figure 4. **RESULTS OF DIFFERENT PERSONNEL PIS WITHIN OVGW BENCHMARKING (BOXES SHOW 25 TO 75 % QUANTILES) (NEUNTEUFEL ET AL. 2006, AMENDED)**

2.2 Why to choose which denominator for PIs?

Following the above mentioned, the calculation of PIs with different denominators in parallel leads to more transparency regarding their interference with the structural conditions of the water supply services. On the other hand, the practical world requires lean systems of performance assessment. Thus, some general issues on the role of denominators shall be pointed at.

As it is clearly mentioned in the IWA Manual of Best Practice for performance indicators (Alegre et al. 2006), *"denominators of variables which may vary substantially from one year to another, particularly if not under the control of the undertaking, should be avoided (e.g. annual consumption), unless the numerator varies in the same proportion"*.

Especially when looking at the key financial PIs unit total costs (IWA code Fi4), unit running costs (Fi5) and unit capital costs (Fi6), the cost figures are referred to the authorised consumption. The configuration of these PI formulas is of course based on the argument that costs shall be covered by the primary cost object, therefore by the cubic metres. For temporal comparisons of one water supply service, these PIs can make the development of the overall business transparent, interpreting the relationships of revenues and costs in different years. Also, the assessment period might be extended to a longer time than one year to gain a better average value within one utility.

But for benchmarking activities, the figures shall be accounted to one year. Otherwise the efforts for data acquisition would go beyond the scope of the project. Moreover, the reference to the cost drivers should be strengthened in benchmarking projects in order to gain comparable results. As experienced in the Austrian project, physical assets are the cost drivers, especially mains length (IWA code C8) which fulfils the criterion of being a more stable variable, too. Due to the limited space in this contribution, charts for the PIs total / running / capital costs per km are presented in Valencia orally.

3 HOW TO COPE WITH STRUCTURAL HETEROGENEITY OF PARTICIPANTS?

Within the Austrian benchmarking project, these circumstances were considered by a diligent approach of incorporating as many influencing aspects as possible. Whereas in other performance comparison projects the participants are – if at all – grouped by the same criteria for the whole set of PIs, eg. by the utility size, more than twenty different grouping factors were used for the PI charts of the individual reports (Neunteufel et al., 2006). For instance, PIs of water losses, mains failures and rehabilitation rates have been grouped by the specially developed network age index factor NAX (Figure 5; Theuretzbacher-Fritz et al. 2007 in general and Neunteufel et al. 2007 in detail). Personnel indicators were grouped by the index factors of outsourcing ratios (external / in-house / total) as mentioned above as well as by the urbanity. Besides this classification for distribution structure, an assessment of the water production conditions was carried out, too. The efforts for water production depend on the local abilities of abstracting water from the natural sources and differ widely. The annual average values for abstraction per source range from 0.1 to 90 litres per second. So, some utilities gain their water from just a few, but large physical assets, others have to build, operate and maintain up to 130 small wells and springs. On the other hand, raw water treatment is more an exception than the rule; only 7 % of the benchmarked water volumes need to be purified due to the high standards of environmental and water legislation in Austria (Theuretzbacher-Fritz et al. 2006).

Furthermore, charts of core PIs were visualised by using several denominators simultaneously; e.g. personnel figures were referred to the amount of service connections, to the cubic metres of system input and to the kilometres of network length as well.

3.1 Assessing overall performances on comparable PI values

Based on this thorough procedure of comparing the participants only within peer groups, an overall performance assessment could be derived. A first draft on this approach was presented at the Salzburg meeting of COST C 18 (Theuretzbacher-Fritz, 2006). Its realisation and the gained experiences can now be described.

For each of the five target categories (reliability, supply quality, customer service, sustainability and efficiency) key performance indicators were selected from the whole PI set. Then the positions of the participating utilities within their peer group were evaluated for each selected key performance indicator. Four points were given for a PI value in the best quartile, three points for a score in the quartile better than the median value (in the better half of the box), two points for a position in the quartile worse than the median value and one point for a PI value in the worst quartile of the peer group distribution. Additionally if there were explicit standards by law or technical standards available for some PIs, the fulfilment was counted as 4 points, an almost fulfilment as 3 points and failing the target values as one point.

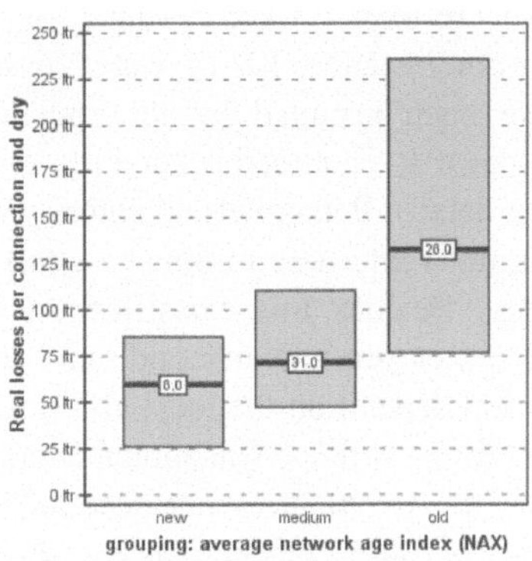

Figure 5. REDUCED BOX CHARTS FOR PIS (THEURETZBACHER-FRITZ ET AL., 2006)

Figure 6. OVERALL PERFORMANCE WITHIN OVGW PROJECT (NEUNTEUFEL ET AL. 2006)

The points of the key PIs performance figures could be calculated by certain weightings for the five target categories of OVGW benchmarking (Figure 6). Moreover, the performance values of the five target categories were summed in order to derive an overall performance value for each participant. Then the participants were divided into peer groups again for comparing their overall performance values. The criterion chosen for this grouping was the urbanity serving as the main explanatory factor of different structural conditions. Thus, the overall performance values were compared within the groups of metropolitan, urban and rural suppliers. Moreover, the special groups of bulk suppliers and water co-operatives were formed separately as well. Within each group one "best practice utility" was gained being ranked at the top and fulfilling the marginal condition of having a well-balanced performance in each of the five target categories. Nevertheless, other utilities can even perform better in some of the five categories (see e.g. sustainability in Figure 6) which corresponds with the fact that an ideal performance within all areas does not exist in practice.

The experiences with the spider chart displaying the overall performance were twofold. Naturally, the chosen procedure can only serve as a fuzzy approach as the results depend on the choice of the key PIs and their weightings to each other. Generally, the calculation of overall performance values proved to work well while auditing the results. Nevertheless, the used model represents a kind of global calibration in order to carry out comparisons between the benchmarked water services. In fact, each enterprise is targeted to individual company policies and strategic aims and would require a local calibration, by means of specific weightings within the assessment model.

Secondly, there have been other participants with overall performance values which were pretty close to the derived best practice utility. In such cases, their high performances were appreciated in the management summaries of their confidential individual reports. Whereas the spider chart shall mainly give a first overview and shall "make hungry" for 140 pages of detailed individual report, it also emphasises the nature of the quasi-competition by motivating the participants to perform even better in the future.

4 CONCLUSIONS

Experiences from the metric benchmarking activities in the Austrian water supply sector have shown that some methodological aspects have to be taken into account in order to derive and visualise comparable results. Besides efforts for homogeneous data collection (clear and well-defined data system, company visits, plausibility checks), the participants must be allocated to peer groups with best-possible similar conditions, separately for each PI and its influencing factors. This led to the high number of more than twenty grouping factors in the OVGW project, but served the purpose of a high-quality approach.

On the other hand, it is recommended to analyse the interference of the structural set-ups of the water supply services with the denominators of the PI formulas. If the PI values for the same numerator, but different denominators vary heavily, a combined view on these neighbouring PIs shall be facilitated.

Once these aspects are considered, overall performance values can be derived as it has been carried out in the Austrian project. Anyhow, their main purpose shall not end up in pointing out who is the best and who is worst. In fact, it shall serve as a starting point to first gain a synopsis and then drop into a plenty of single strengths and weaknesses in order to analyse the reasons for the results and to derive measures for further improvements.

5 REFERENCES

Alegre, H., Baptista, J.M., Cabrera, E., Cubillo, F., Duarte, P., Hirner, W., Merkel, W. and Parena, R. (2006). Performance Indicators for Water Supply Services. Second Edition, Manual of Best Practice. IWA Publishing, London, UK.

Neunteufel, R., Perfler, R., Theuretzbacher-Fritz, H. and Kölbl, J. (2007): Explanatory Factor "Average Network Age Index" (NAX) for Mains Failures and Water Losses. In *Proc. of LESAM 2007, 2nd Leading Edge Conference on Strategic Asset Management*, Lisbon, Portugal.

Neunteufel, R., Theuretzbacher-Fritz, H., Kölbl, J., Perfler, R., Unterwainig, M., Krendelsberger, R. and Mayr, E. (2006). Benchmarking und Best Practices in der österreichischen Wasserversorgung - Stufe B. 71 confidential individual reports, Vienna / Graz, Austria.

Theuretzbacher-Fritz, H. (2006). Performance Indicators and Benchmarking Projects in Austria. - COST C18 meeting on "Performance Assessment of Urban Infrastructure Services: The Case of Water Supply, Wastewater and Solid Waste". Salzburg, 3rd April 2006.

Theuretzbacher-Fritz, H., Kölbl, J., Kainz, H., Neunteufel, R., Perfler, R. and Haberl, R. (2007). New PIs and Explanatory Index Factors in The Austrian Water Supply Benchmarking System. In *Proc. of EFFICIENT 2007. The 4th IWA Specialist Conference on Efficient Use and Management of Urban Water Supply*, Jeju, Korea, 1143–1150.

Theuretzbacher-Fritz, H., Neunteufel, R., Kölbl, J., Perfler, R., Unterwainig, M. and Krendelsberger, R. (2006). Benchmarking und Best Practices in der österreichischen Wasserversorgung – Stufe B. Final Report of OVGW project 2005/06, Graz-Vienna-Wiener Neustadt, Austria.

The canadian national water and wastewater benchmarking initiative: Using process to drive improvement. Strategic Management of Water in Urban Areas

D. Main[1], *L. Ng*[2] *and A. North*[3]

[1] *David Main, 1901 Rosser Avenue 6th Floor, Burnaby, BC V5C 6S3 Canada, david.main@earthtech.ca*
[2] *Leslie Ng, 1901 Rosser Avenue 6th Floor, Burnaby, BC V5C 6S3 Canada, leslie.ng@earthtech.ca*
[3] *Andy North, 47620 UEP Subang Jaya, Selangor Darul Ehsan, Malaysia andy.north@earthtech.ca*

 Performance Assessment of Urban Infrastrucutre Services Edited by Enrique Cabrera Jr and Miguel Ángel Pardo. ISBN: 9781843391913.
Published by IWA Publishing, London, UK.

Keywords: Benchmarking, Continuous Improvement, Performance Measurement, Asset Management.

ABSTRACT

Earth Tech has been successfully benchmarking Canadian municipal water, wastewater and stormwater utility operations since 1997. While the fundamental purpose of this project was metric benchmarking for the purpose of making performance comparisons to guide continuous improvement, the project is now serving as a dynamic platform to consider, examine, and implement a broad range of utility best practices that have resulted in superior performance where they have been implemented. The keys to success however were based more on a process that emphasizes communication, teamwork and collaboration rather than the trend to push computerized data management systems to their fullest potential, and most importantly in recognizing the importance of "hard work". With these success factors now well understood and documented, it is feasible to benchmark almost any public infrastructure amongst agencies that are willing, regardless of their level of technological development. Finally by sharing this methodology, the performance measure descriptions and detailed definitions, it is also feasible to make international comparisons in a simple and cost effective manner, thus opening the door to the broad exchange of international best practices.

INTRODUCTION

In 1997, a group of four municipal wastewater utilities, Earth Tech Canada, Inc, and the National Research Council of Canada met with the purpose of looking for a framework to answer four seemingly simple questions that had been posed by each of the utility's boards:

- "How well are we doing?"
- "How do we compare with similar organizations?"
- "Are we getting value for money?"; and
- "How can we get better at what we do?"

Benchmarking, defined as " the continuous process of measuring products, services and practices against the toughest competitors or those companies recognized as industry leaders" was pioneered by Robert C. Camp at the Xerox Corporation in the 1970s. Because of its widespread use in all leading corporations through the 1990s, it was agreed that benchmarking could be effectively utilized to provide the answer to these important questions. What seemed straightforward and intuitive through many Best Practice publications, was in fact very challenging. The difficulty appeared once the practice was recommended for use in the public sector. In the private sector, competition results in ruthless selection of the "fittest". Whereas virtually all private sector corporations ultimately roll all performance metrics into only one final performance measure (total return to shareholders as measured by profit and shareholder equity); public agencies have always had a much broader set of objectives, including performance on social, environmental, as well as financial matters. Benchmarking was assumed to be a well documented process, but in fact had to be redesigned from the ground up in order to develop a management process that reflected the attainment of goals that by definition will always compete with one another (for example, improved environmental performance will be at the expense of bottom line financial results).

An extensive literature search showed that benchmarking in the public sector had been conducted in the past, even within the municipal water and wastewater utility industry. Further analysis however, showed that past exercises tended to be short lived, usually terminating after one or two iterations. Only in exceptional cases, has benchmarking appeared to deliver remotely satisfactory results. Weaknesses within all examined examples included:

- Inability to provide accurate and comparable data;
- Lack of agreement regarding what to measure;
- Lack of patience within participants to optimise the benchmarking process; and
- Few if any tangible improvements in participating organizations

By focusing efforts at the early stage of the project on each of the above, the National Water and Wastewater Benchmarking Initiative was able to design a robust methodology for conducting metric benchmarking within the public sector successfully and continuously. The project has found success not only in its fundamental purpose of metric benchmarking, but also as a dynamic platform to consider, examine, and implement a broad range of utility best practices that have resulted in superior performance. As well, the project has developed into a highly effective network of peers from which to share and exchange ideas about utility management.

What is Benchmarking? (And More Importantly, What is it Not?)

For the National Water and Wastewater Benchmarking Initiative (NWWBI), benchmarking encompasses regular annual comparisons regarding the attainment of common goals for the purpose of identifying performance gaps.

Equally important is implementing the improvements, monitoring the progress, and reviewing the benefits of the implemented changes. What differentiates this program from the traditional consultancy-lead private sector programs aimed at competitor analysis is that the NWWBI has been undertaken collaboratively, through the willing sharing of performance data, to learn about the circumstances and processes that underpin superior performance. In tandem with this, is the realization that no single organization has all the answers, and that success is measured through a wide range of criteria that may include financial-; sustainability-; reliability-; environmental- and customer service criteria.

Participation is National in Scope

Launched in 1997 as a pilot project that included four participating cities as well as team members from Earth Tech (as the Program Manager) and the National Research Council for technical advice, the Benchmarking Initiative has grown to the point where it serves as a national standard. Today, the National Water and Wastewater Benchmarking Initiative include 37 of Canada's leading municipal and regional utilities. The partnership now represents the Canadian urban centres that account for over 60% of Canada's population.

This wide participation base is not without its challenges. Canada is the second largest country in the world. It spans 4 ½ time zones, and includes climates that range from near Mediterranean to Arctic (including permafrost). In terms of precipitation, it ranges from desert to temperate rain forest. In the past, comparisons amongst such a diverse base were discounted. But by taking local factors that may affect cost, performance, or the selection of practices into account in the methodology, participants are rewarded by being able to evaluate an enormous range of practices in use. The fact remains, regardless of where we operate, we are all in the same business of operating public water and wastewater utilities. Figure 1 shows a map of the distribution of participating utilities.

Figure 7. PARTICIPATING CITIES IN THE NATIONAL WATER AND WASTEWATER BENCHMARKING INITIATIVE

METHODOLOGY: ITS NOT ROCKET SCIENCE, BUT IT IS HARD WORK

Figure 8. ANNUAL METHODOLOGY

The NWWBI utilizes a non proprietary benchmarking methodology that follows a conventional annual cycle as shown in Figure 2. Great efforts are placed on keeping the methodology simple and focusing on the effectiveness of each task within the overall process. The management premise is that if even one link is broken; the overall program has no chance for success. The following continue to be critical in the program's ongoing success:

The Utility Management Model: What Should We Measure?

The first task of the project was to develop a standardized "Utility Management Model" that would provide the framework for the selection and definition of performance measures. This single step was proven to be critical to the foundation of the entire project. Figure 4 illustrates the Utility Management Model that was adopted, with the Utility Goals and the relationship to the many Performance Measures. The Utility Management Model defines a framework of high level goals to which all water and wastewater utilities in Canada aspire to. Success therefore, is literally based on the attainment of these goals. The Model provides the basis for selecting practical and relevant performance measures to measure goal attainment.

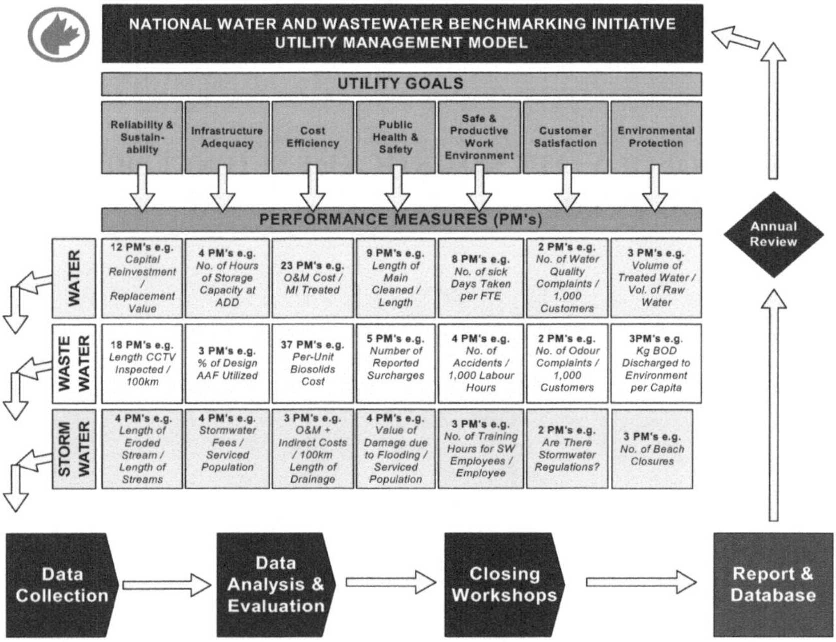

Figure 3: UTILITY MANAGEMENT MODEL

Complete Utility Management Goals are presently stated as:

1. Provide reliable and sustainable infrastructure.
2. Provide accessible and sufficient infrastructure (capacity).
3. Meet service and performance requirements at minimum sustainable cost.
4. Protect public health and safety.
5. Provide a safe and productive work environment.
6. Have satisfied and informed customers.
7. Protect the environment and minimize environmental impacts.

With the Utility Management Model in place, a comprehensive set of performance measures were formulated to measure a utility's success in attaining each individual goal. A very detailed definition for each performance measure numerator and denominator was also documented so that comparable data could be collected to assure valid comparisons. In most cases, each performance measure includes a numerator (expressing the level of goal attainment) and a denominator (normalization factor to enable comparisons, amongst agencies of differing sizes). Table 1 below shows an example performance measure and its corresponding definition:

Table 1: EXAMPLE OF PERFORMANCE MEASURE AND DEFINITION

Performance Measure	Numerator Definition	Denominator Definition
Total Operations & Maintenance Cost / km Length of Distribution system	Sum of the actual O&M costs incurred in the operation of the distribution/transmission/ integrated system (excludes capital costs, indirect costs, transfers to reserves and debt/interest charges). Includes O&M costs for both linear (pipes, meters etc) and non-linear (pump stations, reservoirs etc) infrastructure. Revenues are only included where they are recoveries for work done by water distribution staff that is extraneous to the utility (for example, lab tests for other utilities).	Total length of mains in the distribution/transmission/integrated system (i.e. excluding length of service connections). For the distribution system length include all connecting pipes between pump stations, rechlorination facilities and storage facilities if these are located within the distribution system. For the transmission system length include all connecting pipes between pump stations, rechlorination facilities and storage facilities when located between the source and the treatment plant or between the treatment plant and the distribution system.

In all, the NWWBI conducts benchmarking on about 50 performance measures for each of water and wastewater treatment, water distribution and wastewater collection. There are presently about 15 performance measures for the stormwater drainage function. In keeping with our approach to an open methodology, all the NWWBI performance measures and the definitions are published and available to any interested party on the project website (www.nationalbenchmarking.ca). Other benchmarking practitioners are encouraged to leverage their efforts from NWWBI published standards. It is our view that the more international agencies that use these same measures, the easier it will be to make international comparisons if the parties are so interested

Data collection: Comunicationa and team work; Not data processing

Data collection is the most difficult and time consuming aspect of benchmarking. It is also vitally important, for without accurate and reliable data, it is impossible to make performance observations. Earth Tech's early research into past exercises demonstrated that the data collection effort was severely underestimated and the resulting data collection flaws resulted in the termination and/or failure of the examined benchmarking examples. Our experience is that no less than 30% of the benchmark effort must be dedicated to this single activity at least in the early stages before quality information can be assured. A key distinguishing factor of this benchmarking project over similar benchmarking efforts is that data is collected through on-site data collection visits with the use of trained data collection staff from Earth Tech in close association with key utility staff.

In addition to effort, participants must be willing to be patient in the development of a robust benchmarking database. OFWAT, the government regulator for the UK and Wales water industry, predicted that it would take two years (two complete iterations) before reliable data could be assumed in their benchmarking efforts . In our experience, two years has been an optimistic target that only some utilities were able to achieve. For the NWWBI, it has taken about five years to develop reliable, useable data that is statistically significant, historically sound, and accurate for historic trending. However, this longer term has allowed the NWWBI to document best practices for the collection and management of utility data. Unlike other best practices that rely on computer-based data management system, the key to success is communication and team work within each utility department. Computerized data management projects must be designed only after business processes have been developed and installed.

Because of the need for communication and cooperation within utility agencies, a personable approach to the collection of information has helped to develop greater insight into a utility's local factors and operations. The personal involvement of Earth Tech staff in collecting the information also helps to ensure that quality assured data is collected and is comparable. The unexpected benefit to many participating agencies is that benchmarking may be the first time many utility staff have been asked for very specific information. It is not until the data collection task begins that utilities begin to realize both how much information they actually have about their system, and then whether they can locate it. Many utility organizations still rely on the memories of veteran utility operators or foremen, with intimate knowledge of each valve, pipe, and pump. This complex knowledge is in danger of disappearing with the impending retirement of long-time operators. We try to maintain a personal aspect to the data collection process to ensure that someone talks to this operator (and documents the detail) if the information is not kept elsewhere.

The well documented and rigid structure required of the NWWBI data collection provides a detailed framework to begin the difficult process of documenting operational and management information. The ultimate aim is to get this information into key management systems, (such as CMMS, GIS, etc.) and therefore, ensure that valuable knowledge will not be lost with the retirement of that "one" utility operator.

The results are worth the effort and investment. Because the participants are able to use the results with confidence, the project has been able to tackle a wide range of performance related discussions productively. Each year, the project has added valuable content that make the results more and more accessible to municipal utility practitioners.

Results from data collection to continuous improvement

The usefulness of benchmarking data only becomes apparent when continuous improvement actions are formulated. To start this process, the NWWBI produces a vast array of pictorially based graphs (approx. 5,000) that depict overall group performance, and personalized historical trending graphs. These results are published in an annual report, and are always available online through the project website. Earth Tech even provides a "help desk" service to customers to assist in customizing the data to meet a wide variety of reporting needs. All data is generated from a relational database management system, but the MS Access-based application tends not to be that intuitive for the occasional user, therefore assistance is always available.

Group Comparison Graphs: How do we compare with similar organizations?

First, group graphs are produced which compare results within similar groups. Performance measures are calculated from the compiled data in each of Water Distribution, Water Systems, Wastewater Collection, Wastewater Treatment and Stormwater. In comparing the data in group graphs, we compare similar systems for example, only conventional water treatment plants less than 60 ML/day are compared for measures such as O&M cost to avoid the large gap from economies-of-scale and because clearly conventional treatment plants cannot be compared to unfiltered or even direct or membrane filtration plants. As we make further comparisons and interpretations of the graphs, our personal knowledge of the systems help to evaluate what local factors may be considered as well as regional issues that may be of importance. Despite these unique characteristics, there are always comparable issues and factors. The following are example group performance graphs

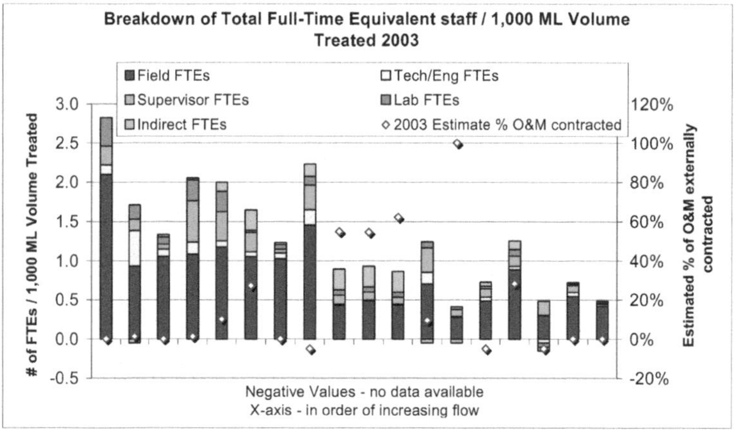

Figure 4: EXAMPLE OF A GROUP COMPARISON GRAPH

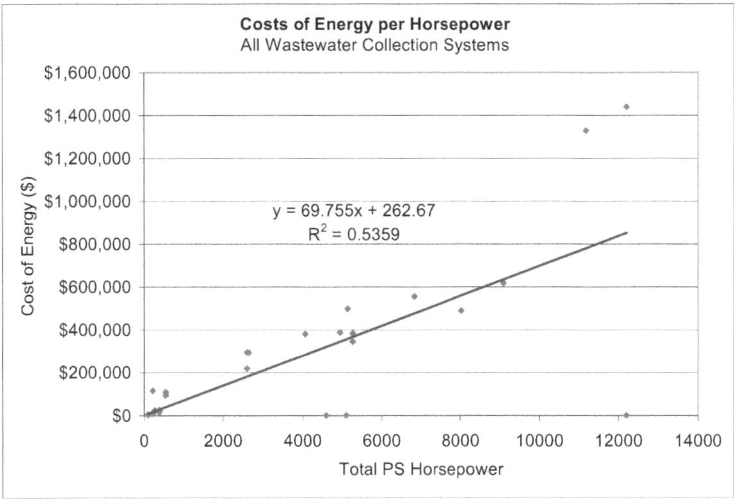

Figure 5: CONTINUOUS IMPROVEMENT EXAMPLE OF A TREND FOR ENERGY COSTS NORMALIZED BY TOTAL HORSEPOWER OF ALL PUMPS IN WASTEWATER COLLECTION SYSTEMS

Minimum, maximum and average graphs: How are we doing

One of the other types of graphs that are produced is the personalized minimum, maximum and average trending graphs. The group average, minimum and maximum are shown over three years as well the utility's own data for three years. In this way, an easy visual comparison can be seen for overall improvement, comparability and performance gaps within a comparable group.

For example, Utility A is clearly above average in terms of Pump Station O&M Cost per Total Pump Station Horsepower. To realize the full benefits in identifying this performance gap, the personable approach to a utility's data collection and operations lends insight to the unique local factors that are affecting the results from this utility. Perhaps energy costs are higher in this area; or perhaps they have oversized pumps. The key to improved performance is to understand what the performance driver is and then be able to respond to it. The graphs only help to identify performance gaps; the personal knowledge of the utility allows the information to be applied into a relevant continuous improvement action plan

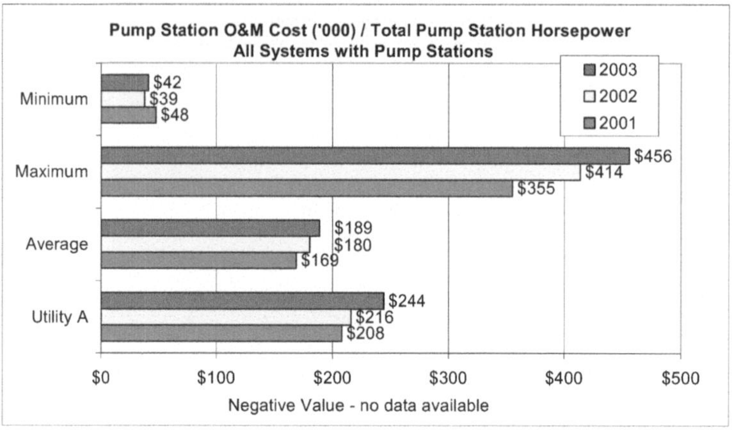

Figure 6: EXAMPLE OF A MINIMUM/MAXIMUM/AVERAGE PERSONALIZED GRAPH FOR WATER DISTRIBUTION SYSTEMS

Radar Charts: How are we doing overall?

When asked how a utility was doing on a whole, we did not initially have a visual representation of the utility's total performance. The radar chart, as observed in the Environmental Sustainability Index project (Yale Center for Environmental Law and Policy, and the Center for International Earth Science Information Network (CIESIN) at Columbia University), was configured as a roll-up of the various Goals and Performance Measures to help utility managers and operators get a quick view of their utility's performance amongst competing goals, where each goal forms its own axis on the radar chart.

Both high and low targets for performance had to be determined to calculate a utility's performance based on their achievement of the targets. To quantify each utility's performance on the goals, representative performance measures were used. The logic for quantifying the performance measures for the radar charts is based upon discussion, assumptions, and conclusions that were agreed upon at the annual Workshop.

A radar chart does not eliminate the need for bar charts but instead summarizes the information provided in the bar charts. A benefit of the radar charts is that it identifies where to start focusing energy on improving the performance of the utility. For example in Figure 8, it is clear that on Goals 3, 4 and 5 the utility is performing on average with the group but on Goals 1 and 2 work needs to be done. The utility would now have to look at the bar charts for the relevant performance measures under Goals 1 and 2 to assess what can be done to improve their performance.

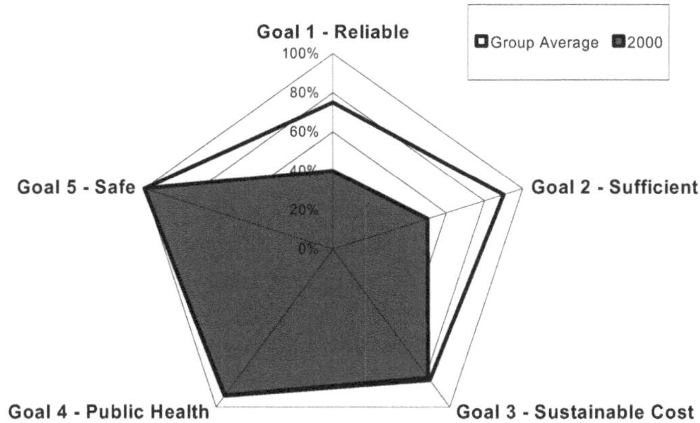

Figure 7: EXAMPLE OF A RADAR CHART

Radar charts can provide annual comparisons for a utility as well as comparisons to "best practices" or targets established by the group. They allow utilities to compare their overall management to similar utilities. Radar charts are very useful as a method to analyse and summarise large amounts of data. They do not replace the analysis of each bar chart and the details behind each performance measure but will enhance the utility's ability to take a quick "snap-shot" of their performance which can be used to communicate performance to all the stakeholders in the utility.

Comparing graphs is one thing, but benchmarking cannot end here. In order to improve any function, you must use the information to assist your overall strategic management program. While this is ultimately the responsibility of each participant, the NWWBI is directly involved in the process with the creation and facilitation of an enormous peer to peer network, to consider, plan, and execute a wide variety of continuous improvement initiatives

Peer to peer network and annual workshop

From the outset of the NWWBI, each benchmarking iteration was thoroughly debriefed in a multi-day workshop forum dedicated entirely to understanding the results of benchmarking. Early workshops were informal events, usually located in meeting rooms of a volunteer host utility. But as the partnership grew, and the cross-departmental interest in the results in each participant also increased, these modest facilities proved inadequate for a productive workshop.

The summary workshop is now a major annual event that has total attendance of up to 150 participating utility staff. The key to success remains the event's workshop orientation, and does not duplicate the professional association "conference" format. It is also the major networking opportunity for all participants. After benchmarking for over five years, the partnership has grown and made significant strides towards continuous improvement through various strategic partnerships, task forces, identifying utility gaps and, not least of all, the development of a network of industry operators and managers collaborating and sharing their information. The network is unique in that a similar private sector partnership could never be expected to share this type of information with their competitors. An atmosphere of familiarity, comradeship, and trust has developed through adherence to the group confidentiality protocol by all partners that encourages individual and organizational learning, and ultimately increases public sector efficiency and accountability. There is no doubt that one of the greatest benefits and results of the Canadian National Water and Wastewater Benchmarking Initiative is the network of peer contacts that has emerged.

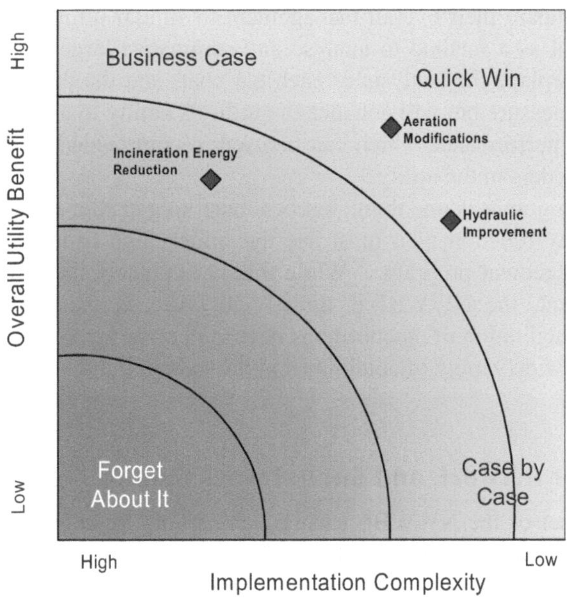

Figure 8: AN EXAMPLE EFFORT AND BENEFITS DIAGRAM

While many of the participants have developed confidence in their data and more accurate methods of data collection, the NWWBI is now recommending the Effort and Benefits diagram as a tool to document Continuous Improvement priorities. The Effort and Benefits diagram illustrates a scale of ease-of-implementation and benefit-of-action that helps to prioritize a Continuous Improvement action. To gain the most benefit for the ease of implementation it is important to conduct an effort and benefits analysis for the top five Continuous Improvement initiatives identified through a high level review. By rating each Continuous Improvement initiative identified for each utility along those two scales, the effectiveness of each action can be seen relative to the others. Those performance improvement initiatives in the upper right hand corner are the initiatives that deserve more focus in the next year as they have been identified as being not very complex to implement and achieving a high level of benefit.

Utility task forces

World-wide, benchmarking has been recognized as an essential tool for effective and efficient public services that can assist with continuously improving these services. In particular, benchmarking can assist with identifying examples of good practice from organizations in both the public and private sectors, within the country and abroad. Benchmarking is a tool for the monitoring of progress brought about by improvement initiatives, ultimately contributing to individual and organizational learning. As such, one of the greatest benefits of the National Benchmarking initiative has been the sharing of insights and knowledge between utilities across the country in overcoming common problems. Accordingly, the annual Water and Wastewater Benchmarking Workshop is the ideal forum for using the benchmarking information to uncover areas of poor performance, and to collectively devise strategies to address these shortcomings.

In the months preceding the May 2002 benchmarking workshop, a number of wastewater "Utility Task Forces" were convened to discuss some of the more pertinent issues that utility managers are dealing with. Utilities could select to participate in one or more of the task forces based on their individual needs. Regular conference calls were convened during which utility managers could contribute to identifying examples of good practices from their organizations, as well as difficulties or pitfalls experienced in implementing certain strategies. An important part of the feedback into the benchmarking process was the identification of new performance measures to better measure a utility's success in attaining the goals, or to refine existing performance measures to improve the accuracy of comparisons.

These task force discussions were used as a basis for further discussions during the benchmarking workshop, where a larger group of utility managers made their contributions. The experience gained from the wastewater benchmarking task forces is now leading to the establishment of task forces for water utilities, where issues such as restoring confidence in public water supply, water security, staff retention and training and water infrastructure reinvestment will be discussed.

To summarize, participation in the utility task forces are as follows:

Step One: Volunteer to Sit on a Task Force

If a benchmarking participant has identified specific issues or challenges as a priority in their utility, the utility can select to participate in one or more of the task forces. In addition to management, staff members beyond those participating in the benchmarking project are encouraged to take part in the process. This way utility managers can broaden their utility's involvement without committing too much of their own time.

Step Two: Task Force Kick-Off with a Conference Call

Once a task force issue has been identified, Earth Tech organizes a kick-off conference call between the interested individuals to consider task force objectives, scope and other planning details. The task force may then direct the selected initiatives.

Step Three: Assist the Task Force in a Manner that You Select

It is not intended that the task forces become an onerous time commitment for its members, but rather that members act as a steering committee to guide the initiative. In most cases, the task forces will need to think through the issues, determine what value utilities need to get out of the initiative, and assist in setting scope and timeframes. Further involvement is on a voluntary basis.

Step Four: Attend and Present Findings in a Dedicated Break-Out Session at the Spring Benchmarking Workshop

Findings from all the task forces are presented at the benchmarking workshop in a round-table setting. A task force chairman is chosen to lead the discussion, with Earth Tech staff members acting as facilitators. With the larger group, each task force can then set a strategy for the next benchmarking iteration. In some cases, the task force can decide to define the requirements for more specific data collection. Other instances may require the undertaking of specialized research, or to conduct a review of practices.

Figure 2 presents a summary of some of the performance improvement initiatives that the National Water and Wastewater Benchmarking Initiative has been examining over the past two years.

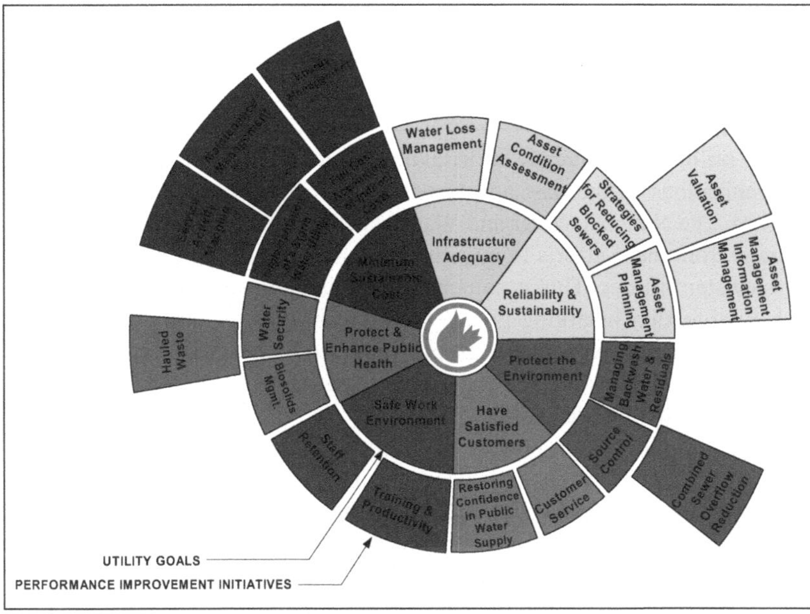

Figure 9: BENCHMARKING GOALS AND CONTINUOUS PERFORMANCE IMPROVEMENT INITIATIVES

Water and Wastewater service activity mapping

To date, the National Water and Wastewater Benchmarking Initiative has focused on establishing an approach to evaluate the effectiveness of utilities in meeting their high level organizational goals. This has formed the basis of the "metric" benchmarking process that is now well established and is generating performance related data on key performance indicators.

But how do water and wastewater utilities meet their "goals"? In short: they perform processes. How well a utility organizes and conducts their various processes will determine how successful they will be in attaining their goals. The success of a utility therefore depends entirely on:

- How effectively a utility performs processes.
- How their processes tie to one another in a logical manner that facilitates the attainment of goals.

Business processes can be likened to a relay race, with the baton passing from person to person within the organization. As in the relay race, the problems occur at the changeover points when the baton is dropped. To illustrate this analogy, a wastewater utility that is highly effective in conducting CCTV sewer inspections, but does not convert the data to information to feed into its sewer main rehabilitation and replacement program has in fact "dropped the baton". A complete and thorough process mapping exercise will address such discontinuities in activities.

Many utilities have already undertaken various levels of process analysis in their organizations in the past and have adopted a range of process mapping approaches. The National Water and Wastewater Benchmarking Initiative hopes to leverage from these efforts to map the various processes performed by water and wastewater utilities. This is somewhat of a daunting task when, in the case of such a large Benchmarking group, there are many organizations that want their organizational activities to be exactly reflected in the chosen framework(s). With these constraints in mind, the following general approach has been adopted, and is proving successful:

1. Start at the highest level of the utility and begin drilling down to the more detailed levels as needed. At the highest level there must be a commonly accepted view. Refer to Figure 3 for a high-level process view of the Maintenance function of wastewater treatment and collection.
2. Identify the inventory of all utility activities. Break the activities down into the most fundamental tasks, and "map" the tasks in a chronological order. Refer to Figure 4 for an example of such a service activity map.

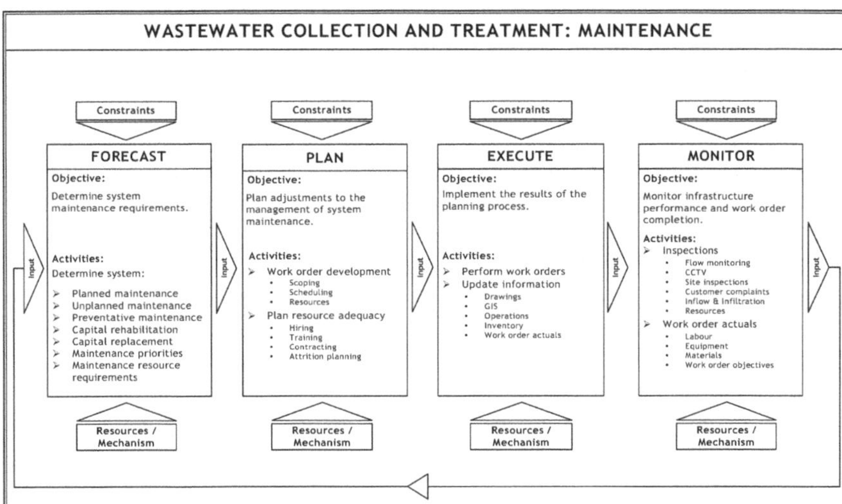

Figure 10: HIGH LEVEL VIEW OF THE MAINTENANCE PROCESS

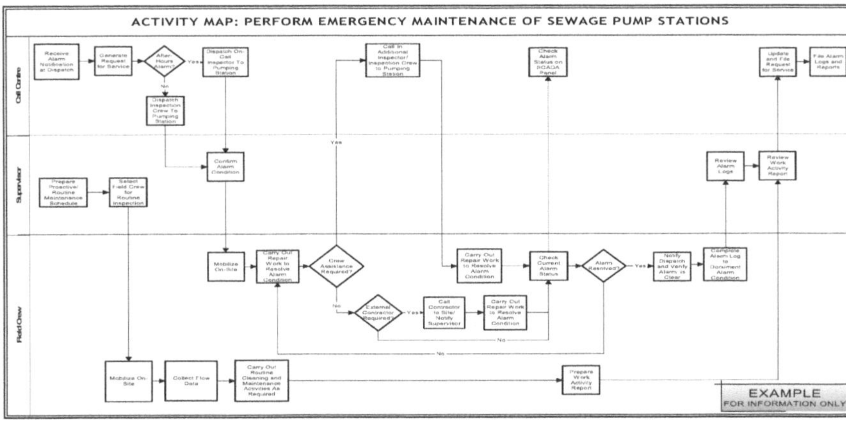

Figure 11: EXAMPLE OF A SERVICE ACTIVITY MAP

The method of constructing service activity maps ensures that utility managers understand all the activities and tasks that make up a process - their sequencing, the resources required to complete the process successfully, and the standards or performance targets to be met. By producing service activity maps, utility managers are encouraged to ask the following questions:

1. Why do we do this?
2. Why do we have to do it this way?
3. Why do we have to do it in this order?
4. Why don't we do it differently?

Service activity maps challenge the notion that "everyone knows how to do that activity properly and efficiently", and encourages the search for continuous performance improvement

INDIVIDUALIZED MANAGEMENT LEVEL BENCHMARKING SUMMARY REPORTS

Each year after the Final Benchmarking report is distributed, Earth Tech offers an individualized management level benchmarking summary report to those utilities that to do not have the in-house resources or time to perform the analysis on their own, or require an independent assessment of their progress. The objective of this report is to summarize a utility's individual results in a concise and accessible format that pieces together the utility's specific information from the wealth of information supplied in the Final report.

A large number of utilities have in the past made use of the Management Level Summary Report to ensure that the benchmarking information that

pertains specifically to their utility is available to department staff, management team and Council in a succinct manner.
Because the report is specific to a utility, the report is be deemed "Commercial Confidential" (and not subject to FOI requests). The format of the report is such that a utility may share the information with any external group that it wishes to without jeopardizing the confidentiality of the benchmarking partnership. As such, the information compiled in the report is suitable for incorporation into a utility's annual report.

Due to the complexities involved in comparing different systems or treatment plants with one another on an "apples-to-apples" basis, each analysis contains a section on the utility-specific factors that define the operating environment and the extent of the operations. With this information as a backdrop, the utility performance is compared according to a wide range of key performance indicators that make up the utility management model. Beyond a mere comparison of averages, the analysis focuses on how a utility performed relative to the distribution of the group results, enabling sound judgment on those areas where the utility excelled compared with its peers, as well as areas where there is room for improvement.

<table>
<tr><th>CITY OF METROPOLIS 2000 PERFORMANCE OBSERVATIONS</th><th>MIN MAX AVERAGE GRAPHS</th><th>DISTRIBUTION SHOWING CITY OF METROPOLIS PERFORMANCE</th></tr>
<tr><td colspan="3">GOAL 3.1 Field FTE's / 100Km length</td></tr>
<tr><td>➢ The City of Metropolis had the group maximum number for FTE's for the wastewater collection system for 1999 & 2000, and had the second highest value for 1998.
➢ The FTE performance measures are not an indicator of total labour since they do not include contracted staff. During the past iteration of data collection, we attempted to obtain data on contract staff but found this difficult, especially in circumstances where only a portion of the utility's overall operation is contracted out</td><td>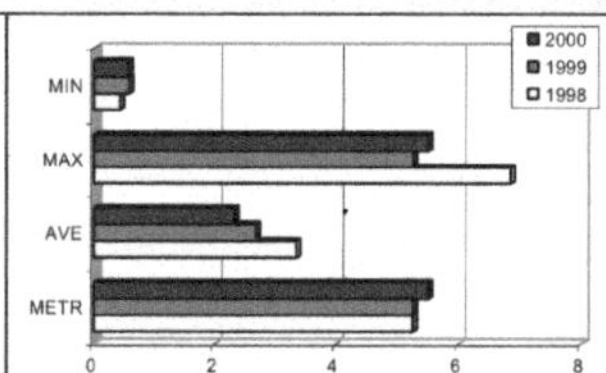</td><td>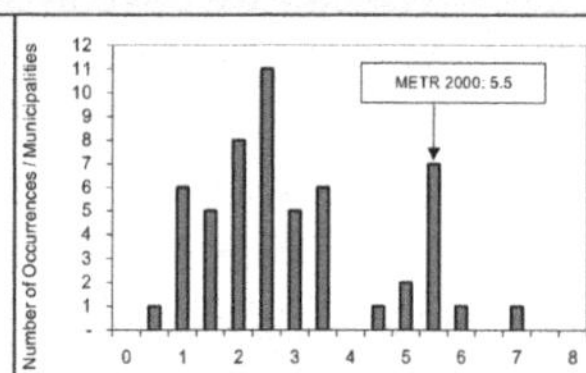</td></tr>
<tr><td colspan="3">GOAL 3.5 Total operating cost ('000)/ km length</td></tr>
<tr><td>➢ The City of Metropolis had the second highest total operating costs / km length than the group for 2000 and the third highest for 1998 & 1999
➢ This value has remained relatively constant over the last three years.
➢ This measure is the sum of both the direct and indirect operating costs.</td><td>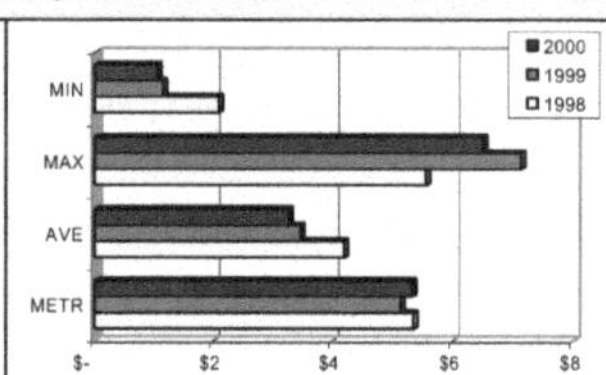</td><td>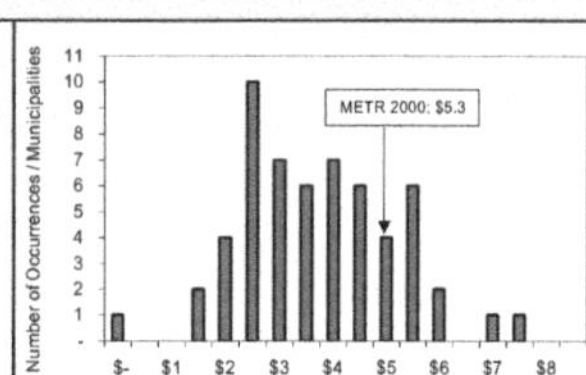</td></tr>
</table>

Figure 12: EXAMPLE OF THE PRESENTATION OF INDIVIDURALIZED BENCHMARKING RESULTS IN A MANAGEMENT

The report makes use of charts, diagrams and graphs to communicate the findings, and allows a utility's senior management team to quickly see the results of the benchmarking efforts, and allow them to focus on improvement strategies and action plans to address problem areas. Figure 5 shows an example of the presentation of Benchmarking results in the Management Summary report.

The Management Level Benchmarking Summary Report is yet another tool that Benchmarking offers to continuously improve Canadian water and wastewater utilities.

CONCLUSIONS

For organizations that are interested in utilizing benchmarking to assist continuous improvement, the above methodology will provide answers to the first three of the four questions originally posed in this paper's introduction:

- "How well are we doing?";
- "How do we compare with similar organizations?";
- "Are we getting value for money?"

But if you look at benchmarking in isolation, you will not have addressed improvement. To address the final, and ultimately most important question (How can we get better at what we do?), the organization must be willing to accept change. Accepting and embracing change is by far the hardest thing that any organization faces, and there are no easy answers, since human beings are creatures of habit. But if you approach benchmarking not as a numerical exercise that relies on data, automation, and computerization, but rather as a process to expand communication, teamwork, and collaboration, the door to performance improvement will suddenly open, and you will be surprised at the result.

As the National Water and Wastewater Initiative matures, we are now starting to benchmark the final result: success in implementing change. To state it another way, the final "question" becomes "Did you do what you said you were going to do?" By answering this question with "Yes; and here is the proof" we now see organizations focusing their efforts on what is truly important. Only now, the leading Canadian municipal utilities that participate in the National Water and Wastewater Benchmarking Initiative know that hard work, communication and teamwork will once again be the keys to success

REFERENCES

Alegre, H., Baptista J.M., Hinir, W., and Parena, R. (2000). *Performance Indicators for Water Supply Services.* Published by IWA Publishing in its Manuals of Best Practice series. London, UK.

Performance assessment of urban infrastructure services

Arbour, R., and Kerri, K. (1998).*Collection Systems: Methods for Evaluating and Improving Performance*. Office of Water Programs, California State University, Sacramento, USA.

Bain Consulting's (2003). *Management Tools & Trends* Published in the Globe and Mail, June 2, 2003, Toronto, Ontario.

Camp, R. C. (1989). *Benchmarking: The Search for Industry Best Practices that Lead to Superior Performance*, Manager of Benchmarking Competence Quality and Customer Satisfaction, Xerox Corporation.

Eisenhardt, P. and Waltrip, G. D. (1999). *Improving Wastewater Treatment Plant Operations Efficiency and Effectiveness.* Water Environment Research Foundation, USA.

Environmental Sustainability Index: An Initiative of the Global Leaders for Tomorrow Environment Task Force, World Economic Forum (2002). Yale Center for Environmental Law and Policy and the Center for International Earth Science Information Network (CIESIN) at Columbia University. http://www.ciesin.columbia.edu/indicators/esi/

Stahre, P. and Adamson, J. (2001). *Performance Benchmarking: A Powerful Management Tool for Water and Wastewater Utilities.* Sweden.

The Hungarian Inception Project on Benchmarking

D. Laky[1], *R. Neunteufel*[2], *R. Perfler*[2] *and P. Darabos*[1]

[1] *Budapest University of Technology and Economics, Department of Sanitary and Environmental Engineering, Budapest, Műegyetem rkp. 3, 1111 Hungary, laky@vkkt.bme.hu, Tel: +36 1 463 15 30, Fax: +36 1 463 37 53*
[2] *Institute of Sanitary Engineering and Water Pollution Control, University of Natural Resources and Applied Life Sciences Vienna, Muthgasse 18, A-1190 Vienna, Austria; roman.neunteufel@boku.ac.at, Tel: +43 1 36006 5800, Fax: +43 1 368 99 49*

Keywords: Austria, Hungary, inception project, key performance indicators, water supply benchmarking

 Performance Assessment of Urban Infrastructure Services
Edited by Enrique Cabrera Jr and Miguel Ángel Pardo. ISBN: 9781843391913.
Published by IWA Publishing, London, UK.

ABSTRACT

Today many countries already have highly-developed benchmarking systems in the water infrastructure sector. Based on the experience of some existing benchmarking projects (e.g. the Austrian OVGW project, and the Bavarian EffWB project), key performance indicators have been chosen and will be applied at some selected Hungarian water utilities. The comparison of the selected PIs with the Austrian and German values shall help for the Hungarian utilities to see other practices. In the future, the results of this inception project together with similar projects from new EU countries might be the basis of the "Central and Eastern European Benchmarking Initiative for Water Supply Utilities" (CEEBI) project. The strategic and long term goal of CEEBI is to introduce high quality benchmarking similar to OVGW benchmarking as a continuous management tool among the new EU countries.

1 INTRODUCTION

Many countries have started benchmarking studies with the aim of improving performance in their water sectors. Metric benchmarking studies use standardized performance indicators like e.g. unit costs, water losses, renewal rates, supply safety, customer services. Benchmarking is used as a continuous process of comparing performance between different utilities but also over the timescale to track enhancements of each firm. The core idea of benchmarking is to provide and share information on best practices to achieve improvements in the company's performances.

Aim of the Hungarian Inception Project on Benchmarking is to see whether utilities are willing to join in a benchmarking study and whether they are able to deliver the data requested. In future, different inception projects of some new EU-countries shall be merged with the existing benchmarking project of Austria and some German benchmarking projects. This merger is planned under the lead of the University of Natural Resources and Applied Life Sciences Vienna, and shall be called "Central and Eastern European Benchmarking Initiative for Water Supply Utilities" (CEEBI, 2007) Project.

Aim of the CEEBI project is to determine some key PIs out of different benchmarking studies like the Austrian OVGW benchmarking (Theuretzbacher-Fritz et al., 2006) and some German Projects e.g. the Bavarian EffWB benchmarking (Kiesl and Schielein, 2005) according to the IWA benchmarking PI set (Alegre, 2006). Some Hungarian utilities have already participated in the IBNET benchmarking project (Kiss, 2006), which provides performance assessment through the comparison of many performance indicators. By using experiences of existing projects together with standardised definitions of variables, context information and performance indicators, a high standard benchmarking shall be established for introduction to new EU-countries.

2 METHODS

The inception project should provide a simple benchmarking platform that helps utilities gathering structured and consolidated data. A crucial step during data processing is to determine the most relevant influencing factors on performances for the appropriate definition of peer groups or at least as explanatory factors respectively. Thus an optimal validity of comparisons should be achieved.

The inception project as well as the CEEBI benchmarking is a voluntary and anonymous comparison of company performances. Strong focus is laid on data security and data confidentiality. All company-data are governed and interpreted confidentially by a neutral and objective project team.

The arrangements and preparations of the Hungarian inception project started within the scope of a bilateral cooperation (Asset Management for Water Supply and Waste Water Infrastructure – Approach, Status quo and Potentials) between the Budapest University of Technology and Economics and the University of Natural Resources and Applied Life Sciences Vienna in September 2007. The determination of PIs to use and setup of a simplified survey form was finished in September 2007.

First an English data-gathering form was built up, which consists of the following data sheets:

- a detailed water balance (as this is the basis for many PI's)
- assets data (length, age, failure data, operating pressure, service interruptions, etc.),
- operation data (regarding quality, number of personnel and electricity costs) and
- simplified economic data.

The gathering form also provides the detailed definition of each variable and context information. The unequivocal definitions are of key importance in order to achieve comparability among utilities, and also among the different datasets from the different countries.

As a next step, the English form was translated to Hungarian language and sent out to 25 water utilities in Hungary.

3 RESULTS AND DISCUSSION

3.1 Expected benefits

Based on the feedback of participating utilities of the Austrian benchmarking project, the following three benefits can be expected:

- One of the first outcomes of data collection is the future availability of structured and consolidated data.
- The positions of the utilities within the sector will be determined and the performance of each firm can be documented to the stakeholders as e.g. customers, shareholders, municipality and authorities. This creates transparency and reduces mistrust.
- Gaps or deficits in a firm's performance can be identified and measures for improvement can be derived.

Additional expected benefits from the Hungarian Inception Project is that the comparison with existing data of Austrian and German projects shall help for the Hungarian water utilities to see other practices as well.

3.2 Basis of the project

The definitions of the CEEBI project are mainly based on the IWA Performance Indicator System (Alegre et al, 2006). These definitions together with the experiences from the existing benchmarking projects, namely the Austrian OVGW Benchmarking Project with 70 Participants (Theuretzbacher-Fritz et al., 2006) and the Bavarian EffWB benchmarking project (Kiesl and Schielein, 2005) give the basis of the Hungarian Inception Project. In particular the experiences showed that some of the IWA definitions are not appropriate to small and medium scale water works and therefore had to be changed slightly to enable all participants to fill in the gathering form completely.

3.3 Key performance indicators

The inception phase of the CEEBI project shall only use a few key performance indicators applied among some selected water supply utilities. Nevertheless some more sophisticated PIs shall be used and some recently developed Explanatory Factors like "Average Network Age Index (NAX)" (Neunteufel et al., 2007b) shall be tested as well in the Hungarian inception project. Therefore the inception project is planned to be an enhancement for utilities, which have already participated in the IBNET benchmarking.

Table 1. THE LIST OF THE KEY PERFORMANCE INDICATORS USED IN THE HUNGARIAN INCEPTION PROJECT

Key-Performance Indicators	IWA	OVGW	IBNET
Economic efficiency			
• Unit total costs (costs/m^3), divided in:	✓	✓	✗
o Unit running cost	✓	✓	✗
o Unit capital cost	✓	✓	
• Total personnel divided in:			
o Employees per 1,000 connections	✓	✓	✓
o Employees per mains length		✓	
o Employees per 1,000 people supplied	✓		✓
o Employees per system input		✓	
• Non-revenue water by volume	✓	✓	✓
• Electricity costs			
o Total energy consumption (kWh/m^3)	✓	✓	
o Standardised total energy consumption (kWh/m^3/1000 m)	✓	✓	
Quality			
• Water losses (%)		✓	✗
• Water losses per mains length (m^3/km/day)	✓	✓	✗
• Water losses per connection (m^3/connection/year)	✓	✓	✗
• Infrastructure Leakage Index	✓	✓	
• Mains failures (no./100 km/year)	✓	✓	✓
• Mains failures without transmission pipes (no./100 km/year) *(to enhance comparability)*		✓	
• Network Average Age Index (NAX)		✓	
• Quality of supplied water (% of total number of treated water tests complying with the standards)	✓	✓	
Sustainability			
• Mains rehabilitation (%/year)	✓	✓	
• Mains rehabilitation (%/year) without transmission pipes (to enhance comparability)		✓	
• Water tariffs for a residential consumption of 150 m^3	✓	✓	✗
• Total cost coverage ratio (%)	✓	✓	✗
Reliability			
• Total reservoir capacity (within transmission and distribution system) (days)	✓	✓	
• Peak Supply Safety (%)		✓	
• Service interruption (%)	✓	✓	✓

The selected key performance indicators are listed in Table 1. according to four of the five IWA categories:

- Economic efficiency
- Quality
- Sustainability
- Reliability

The fifth area – customer service – was omitted in the inception phase yet as the IWA definitions and the experiences of the existing project diverge too much on that topic and a common appropriate standard has yet to be defined for future applications. The IWA PI values regarding the customer service are strongly influenced by the way recording of the complaints. If there are no uniform complaints registration systems, the costumer indicators will be not comparable.

The ticks in Table 1 indicate that the PI is used in other benchmarking projects as well, and "x" means that similar, however not exactly the same PI is used in the other project.

3.4 Context information

Grouping utilities according to various influencing factors is essential to achieve high quality benchmarking. On one hand, the formation of peer-groups enhances comparability across the different utilities. On the other hand, differences of median values between different groups quantify the influence of structural parameters (Neunteufel et al, 2007a). Context information and aggregated index values are used as explanatory factors (e.g. urbanity, NAX, network delivery rate).

The following context information was asked from the Hungarian participants:

- Type of asset ownership (public, private, mixed)
- Type of operation (public, private, mixed)
- Type of system (bulk water supply, direct distribution, bulk + direct distribution)
- Type of water sources used (groundwater, springs, surface water, deep groundwater, karstic spring, bank filtrate)
- Type of treatment (disinfection only, conventional methods, advanced methods)
- Number of water meters
- Percentage of metered water in each category (direct distribution, bulk water, unbilled authorised consumption)

- Number of reservoirs
- Number of pressure zones
- Electricity costs
- Estimated service life according to material

3.5 Further work

At the moment, the data gathering and validation is outstanding but planned for November and December 2007. One aim of the project is to see whether the utilities are able to deliver the data requested.

The analysis of influencing factors, reflections on the assembling of peer groups and comparisons of PI results will be carried out within December 2007 and January 2008. The outcomes of the Hungarian inception project on benchmarking will available by February 2008 and will be presented at the PI08 conference in Valencia.

4 CONCLUSIONS

CEEBI shall induce an information sharing network among utilities that leads to the derivation of measures to enhance performance. The comparison with existing data of Austrian and German projects shall help to see other practices.

The strategic and long term goal of CEEBI is to introduce high quality benchmarking similar to the Austrian OVGW benchmarking as a continuous management tool among the new EU countries.

The participants will benefit by receiving a feedback report on their own performance in comparison with the anonymous results of other comparable utilities.

ACKNOWLEDGEMENTS

The project is carried out within the frame of the bilateral Science & Technology co-operation between Austria and Hungary financed by the ÖAD (Austrian Exchange Service), and TéT (Hungarian Science and Technology Foundation).

5 REFERENCES

Alegre H., J.M. Baptista, E. Cabrera Jr, F. Cubillo, P. Duarte, W. Hirner, W. Merkel, R. Parena (2006): Performance Indicators for Water Supply Services - Second Edition; ISBN: 1843390515

CEEBI (2007) Central and Eastern European Benchmarking Initiative for Water Supply Utilities, available at: http://ceebi.boku.ac.at/ (November 2007)

Kiesl, H. & J. Schielein (2005): Effizienz- und Qualitätsuntersuchung der kommunalen Wasserversorgung in Bayern (EffWB) 2004. – Final Public Report on Bavarian project 2004/05) [german] (http://www.effwb.de/)

Kiss A., Ungvári G. (2006) IBNET teljesítményértékelés hazai víziközmű-szolgáltatók körében (IBNET performance assessement for Hungarian water utilities, *in Hungarian*). Vízmű Panoráma **14**(4), pp 7-9.

Neunteufel R., Kiesl H., Perfler R., Haberl R., Schielein J., Theuretzbacher-Fritz H., Kölbl J. and Kainz H. (2007a): Water supply benchmarking – a cross-border comparison of outcomes and benefits in Austria and Southern Germany, *4th IWA Specialist Conference on Efficient Use and Management of Urban Water Supply*, 21.-23. May 2007, Jeju Island, Korea

Neunteufel R., Perfler R., Theuretzbacher-Fritz H., Kölbl J. (2007b): Explanatory Factor "Average Network Age Index" (NAX) for Mains Failures and Water Losses, *IWA leading edge conference on strategic asset management 2007*, 17.-19. October 2007, Lisbon, Portugal

Theuretzbacher-Fritz H., Neunteufel R., Kölbl J., Perfler R., Unterwainig M. & R. Krendelsberger (2006): Benchmarking und Best Practices in der österreichischen Wasserversorgung – Stufe B - Final Public Report on Austrian project 2005/06 [german] (www.trinkwasserbenchmarking.at)

Benchmarking the Performance of Portuguese Water Utilities

T. Correia[1], *A. Brochado*[2] *and R.C. Marques*[3]

[1] *Center for Urban and Regional Systems, DECivil-IST, Technical University of Lisbon.* taniamcorreia@gmail.com, Av. Rovisco Pais, 1049-001 Lisbon, Portugal, *Tlf +351 938266968, Fax: +351 218409884*
[2] *Center for Urban and Regional Systems, DECivil-IST, Technical University of Lisbon.* abrochado@civil.ist.utl.pt, Av. Rovisco Pais, 1049-001 Lisbon, Portugal, *Tlf +351 218418318, Fax: +351 218409884*
[3] *Center for Urban and Regional Systems, DECivil-IST, Technical University of Lisbon.* rcmar@civil.ist.utl.pt, Av. Rovisco Pais, 1049-001 Lisbon, Portugal, *Tlf +351 218418319, Fax: +351 218409884*

Keywords: Benchmarking, efficiency, stochastic frontier analysis, water and sewerage services

 Performance Assessment of Urban Infrastructure Services
Edited by Enrique Cabrera Jr and Miguel Ángel Pardo. ISBN: 9781843391913.
Published by IWA Publishing, London, UK.

ABSTRACT

This study applies the stochastic frontier analysis (SFA) as a methodology of benchmarking to the water sector in Portugal. It estimates a cost function in order to determine efficiency levels by using a sample of 66 water utilities. The sample represents around 60% of the Portuguese population, approximately 6.3 million inhabitants. The benchmarking results corresponding to the year 2005 reveal high inefficiency levels, pointing to potential earnings of scale inefficiencies and densities economies. There were no meaningful differences in efficiencies between private and public water utilities.

1 INTRODUCTION

Benchmarking can be defined, in a straightforward way, as the process of seeking excellence through a systematic comparison of performance measures with reference standards (Marques and Witte, 2008). Among the several methodologies that exist to apply benchmarking there are different techniques that enable the computation of efficiency (Coelli *et al.*, 2008), such as index numbers, data envelopment analysis (DEA), stochastic frontier analysis (SFA) and regression models (e.g. OLS and COLS).

One of the benchmarking methodologies generally used by the regulators is the SFA. It is a parametric frontier method usually used to determine productive efficiency which was introduced by Meeusen and Van den Broeck and by Aigner, Lovell and Schmidt in 1977. The main feature of this technique is the error term decomposition (ε_i) in two parts, a non-negative random variable which represents inefficiency u_i, and a random error term v_i. It is assumed that v_i is an independent identically distributed (i.i.d.) variable with normal distribution with a mean equal to zero and a standard deviation σ_v^2, which is independent of u_i, a variable similarly i.i.d., which follows a distribution equally half-normal. The productive efficiency of the utility i is obtained by the expression $\exp(-u_i)$.

One of the greatest difficulties of efficiency measurement through SFA is related to the high number of options, namely the model's choice, if cost function or production function and their functional forms (e. g. Cobb-Douglas or translog) and also the option for calculating the error (Marques, 2005).

The water sector in Portugal ascribes the responsibility for the water and wastewater services to the municipalities, which can opt for five different management frameworks, such as municipal services, municipal services with autonomy, municipal companies, public companies and companies operating under concession arrangements. The first four are publicly managed by the municipality whereas the last one is privately managed. Regulation in this sector is almost non-existent, mainly in the case of the services under direct municipal management. However, there is a sector specific regulator, the Institute for the Regulation of Water and Solid Waste (IRAR), whose main roles are regulating the water supply, wastewater and urban solid waste management but it only acts in the scope of the private services (concessionary services). This study employs the parametric frontier benchmarking method of SFA to the water sector in Portugal. After this introduction section 2 focuses on the literature revision. In section 3 some considerations about SFA methodology are made. Section 4 presents the case study and section 5 gives the concluding remarks.

2 REVIEW OF RELEVANT STUDIES

More and more studies are carried out to evaluate the levels of efficiency in the water sector. Most of them deal with the comparison between efficiency levels presented by the available management frameworks, either public or private. It is also common to find studies about the existence of scale and scope economies. Next, we will present some of them.

Lynk (1993), through the application of SFA, compares the efficiency of public and private water utilities in the UK in the period immediately before the privatisation. It adopts a translog functional form and the half-normal distribution for the variable inefficiency of the error term. These studies provide evidence for the existence of scope economies in the water sector. The inefficiency level observed in the private water utilities is considerably higher than that of the public ones.

Bhattacharyya *et al.* (1995) use SFA, through a translog cost function, to describe the inefficiency cost of 221 water utilities, both public and private, operating in the USA in 1992. The variable inefficiency of the error term u_i follows a truncated normal distribution. The estimation was done by two steps and unlike the standard frontier models does not need to assume rigid distributions. The results show that the efficiency level does not only depend on the management framework but also on the utility's size. In the small utilities the private companies are more efficient than the public ones and with regard to the public utilities the most efficient ones are those with larger size.

Antoniolli and Filippini (2001), through a variable cost function presented the cost structure of 32 water utilities that operate at the municipal level in Italy during the period 1991-1995. The variable cost function was obtained according to two methodologies, ordinary least squares (OLS) and random-effects model (GLS), where the Cobb-Douglas functional form was applied to avoid multicollinearity problems. The results show the importance of the explanatory variables such as the price of work, the water losses and the characteristics of the area where the service is provided as well as the existence of economies of scale and economies of density.

Estache and Rossi (2002) analyse the difference between public and private water utilities by applying a Cobb-Douglas cost function for the approximation to the SFA in a sample of 50 public and private operators in 29 regions of Asia and the Pacific in the year 1995. It uses two functions for the measurement of efficiency, error components (EC) and technical efficiency effects (TEE), by means of three estimation methods, OLS, COLS and ML. The results show that there is no significant difference between public and private water utilities.

Filippini, Hrovatin and Zoric (2007) estimate the cost inefficiency and the economies of scale of 52 Slovenian water utilities, between 1997 and 2003, applying four stochastic frontiers methods [Pooled (ML), random-effects (GLS), random-effects (ML) and technical efficiency effects (ML)] and following the translog function as functional form. The outcomes point to a significant cost inefficiency and although the inefficiency scores obtained by the different models are not consistent between them there is a clear presence of economies of scale and of density.

3 MODEL SPECIFICATION AND METHODOLOGY

The study applies the SFA to the Portuguese water and wastewater services with the aim of determining their efficiency. It is more and more frequent to find the application of this methodology in this sector due to its specific characteristics and owing to the fact that it considers the influence of the random factors that cannot be controlled by the utilities themselves.

For the cost function specification, four inputs were considered in the water utilities, namely staff costs, costs of goods sold, capital and current costs, obtaining the output distributed water volume, the number of customers and size of the utility. It is given by

$$CT = c(W, L, S, K, E, CU, N, LO, dP, dO, dC) \tag{1}$$

where CT represents the total cost of water production and W the distributed water volume. L is the unit staff cost, the staff cost divided by the total number of employees. S is the cost of imported water, obtained by the ratio between the cost of goods sold and the volume of distributed water. K is the capital cost, calculated by the division between the CAPEX and network length. E is the energy cost, found by the ratio between current costs and the distributed water volume. CU indicates the number of customers, N the mains length and LO the water losses value. dP, dO and dC are the dummy variables regarding respectively: the type of management, 1 if private, 0 in the opposite case; the scope of the services, 1 when the utility provides other services besides the water supply, 0 if this does not happen; the existence of water abstraction facilities, 1 if they exist and 0 if it they do not exist. The existence of water abstraction facilities is assumed if the utilities provide water exclusively from their own sources.

It is necessary to define a functional form for the calculation of the cost function. The most frequent are the translog (transcendental logaritmic) function and the Cobb-Douglas function (Kumbhakar and Lovell, 2000). We opted for the Cobb-Douglas function in spite of the limitations it presents, such as the constant elasticity of inputs which is equal to one and the fact that it does not enable us to assess the existence of scope economies (although it can be suggested to uniproduct companies). If we apply logarithms, we obtain:

$$\ln CT = \alpha_0 + \alpha_W \ln W + \alpha_L \ln L + \alpha_S \ln S + \alpha_K \ln K + \alpha_E \ln E + \alpha_{CU} \ln CU$$
$$+\alpha_N \ln N + \alpha_{LO} \ln LO + \alpha_{dP} dP + \alpha_{dO} dO + \alpha_{dC} dC + v_i + u_i \quad (2)$$

The variable inefficiency of the error term u_i can admit different types of distribution. We applied two models in this study which differ in the kind of distribution and in the units of the variable *LO* as presented in Table 1.

Table 1. DESCRIPTION OF THE MODEL USED

	Distribution	*LO*
Model I	Gamma	m^3
Model II	Exponential	%

The data allow for the computation of scale and density efficiencies but unfortunately they do not enable us to determine scope efficiencies. In order to do so we should have applied the translog functional form.

The efficiency of density output, E_{OD}, measures the relationship between costs and an increase in the output, keeping constant the number of customers and the size of the area provided. The existence of economies of output density ($E_{OD} > 1$) implies that the average cost of the water utility decreases as a physical output increases.

$$E_{OD} = \left(\frac{\partial \ln CT}{\partial \ln W} \right)^{-1} \quad (3)$$

With regard to the density efficiency, if the average cost decreases as the output and number of customers proportionally increase there will be economies of customer density ($E_{CD} > 1$). Economies of customer density are calculated as:

$$E_{CD} = \left(\frac{\partial \ln CT}{\partial \ln W} + \frac{\partial \ln CT}{\partial \ln CU} \right)^{-1} \tag{4}$$

Finally the scale efficiency E_S, is an important index to determine the optimal size of the utilities. The economies of scale measure the change of costs when the output, the number of customers and the area size increase proportionally. Economies of scale are obtained as follows:

$$E_S = \left(\frac{\partial \ln CT}{\partial \ln W} + \frac{\partial \ln CT}{\partial \ln CU} + \frac{\partial \ln CT}{\partial \ln N} \right)^{-1} \tag{5}$$

4 CASE-STUDY

4.1 Data

This research applies the SFA methodology to 66 water utilities in Portugal which represent about 60% of the Portuguese population, approximately 6.3 million inhabitants. The sample includes all the concessionaire companies, municipal companies, the public company of Lisbon (EPAL) and the municipal services with autonomy. It was impossible to include in the study the remaining utilities operating under direct management by the municipalities (municipal services without autonomy) as it was very difficult to collect the data necessary to the database. Figure 1 shows the number of water utilities from the sample according to their management framework.

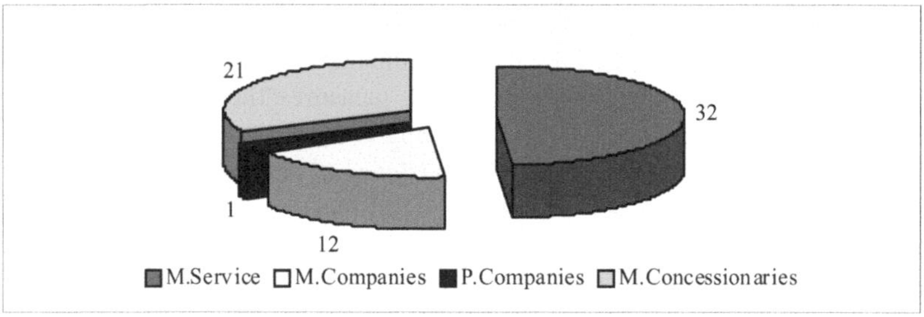

Figure 1. SAMPLE MARKET STRUCTURE

All the data used were collected by means of the utilities annual reports and accounts for the year 2005. However, in some cases it was necessary to complete these data with information from other sources, such as IRAR (2006) and APDA (2006). Table 2 presents the statistical summary of the variables used in the models.

Table 2. SAMPLE DESCRIPTIVE STATISTICS

Variable Description	Variable	Mean	Median	Std. Dev.	Minimum	Maximum
Total Cost (€)	*CT*	7996151,33	4620466,39	9407135,35	460614,00	55280022,77
Water Supply (m^3)	*W*	8.984.410,2	4519716,5	13.896.358,6	530.710,0	98.952.990,0
Customers	*CU*	45188,0	27362,5	55363,6	4364,0	341764,0
Labour Costs (€)	*L*	13900,13	14059,07	4258,04	2247,18	30319,11
"Bulk" Water Costs (€)	*S*	0,18	0,11	0,17	0,00	0,54
Capital Costs (€)	*K*	2915,72	2295,26	2532,61	117,39	13534,63
Energy Costs (€)	*E*	0,33	0,28	0,25	0,06	1,84
Network Lenght (km)	*N*	633,894	498	458,162	87	2789
Water Loss (m^3)	*LO*	2897520,6	1327934,5	5291932,3	186810	36655050
Water Loss (%)		0,310	0,303	0,080	0,148	0,541
dummy Private	*dP*	0,318	0	0,469	0	1
dummy Others	*dO*	0,833	1	0,376	0	1
dummy Sources	*dC*	0,424	0	0,498	0	1

4.2 Results

The results of cost function estimation for both models, obtained with the software *Limdep*, are presented in the Table 3. Figure 2 displays the results by water utility.

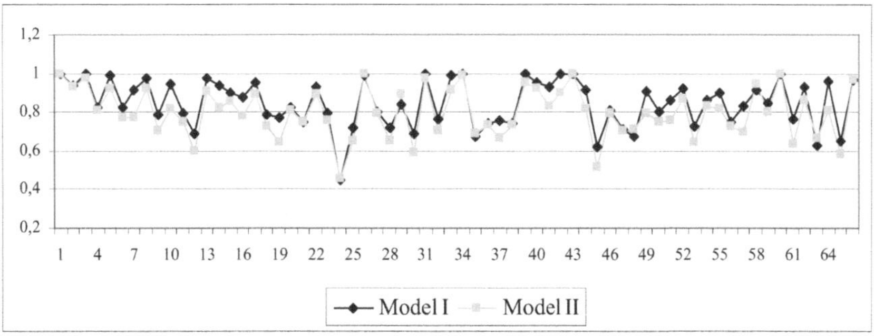

Figure 2. PORTUGUESE WATER UTILITIES EFFICIENCY

Table 3. PARAMETER ESTIMATES

Variable	Model I	Model II
Constant	1.02*** (4.648)	2.343*** (7.386)
ln*W*	0.515*** (23.917)	0.623*** (8.231)
ln*CU*	0.277*** (12.921)	0.344*** (5.6)
ln*L*	0.083*** (3.895)	0,0004 (0.012)
ln*S*	0.062*** (9.986)	0.103*** (9.36)
ln*K*	0.21*** (27.972)	0.203*** (10.756)
ln*E*	0.145*** (9.074)	0.142*** (5.261)
ln*N*	0.199*** (12.107)	0.198*** (5.556)
ln*LO*	-0.212*** (-30.684)	-0.17*** (-3.893)
dP	-0.047*** (-5.022)	-0.139*** (-4.824)
dO	-0.055*** (-4.369)	-0.2*** (-7.976)
dC	-0.177*** (-8.314)	-0.102*** (-3.626)
$\lambda=\sigma_u/\sigma_v$	NA	NA
$\sigma=(\sigma_u^2+\sigma_v^2)^{1/2}$	NA	0,082
θ	3,703	12,251
P	0,172	NA
σ_u	0,112	0,082
σ_v	0,007	0,000
Log L	249,403	240,477

*Significant at 10%; **significant at 5%; ***significant at 1%.

The values obtained for the variables show very good significance levels, of 99%, and the expected signal. The exception is the variable *LO*, which shows an opposite signal to what was supposed. This fact can be explained by an increase in costs regarding maintenance works and network rehabilitation necessary to reduce the water losses volume. Furthermore, the software used has the capability to provide simultaneously the efficiency levels of the different utilities obtained by means of the exp($-u_i$). Table 4 shows the statistical results.

Table 4. ESTIMATED COST EFFICIENCY SCORES

Statistics	Model I	Model II
Mean	0,847	0,796
Median	0,854	0,800
Std. Dev.	0,122	0,126
Minimum	0,450	0,453
Maximum	1,0	1,0

As Figure 2 shows that the levels of efficiency computed by the models are somewhat similar. Table 5 illustrates the correlation matrix of two models. The utilities can be grouped according to efficiency levels intervals. Figure 3 enables us to highlight the higher efficiency levels with the application of Model I.

Table 5. CORRELATION BETWEEN EFFICIENCY SCORES

		Model I	Model II
Model I	Pearson Correlation	1	0.925 **
	Sig. (2-tailed)		0.000
	N	66	66
Model II	Pearson Correlation	0.925 **	1
	Sig. (2-tailed)	0.000	
	N	66	66

**. Correlation is significant at the 0.01 level (2-tailed).

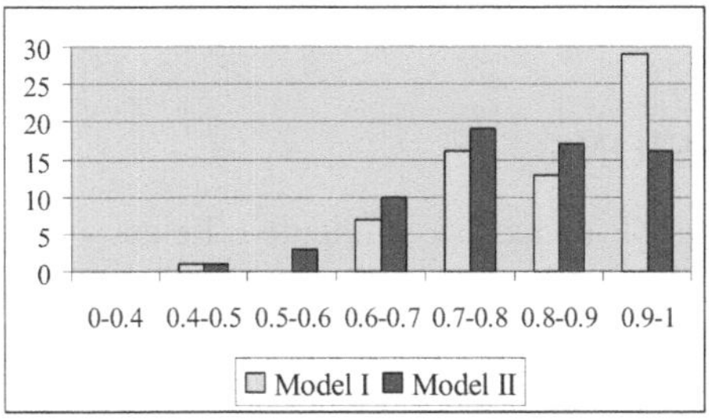

Figure 3. CLASSES OF EFFICIENCY LEVELS

Against the expectations, it is in the private utilities that the efficiency level is lower as we can see in Table 6. Most private utilities operate with efficiency levels above the average whereas the public utilities show the opposite, since most of them operate with efficiency levels below the average, underpinning the idea that the sample could contain an outlier.

Table 6. ESTIMATED COST EFFICIENCY SCORES BY OWNERSHIP

	Model I		Model II	
Statistics	*Public*	*Private*	*Public*	*Private*
Mean	0,851	0,837	0,808	0,770
Median	0,836	0,907	0,806	0,779
Std. Dev.	0,109	0,149	0,112	0,153
Minimum	0,627	0,450	0,601	0,453
Maximum	1,0	0,9999	0,996	1,0

The estimation of economies of output density (E_{OD}), customer density (E_{CD}) and economies of scale (E_S) obtained, respectively, by the formulas 3, 4 and 5 are presented in Table 7. The indicators of E_{OD} and E_{CD} are both higher than the unity, which means that the water utilities operate at an inappropriate low-density level. The indicators of E_S are not inferior to the unit but with a value higher than this we admit to be in the presence of a situation of diseconomies of scale. The existence of diseconomies of scale in the water sector be justified by the fact that the larger companies usually have higher staff costs, do not need specialized manpower or because the features of the services lead to an increase in production that requires a new source (and distant) of raw material.

Table 7. ECONOMIES OF OUTPUT AND CUSTOMER DENSITY AND SCALE

Economies	Model I	Model II
E_{OD}	1,940	1,605
E_{CD}	1,262	1,034
E_S	1,009	0,858

5 CONCLUSIONS

This study uses the SFA model to compute efficiency levels of the Portuguese water utilities. It shows that for the sample studied there are some companies with good efficiency levels, but most of them still need to walk a long way until they reach the best practices.

The values obtained do not allow us to conclude which is the best management model. Despite the fact that the efficiency levels are higher in utilities of public management, most of them operate at levels below the average. However, while the opposite is true for the utilities under private management, on average there are more private utilities operating over the average efficiency levels.

Finally, as far as the economies of scale and density are concerned, we concluded that the water utilities operate at density levels below what is recommended and that diseconomies of scale are associated with the sector.

8 REFERENCES

Antonioli, B. and Filippini, M. (2001). The use of a variable cost function in the regulation of the Italian water industry. *Utilities Policy*. **10**(3-4), 181-187.

APDA (2006). Quem é quem no sector das águas em Portugal? Associação Portuguesa dos Distribuidores de Água, Lisbon.

Bhattacharyya, A., Harris, T., Narayanan, R. and Raffiee, K. (1995a). Specification and estimation of the effect of ownership on the economic efficiency of the water utilities. *Regional Science ad Urban Economics*. **25**(6), 759-784.

Coelli, T., Rao, P. and Battese, G. (1998). An introduction to efficiency and productivity analysis. Kluwer Academic Publishers, Boston, Massachusetts.

Estache, A. and Rossi, M. (2002). How different is the efficiency of public and private water companies in asia? *The World Bank Economic Review*. **16**(1), 139-148.

Filippini, M., Hrovatin, N. and Zorić, J.. (2007). Cost efficiency of Slovenian water distribution utilities: An application of stochastic frontier methods. In *XIX Riunione Scientifica Siep, Economia del Capitale Umano - Istituzioni, incentivi e valutazioni*, Pavia, Aule storiche Università, 13 - 14 September.

IRAR. (2006). Relatório Anual do Sector de Águas e Resíduos em Portugal – 2005. Instituto Regulador de Águas e Resíduos, Lisbon.

Kumbhakar, S. and Lovell, K. (2000). Stochastic frontier analysis. Cambridge University Press, New York.

Lynk, E. (1993). Privatisation, joint production and the comparative efficiencies of private and public ownership: The UK water industry case. *Fiscal Studies*. **14**(2), 98-116.

Marques, R. (2005). Regulação de Serviços Públicos. Sílabo Press, Lda., 1st Edition.

Marques, R. (2005). Towards a benchmarking paradigm in the water and sewerage services. Accepted for publication in 'Public Money and Management'.

The development and application of Performance Indicators for sewage system service in Korea

Y. M. Ahn[1], *D. Kim*[2], *H. S. Yoon, M. S. You, Y. J. Kim, Y. C. Yoon and Y. S. Shim*

[1] *Young Mi Ahn, kandari@emc.or.kr,* Research & Development Team, Environmental Management Corporation, Environmental Research Complex, Kyungseo-Dong, Seo-gu, Incheon, Korea, 404-708, *TEL: +82-32-560-2365 Fax: +82-32-560-2288*

[2] *Duk-Jin Kin, duk-jin@emc.or.kr,* Research & Development Team, Environmental Management Corporation, Environmental Research Complex, Kyungseo-Dong, Seo-gu, Incheon, Korea, 404-708, *TEL.: +82-32-560-2686 Fax: +82-32-560-2288*

Keywords: ISO/TC224, Performance Indicator(PI), Sewage utility

 Performance Assessment of Urban Infrastructure Services
Edited by Enrique Cabrera Jr and Miguel Ángel Pardo. ISBN: 9781843391913.
Published by IWA Publishing, London, UK.

ABSTRACT

With rapid expansion of the water market and growth of water companies, the recent trends in international water industry show that standardization of service activities concerning management of water and wastewater system (ISO/TC 224, to be islsued in the first half year of 2008) is expected to have a large influence on the global water market including liberalization of water service market.

To evaluate public service level and establish plans to improve service level of wastewater utilities, we developed PIs for 120 items and applied them to 10 sewage utilities in the consideration of their scales, circumstances and conditions. In this study, we verified the applicability of the developed performance indicators (PIs). So, we concluded that PIs developed for the purpose of this study can be a useful tool applicable to objective evaluation of the service level of sewage utilities. Using the evaluation tool of PIs developed, it can also be used as a benchmarking tool in this industry as well as for enhancing the efficiency of operation and management of the sewage system. It will also contribute to the water resource management through determination of investment priorities based on the evaluation results and to raising international competitiveness of the domestic sewage utilities to meet the future opening of the water market.

1 INTRODUCTION

With rapid expansion of the water market and growth of water companies, the recent trends in international water industry show that standardization of service activities concerning management of water and wastewater system (ISO/TC 224, to be issued in the first half year of 2008) is expected to have a large influence on the global water market including liberalization of water service market. In the Final Draft International Standard (FDIS) of ISO/TC224, Performance Indicators (PIs) are suggested to estimate the service activities relating to water and wastewater treatment facilities. PIs have been developed in many countries including some international organization (World Bank, IWA and etc.) for service benchmarking. Therefore, it is urgently required for wastewater utilities to make efforts to reinforce international competitiveness and to improve the quality of service. For this purpose, it is essential to develop a tool to evaluate wastewater utilities' service quality objectively. To evaluate public service level and establish plans to improve service level of wastewater utilities, we developed PIs for 120 items and applied them to 10 sewage utilities in the consideration of their scales, circumstances and conditions. In addition, we verified the applicability of those PIs. Finally, the method to increase efficiency of domestic wastewater service is suggested.

2 MATERIALS & METHODS

2.1 Development of PI for domestic wastewater service

PI developed in this study reflected the problems and improvements of ME (Ministry of Environment) and Ministry of Government and Home Affairs in Korea, and benchmarked the standards of IWA (International Water Association, 2006), World Bank, JSWA (Japanese Sewage Water Association, 2003), service standard of water and wastewater in France (KWWA, 2006). For the sustainable development of domestic wastewater service, necessary elements were added to develop PI that corresponds to real condition of Korea. On the selection and evaluation of PI, "expert group meeting", which was consisted of academic, wastewater operator, customer, and related professional, developed PIs. These PIs consisted of 120 items, which were 12 for personnel, 14 for facility, 22 for operation, 25 for service quality, 20 for environments, 27 for finance.

2.2 Application on domestic wastewater service

For the actual application of 120 PIs on wastewater service, evaluation on the 2 wastewater utilities in metropolitan city, 4 wastewater utilities in city level, and 4 wastewater utilities in county level was conducted for personnel, facility, operation, service quality, environments, and finance. Table 1 shows the general status of corresponding wastewater utilities.

Total 193 variables were consisting of PIs through relevant equations. 10 wastewater utilities were asked to input the corresponding value of 2003, 2004, 2005 for these variables. For the reliability confirmation of used data in this evaluation, data should reference the corresponding documents, and reliability of data is divided into 5 levels. Table 2 shows the reliability level for variables. Evaluation program for wastewater service was developed, and PI was automatically calculated by the input of variables. After gathering variables and PIs, PI that was impossible to apply was deleted. If PI was incomplete, or showed low reliability, but it was necessary for evaluation, then PI was partially modified for the accurate evaluation.

Table 1. THE GENERAL STATUS OF CORRESPONDING WASTEWATER UTILITIES

Item	A	B	C	D	E	F	G	H	I	J
Served population (1000person)	1,389	1,004	312	392	225	61	42	18	39	26
Percentage of sewered population (%)	98.7	91.6	82.5	86.3	87.6	55.1	40.3	47.2	46.3	31.3
Supplied ratio of sewer (%)	87.5	78.7	75.0	29.5	70.0	47.4	47.0	41.5	60.3	24.5
Capacity (1000m^3/d)	720	597	331	125	150	26	16	11	22	32
Type of operation	C2	C2 D4	C2	C1 D9	D1	C1	C3	C4	D2	C1
Type of accounting	P	P	P	P	P	P	P	G	G	G

C: Commitment, D: Direct , G: General, P: Public

Table 2. THE RELIABILITY LEVEL FOR VARIABLES

Item	Definition	Value
High	Highly reliable data source: data based on sound records, procedure, investigations or analyses that are properly documented and recognized as the best available assessment methods (accuracy≥75%)	3
Medium	Fairly reliable data source: worse than high, but better than low (accuracy≥50%)	2
Low	Unreliable data source: data based on extrapolation from limited reliable samples or on informed guesses (accuracy≥25%)	1
No data	No data and no informed guesses (accuracy≥0%)	0
Not application	All others not apply	-

2.3 Service level comparison for domestic wastewater utilities

To compare service level between 10 domestic wastewater utilities, questionnaires to professionals were conducted. Questions were made for the importance level in each PIs. 97 professionals were questioned, and 86% of them were replied. From the results of questionnaire, AHP (Analytical Hierarchy Process) and Delphi analysis were conducted to calculate weight for PI. In addition, 46 PI with the high importance level among 120 PIs were selected. Among these, the 4 PIs that are difficult to evaluate objectively are omitted, and therefore 42 major PIs were selected, which were 5 for personnel, 6 for facility, 10 for service quality, 7 for environments, 9 for finance. Rank analysis for each PI was conducted on 10 wastewater utilities, and weight and reliability level is applied for comprehensive evaluation results. Table 3 shows key PI and weight through AHP and Delphi analysis.

3. RESULTS AND DISCUSSION

3.1 PI application results and reliability analysis

By adopting developed PIs on 10 wastewater services in local communities, completion ratio on PIs was more than 94% for all cases. PI completion ratio with more than 2 in the reliability level was over 82% excluding J county. For D city, G county, and J county, completion ratio showed 100%, but reliability level (more than 2) was 84~88%. Therefore, not only PI values but also reliability level of related variables have great importance in service evaluation, and the consideration and reaffirmation on the variables must be followed. In addition, wastewater utilities of county level with less than 50% of service ratio excluding Q county (H county, I county, J county) exhibited many instances of "not applicable" and "no record" PIs. Wastewater utilities under general account (H, I, J counties) showed more than 74% of "not applicable" PIs in finance. Among 27 PIs in finance, 19~20 of PIs were not completed, and wastewater utilities with general account should be changed to public enterprise account in near future.

Table 3. KEY PI AND WEIGHT THROUGH AHP AND DELPHI ANALYSIS

	PI	Weight		PI	Weight
P	Rate of technical staffs	1.25	S	Eco-friendly facility rate	0.92
	Rate of qualified staff	0.97		Sewer rehabilitation rate	1.22
	Rate of service staff	0.83		Complaints about sewer blockage	0.76
	Working years in sewage	1.11		Complaints about sewage services	1.07
	Operation staff in sewage	0.83		Complaints about odour, noise, and vibration	0.76
F	Frequency of sewer blockage	0.96		Complaints about pollution accident	0.76
	Renovation rate of sewer	1.09	E	Reuse of effluent	1.01
	Management of sewer	1.09		Inflow rate for generated wastewater	1.16
	Inflow amount per sewer length	0.82		Residue utilization	0.87
	Database rate of sewage register	1.09		Recycling of digestion gas	0.7
O	Rate of untreated wastewater	1.11		Rate of advanced sewage treatment	1.3
	Management of inflow quality	1.25		Supplement rate of deodorization facility	0.87
	Rate of observation of effluent quality	1.11		Observation of raw (Discharge facility of odour)	1.16
	Rate of operation of wastewater treatment plant	0.97	F	Composition rate of equity capital	1.16
	Rate of correction of flow meter	1.11		Debt rate	0.87
	Rate of space of facility	0.97		Rate of total revenue and expenditure	1.01
S	Percentage of sewered population	1.38		Current account ratio	1.01
	Supplied rate of sewer	1.4		Sales account ratio	0.87
	Flood damage rate	0.9		Total cost	1.30
	Risk management	0.76		Total cost coverage ratio	1.16

From the reliability analysis on the each parameter, B city, C city, D city, G county and J county showed lower reliability, but A and E city displayed good reliability (Fig. 1). Therefore, wastewater utilities with low reliability must concentrate on the data management and maintenance.

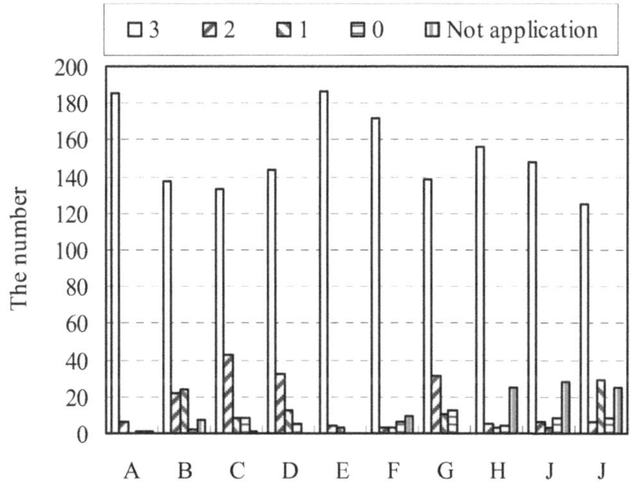

Figure 1. RESULT OF RELIABILITY ANALYSIS ON 193 VARIABLES

On the analysis of completion of PI, occurrence of incompletion and low reliability was higher at facility, service quality, and environments than facility, personnel, operation, and finance. Therefore, to improve the service of domestic wastewater treatment, facility, service quality, environments area should be focused. Especially, sewer system is the important facility in wastewater system, but the fundamental data for sewer was insufficient. Currently, large-scaled sewage improvement projects such as BTL, etc are undergone, and the establishment of basic information on the comprehensive map of sewer must be accompanied. For the long-term, the management of end-of-pipe treatment facility and sewer system should be integrated to enhance its operation efficacy.

Table 4 indicates PIs with low reliability and corresponding wastewater utilities. PIs with low reliability showed less than 1 of reliability level, and PIs with more than 4 occurrence of low reliability in corresponding wastewater utilities were shown. There were several PIs with low reliability in personnel, operation, and finance target, but it was relevant to one or two wastewater utilities. Most of PIs with low reliability was for sewer system. Related data for "emission of GHG" by UNFCCC and "Observation of law (Discharge facility of odour)", which was effectuated on 2006, was not established in many wastewater utilities.

Table 4. PIS WITH LOW RELIABILITY AND CORRESPONDING WASTEWATER UTILITIES

Object	PI	Corresponding wastewater utilities
FACILITY	Frequency of sewer defect	B, D, F, G, I, J
	Frequency of sewer blockage	B, G, I, J
	Deterioration rate of sewer	C, D, F, G, J
SERVICE QUALITY	Improvement ratio of aged pipes and sewers	B, C, F, G, H, I, J
	Service ratio of flush toilet	B, C, F, G, I, J
	Complaint about sewer blockage	B, C, H, I, J
	Time for processing complaint	B, C, F, J
ENVIRONMENT	Protection management for soil contamination from sewer system	B, C, D, F, G, J
	Emission of GHG (CH_4, N_2O)	B, C, G, J
	Observation of law (Discharge facility of odour)	°C, D, E, G, I, J

3.2 Service level comparison between domestic operators

Service level comparison and comprehensive evaluations on wastewater service in 10 wastewater utilities are shown in Table5. By comparing and evaluating general score by each object , wastewater utilities in metropolitan city and city level with high service ratio of sewage showed 1st to 5th place. Wastewater utilities in county level exhibited the lower score. Wastewater service provider in county level lacked the operation and management in facility, and financial investment. Wastewater utilities with general account such as H, I, and J county displayed the 8 to 10th place in rank analysis between the 10 providers on finance area of PIs (19~20 in 27 items). From the results of Table 5, wastewater utilities with general account would receive relatively low score than the one with public enterprise account.

Table 5. SERVICE LEVEL COMPARISON AND COMPREHENSIVE EVALUATIONS ON WASTEWATER SERVICE IN 10 WASTEWATER UTILITIES

Item	Public accounting							General accounting		
	A	B	C	D	E	F	G	H	I	J
Personnel	74.6	61.2	32.2	72.6	61.0	42.4	24.6	62.4	48.9	57.4
Facility	67.7	41.3	66.6	49.4	36.3	69.3	62.8	61.4	62.8	25.7
Operation	72.1	46.7	76.8	74.5	74.0	63.0	62.1	69.4	62.4	58.5
Service Quality	75.0	62.3	56.1	55.9	72.5	61.0	46.4	48.3	76.1	38.3
Environment	67.6	67.0	66.4	68.3	64.2	48.2	41.1	70.1	55.3	37.7
Finance	69.0	67.8	66.0	71.4	60.8	43.9	58.9	50.9	48.2	37.6
Total score (Except for finance)	71.7	57.1	60.0	63.5	63.8	57.1	47.3	60.9	63.0	42.8
Grade (Except for finance)	**1**	**7**	**6**	**3**	**2**	**7**	**9**	**5**	**4**	**10**
Total score	71.1	59.4	61.3	65.1	63.1	54.3	49.8	58.7	59.9	41.7
Score	**1**	**6**	**4**	**2**	**3**	**8**	**9**	**7**	**5**	**10**
Method of calculating score	**{(Personnel score×5) + (Facility score×5) + (Operation score×6) + (Service Quality score×10) +(Environment score×7) + (Finance score×9)}/42**									

Therefore, to compensate this shortcoming, comprehensive evaluation was conducted on personnel, facility, operation, service quality, and environments excluding finance. By considering reliability level of data, A, E, D city occupied 1st, 2nd, 3rd place, respectively. I and H county held 4th and 5th place. For I and H county, service ratio of sewage was lower than B and C city. In addition, facility investment and management was inferior. However, data management and its effort were far superior to the others. From this result, the investment priority should be placed on the low supplement of wastewater facility and relatively low service level to enhance its efficiency during commissioned operation of water and wastewater business.

4. CONCLUSION

We developed 120 performance indicators (PIs) in consideration of domestic circumstance to enhance the competency of water and wastewater utilities, and meet the effectuation of international standard for water and wastewater service (ISO/TC224) on the first half of 2008. Upon the application of developed PIs on 10 wastewater utilities, the following results are acquired.

(1) More than 94% of PI completion ratio was achieved. PI completion ratio with more than 2 of reliability level was more than 82% in 9 wastewater utilities excluding J county. Therefore, the effectiveness of developed PIs is evaluated as high.

(2) From the comprehensive evaluation on the wastewater service level for 10 domestic wastewater utilities, A city positioned 1st place, D city marked 2nd position, G county was 7th place in public enterprise accounting (total 7). In general account (total 3), I county marked the first place. In overall, high service ratio of sewage exhibited the higher position. I county, which lacks the wastewater facility and financial investment comparing to metropolitan city, achieved higher service level.

(3) PI for wastewater service in this study can be utilized as evaluation standard for efficacy in business operations under local administration and private commission. Through developed PIs, local administration can be directly compared for business efficacy and service level with foreign water companies including private ones. This comparison and competition will be very positive stimulation for effective operation and management in water and wastewater business.

(4) As shown in the PI application results, service levels in each areas (personnel, facility, operation, service quality, environments, finance) are compared, and service provider can focus on the weak area. Along with the consultation and effective management, the overall effectiveness in wastewater business operation can be raised.

ACKNOWLEDGEMENT

The authors would like to acknowledge the Ministry of Environments and KIEST (Korea Institute of Environmental Science and Technology)" for their financial support through "development of evaluation criteria on water and wastewater service for intensification of international competency" under technical development project for next-generation core environment technology in 2006.

4. REFERENCES

Ministry of Environment (1993) The study on the development of water and wastewater industry in Korea, 1-2

ISO (2007) Service activities relating to drinking water supply systems and wastewater systems, ISO/TC224

IWA (2006) Performance measurement systems based on IWA performance indicators, 33-68

JSWA (2003) Guide line for improvement of management and operation of sewage, 10-25.

KWWA (2006) Development and application of evaluation standard to reinforce international competitiveness for service activities relating to sewage systems (The 1st report), 33-35

Optimal Incentives in Local Public Utilities: An International Benchmarking Study of the Drinking Water Sector

K. De Witte[1] *and R.C. Marques*[2]

[1] *Center for Economic Studies,* University of Leuven (KUL). kristof.dewitte@econ.kuleuven.be_ HOG 02.161, Naamsestraat 69, 3000 Leuven, Belgium, *Tlf*: *+32 16326850, Fax: +32 16326796*
[2] *Center for Urban and Regional Systems, DECivil-IST, Technical University of Lisbon. rcmar@civil.ist.utl.pt,* Av. Rovisco Pais, 1049-001 Lisbon, Portugal, *Tlf +351 218418319, Fax: +351218409884*

Keywords: Drinking Water Sector, Benchmarking, Incentive Regulation

ABSTRACT

Why does the drinking water sector in most countries present perennial poor performance and low productivity? The absence of regulatory and benchmark incentive schemes in this natural monopoly can at least partially justify that reality. This article uses cross-country comparisons to demonstrate that a governance model of the water utilities based on performance incentives leads to a higher average efficiency. By means of Data Envelopment Analysis (DEA)

Edited by Enrique Cabrera Jr and Miguel Ángel Pardo. ISBN: 9781843391913.
Published by IWA Publishing, London, UK.

and bootstrapping algorithms, the efficiency of the water utilities of the Netherlands, England and Wales, Australia, Portugal and Belgium are compared. We show that after incorporating physical, social and institutional environmental variables the regulatory incentive schemes have a noteworthy positive effect on efficiency.

1 BENCHMARKING AS A TOOL TO COMPARE AND ASSESS PERFORMANCE

Benchmarking can be defined, in a simple way, as the process of seeking best practices through a regular comparison of performance measures. In the drinking water sector, which is endowed with several market failures such as natural monopoly, sunk costs, externalities and asymmetric information, benchmarking obtains a key role. As regional monopolists, the water utilities can provide the service at a high cost and poor service level and sometimes earn unreasonable and unfair profits. The benchmarking is even pointed up as the unique tool to fight the *quiet life* of Hicks (1935) and the *X-inefficiency* of Leibenstein (1966). In this sense, it replaces the competitive incentives by competition by comparison. In Europe, there is even a view that defends a new paradigm for the water sector based on use and abuse of benchmarking (Marques and Witte, 2008).

Several kinds of benchmarking exist and are applied with diverse aims. For example, benchmarking can be classified into bottom up or top down, into metric or process and regulatory or non-regulatory. The methods to benchmark can further be divided into total or partial, parametric or non-parametric, frontier or non-frontier and stochastic and deterministic. They can also be applied by different actors, for instance operators, regulators, customers and financial entities. Benchmarking can be used by the operators as a tool to improve the performance, but it can simultaneously be employed by regulators as a tool to establish tariffs or to enforce the operators to improve the quality of service (labelled as yardstick competition). Even the '*name and shame'* strategy of the performance, known as sunshine regulation, put into practice by some regulators, leads to good outcomes (Marques, 2006). However, independently of the stakeholder, benchmarking has always common principles and goals. All benchmarking methodologies intend to assess performance, measuring the inputs consumed and the results (outputs) produced and, as such to compute the efficiency of activities, processes or organizations.

In this article, we apply benchmarking with a different aim. We assess the performance of drinking water utilities by country, intending to identify the best practices regarding the incentives provided to the water utilities to be efficient and to provide the value for money. By the non-parametric frontier benchmarking technique of Data Envelopment Analysis (DEA), we carry out an international benchmarking research comparing the efficiency of drinking water utilities while taking into account the institutional framework and the operational environment of five countries, respectively, the Netherlands, England and Wales, Australia, Portugal and Belgium (Witte and Marques, 2007a). The model accounts for a-typical observations and measurement errors and investigates if the existence of incentive regulation has a positive influence on the efficiency of the water utilities. In addition, we first commentate on mergers which, although undermining the effectiveness of the benchmark, are often justified on the basis of scale economies and inefficiency reductions.

The paper is organized as follows. Next to this introduction, Section 2 highlights the advantages and problems of carrying out international benchmarking studies. Section 3 displays the selection and main features of the countries and their water sector included in the research. Afterwards Section 4 briefly introduces the non-parametric methodology used to measure the efficiency of the drinking water utilities and discusses the main results. Section 5 concludes the paper.

2 INTERNATIONAL BENCHMARKING IN THE WATER SECTOR

In many countries, the number of comparable utilities is too low to make a valuable benchmark study. This can be attributed to the many mergers in the sector. We start this section by discussing the absence of merger economies.

2.1 Merger economies

Many drinking water utilities are invoking economies of scale and inefficiency reductions when they try to argument a merger with another utility. As it is hard to proof that mergers in the drinking water sector indeed create scale economies, regulators often give the benefit of the doubt to merging companies. However, by merging, the number of comparable utilities in the sample decreases, so that benchmarking becomes more intricate. Indeed, after the last merger, the one remaining utility is efficient by default. However, recent research proofs that mergers in the drinking water sector do not provide merger economies (De Witte and Dijkgraaf, 2007).

The literature distinguishes three possible merger economies (Roller *et al.*, 2000). Firstly, merger economies could arise from the increased market power. In a benchmarking context, this corresponds to the decrease of comparable utilities, so that the remaining companies could more easily invoke different exogenous characteristics. Secondly, merger economies could arise from an increase in scale economies, i.e. by operating at a larger scale, costs per unit decrease. Thirdly, merging companies could benefit from increased incentives to fight inefficiencies. This is particularly the case when stakeholders agreed on the merger on the basis of cost reductions.

By the use of both non-parametric and parametric techniques, De Witte and Dijkgraaf (2007) show for the Dutch drinking water sector that the large efficiency gains in the sector could be attributed to the effect of benchmarking, while the merger economies, both in the sense of scale economies and increased incentives to fight inefficiencies, are absent. Only for very small companies, merger economies are detected. Therefore, they conclude that regulators should be suspicious to mergers and do not give them the benefit of the doubt. However, if scale economies would be present, the regulator can opt for an international benchmark study to compensate for the decrease in comparable units.

2.2 International comparisons

With only few remaining utilities, companies can easily invoke exogenous characteristics. To circumvent this problem, utilities from other countries and regions could be added to the reference sample. An international benchmark study arises. As pointed out in De Witte and Marques (2007a), a cross-country comparison offers some advantages.

Firstly, it offers the possibility to escape the unsteady equilibrium between the claimed economies of scale for merging utilities and a sufficient number of remaining comparators in the sample. Secondly, also the best practices, which are usually the same for a longer period of time, can be benchmarked. This creates additional incentives for those utilities. Indeed, the possibility that a national best practice remains the reference in an enlarged data set decreases. Finally, international benchmark studies allow comparing the incentive schemes in different countries. Therefore, in an international data set, we measure the efficiency of the water utilities by the use of DEA. After correcting bias in the efficiency estimates and after taking into account environmental factors, which are out of control of the firm's management, we evaluate the effect on efficiency of a benchmarking and regulation incentive scheme.

However, these advantages come at a cost as international benchmarking creates some difficulties as well. Especially the intricate comparability of the data sources should attract a lot of attention. Indeed, national regulators and utilities employ different definitions of similar concepts. For example, the definition of medium and large sized customers differs even within countries. Therefore, one should be cautious when merging these samples. Secondly, the data could also suffer from exogenous differences which influence the values.

For example, exchange rate fluctuations or purchasing power differences could give rise to differences in the sample. Thirdly, in some countries a larger extent of the activities are outsourced to specialized companies, so that the total number of employees and staff costs differ. Finally, heterogeneity arising from exogenous characteristics should be taken into account to obtain a fair comparison.

In spite of these difficulties, we think it is possible to perform an international benchmark study. In particular, we account for the intricate lack of comparability of the data by making four specific assumptions. Firstly, we adopt variables in quantities (e.g., the inputs staff and mains length) that are less susceptible to the lack of comparability. Secondly, the major differences among countries are related to taxation issues. However, thanks to the quantities variables, tax heterogeneity does not significantly influence the model. Thirdly, also other heterogeneous factors which characterize the operational and institutional environment are integrated in the second stage analysis of the model. Fourthly, we removed outliers and a-typical observations from the data set. For example, the Australian water services, which may seem very different from the others, correspond only to major cities of Australia, and thus, are more comparable to the other observations.

3 INSTITUTIONAL FRAMEWORKS

In this article, we selected five countries which each have some specific incentive regulatory characteristics: the Netherlands, England and Wales, Australia, Portugal and Belgium. For each of these countries, the data are obtained from the sector associations and regulators for the year 2005 (except for Belgium where we have 2004 data).

After a debate on the possibilities to deregulate monopolistic markets (among which the drinking water sector), the Dutch drinking water utilities started a voluntary benchmarking in 1997. The benchmarking works on the principle of '*naming and shaming*' and is closely observed by the Dutch public opinion.

In England and Wales, the Thatcher administration privatised the water sector in 1989. To avoid the exploitation of the newly established regional monopolists, three regulators were established, from which the Office of Water Services (OFWAT) is the economic regulator. It uses yardstick competition to set drinking water prices and to monitor the utilities.

Australia benefitted from the US and UK regulatory models in that they incorporated the US ideas of transparency, enactment and accountability and the UK ideas of performance incentives. Australia possesses a strict regulatory model which aims at sustainability and corporatization.

Portugal employs regulation for its private utilities, while the public utilities are not triggered by incentive regulation. Only recently, benchmarking was introduced in the sector.

In the federal country of Belgium, the Walloon area started benchmarking in 2006 (hence, not observable in our data yet) while the Flemish utilities are not influenced by incentives. In both regions, the maximum drinking water prices for a particular provision area are determined by a federal pricing commission (although drinking water is a regional competence).

4 NON-PARAMETRIC EFFICIENCY ESTIMATION

Efficiency can be measured by a broad range of tools. A first branch of methodologies are the parametric cost functions. These assume a specific functional form to the data and could easily incorporate exogenous characteristics and measurement errors. In this article, we focus on a second branch of methodologies, non-parametric models which do not assume any *a priori* assumptions on the data and, therefore, let the data speak for themselves. We concentrate on the Data Envelopment Analysis (DEA) model. We briefly explain this approach.

Consider Figure 1 where the horizontal axis presents a one dimensional input and the vertical axis a one dimensional output. Let observation A denotes the evaluated utility with inputs x and outputs y. In an input-oriented model, DEA detects all observations which are using less inputs to produce the same amount of outputs. In an output-oriented model, DEA considers all observations which are producing more outputs for a given amount of inputs. For observation A, this corresponds to all observations in its fourth quadrant. We label these observations as dominant observations relative to A. Some observations are not dominated by any of the reference points. These undominated observations constitute the best practice frontier. Any deviation relative to the best practice frontier is considered as inefficiency. The input-oriented inefficiency for observation A corresponds to the distance AB relative to the distance AC. Its output-oriented inefficiency corresponds to the distance AD relative to DE. Obviously, observations which constitute the frontier obtain an efficiency score of 1, while inefficient observations obtain scores larger than 1.

However, Simar and Wilson (1998, 2000) point out that not the full distance AB or AD could be attributed to inefficiency. A part of this distance is due to measurement errors, a-typical observations which influence the shape of the frontier and noise in the data. Taken together this corresponds to the bias. We account for the bias by bootstrapping (i.e., repeatedly estimating the problem with small disturbance in the input variables). Thanks to these bootstrap estimates, we can induce the properties of the DEA estimates. We refer to Simar and Wilson (1998, 2000) for an extensive introduction in the methodology.

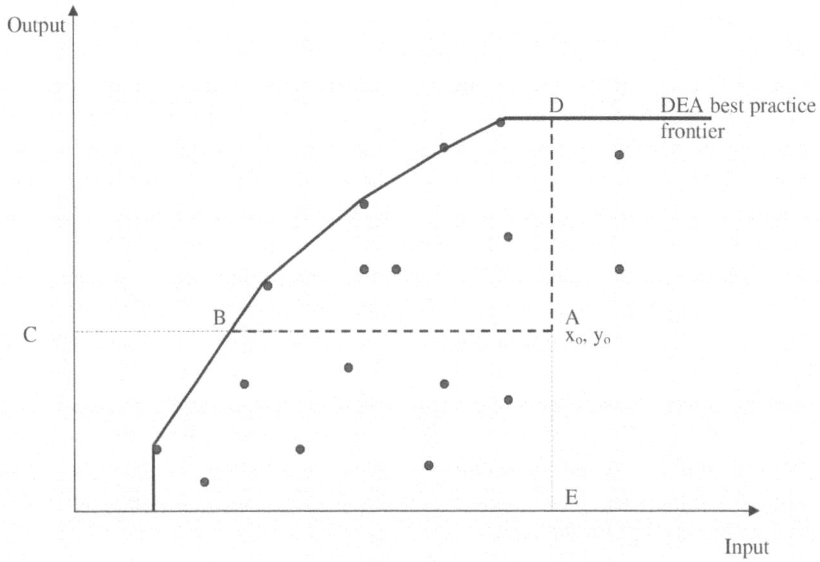

Figure 1. DEA BEST PRACTICE FRONTIER

In the assessment of efficiency, an objective and fair comparison is needed. Therefore, we have to equalize all physical, social and institutional characteristics which are beyond the scope of the firm's managers. We include three physical variables: (1) the percentage of leakage (it captures the geographical relief as a hilly landscape requires more pressure on the mains and the extent of maintenance as more leakages correspond to less expenses with maintenance), (2) groundwater use (the production cost is much lower than the counterparts that abstract superficial water or import water from other utilities) and (3) delivery to medium and large sized customers. The social variables capture the characteristics of the customers. Firstly, gross regional product (GRP) captures the relative wealth of the customers, the difference in skill distribution (see above) and approximates the average productivity of a region. Secondly, water consumption per capita reflects demand side management. By the institutional variables we try to capture (1) the scope economies (i.e., the only activity is the delivery of drinking water), (2) the corporization (i.e., the hard or soft budget constraints), (3) the scale economies (i.e., delivery to more than three municipalities) and finally, the type of incentive regulation, i.e. (4) yardstick competition or (5) benchmarking.

The literature developed several procedures to incorporate environmental characteristics in DEA efficiency scores. Initially, the literature employed truncated Tobit regressions to detect the size and direction of the exogenous influence. However, these models suffer from a separability condition in that the heterogeneity does not influence the shape of the frontier. Therefore, Simar and Wilson (1998, 2000) develop a double bootstrap procedure and Daraio and Simar (2005) introduced the conditional efficiency estimates. The latter procedures are combined in De Witte and Marques (2007a). As in De Witte and Marques (2007b), in this article we use the residuals of the truncated Tobit regression to incorporate the environmental variables. By using the residuals of the Tobit regression, we avoid the above mentioned separability condition.

We first present the results of the bias corrected double bootstrap procedure in Table 1. A positive sign denotes a negative influence to efficiency (i.e., an increase in inefficiency) while a positive sign points to a positive influence to efficiency (i.e., a decrease in inefficiency). We observe that industrial customers, high consumption per capita and corporatization decrease the efficiency of the utilities. Contrary, a higher proportion of leakage, groundwater extraction, GRP, scope economies, scale economies, regulation and benchmarking increase the efficiency of utilities. All of these variables have the expected sign, except for leakage.

By the use of the residuals of the Tobit regression, we correct the efficiency scores for these characteristics. The average results, as presented in Table 2, show the efficiency scores as would all utilities face exactly the same physical, social and institutional environment (indeed, all other influences than the incorporated are captured by the intercept). We observe that in comparison to the uncorrected efficiencies, the Belgian and private Portuguese utilities gain most. The efficiency of the Dutch companies even reduces if exogenous characteristics are taken into account. More interesting is the fourth column of Table 2 where we equalize all physical, social and institutional influences, except for the effect of regulation and benchmarking. The comparison of the latter with the equalized efficiency scores indicate the positive effect of benchmarking and regulation. Indeed, it reveals the effectiveness of the Dutch benchmarking scheme (as the Dutch companies are performing more efficient if benchmarking is taken into account) and the power of the English and Welsh, Australian and (private) Portuguese regulatory models. As there is no clear incentive structure for the Belgian and Portuguese public utilities, their average efficiency falls in comparison to the equalized situation. The Belgian and Portuguese authorities could ameliorate the performances of their drinking water sector by introducing a clear incentive scheme.

Table 2. INFLUENCE EXOGENOUS ENVIRONMENT

	DEA eff.	Lower bound estimate 5 %	Upper bound estimate 5 %
Intercept	4,2216 ***	1,9184	6,6944
Leakage (%)	-0,02258 ***	-0,06406	0,01550
Industry water/household delivery	0,02396 ***	0,01353	0,03404
Groundwater extraction (%)	-0,0001359	-0,007760	0,007383
Gross regional product (GRP)	-0,00006879 ***	-0,0001579	0,000007980
Consumption per capita	0,00005716 ***	-0,00006598	0,0001714
Water unique activity (=1)	-0,2644 ***	-1,0319	0,4602
Corporatization (=1)	1,2254 ***	-0,3341	2,6574
Delivery in one municipality (=1)	-1,3448 ***	-2,4574	-0,3601
Regulator (=1)	-0,9637 ***	-2,1884	0,1741
Benchmarking (=1)	-0,1198 ***	-1,3720	1,1548

*** denotes significance at a 1%-level

Table 3: AVERAGE CORRECTED EFFICIENCY SCORES

Method	Uncorrected efficiencies	All influences equalized	Incentive scheme
The Netherlands	1.216	1.274	1.322
England and Wales	1.358	1.202	1.219
Australia	1.453	1.322	1.331
Portugal public	1.761	1.554	1.524
Portugal private	1.741	1.411	1.372
Belgium	1.600	1.355	1.347

5 CONCLUDING REMARKS

The analysis provides significant evidence for the positive effects of incentives schemes on efficiency. It analysis even demonstrates that in absence of clear and structural incentives the average efficiency of the utilities even falls in comparison with utilities which are encouraged by incentives. The natural monopoly in the drinking water sector leads to the *quiet life* of Hicks (1935) and *X-inefficiency* of Leibenstein (1966). The presence of benchmarking (in the sense of sunshine regulation or yardstick competition) is a key element which replaces competition *in* the market or competition *for* the market by competition by comparison.

6 REFERENCES

Daraio, C. and Simar, L. (2005). Introducing environmental variables in nonparametric frontier models: a probabilistic approach. *Journal of productivity analysis*. **24** (1), 93-121.

De Witte, K. and Dijkgraaf, E. (2007). Mean and Bold? On separating merger economies from structural efficiency gains in the drinking water sector. *Tinbergen Institute Discussion Paper*, 092/3.

De Witte, K. and Marques, R. (2007a). Designing incentives in local public utilities, an international comparison to the drinking water sector. *CES Discussion Paper Series* DPS 07.32.

De Witte, K. and Marques, R. (2007b). Regulering brengt efficiëntie in de drinkwatersector. *Economisch Statistische Berichten*. vol **92**, no. 4506.

Hicks, J. (1935). Annual survey of economic theory: the theory of monopoly, *Econometrica* **3** (1), 1-20.

Leibenstein, H. (1966). Allocative efficiency vs. 'X-efficiency', *The American Economic Review* **56** (3), 392-415.

Marques, R. (2006). A yardstick competition model for Portuguese water and sewerage services regulation. *Utilities Policy*, **14** (3), 175-184.

Marques, R. and De Witte, K. (2008). Towards a benchmarking paradigm in the water and sewerage services. Forthcoming in *'Public Money and Management'*.

Röller, L., Stennek, J. and Verboven, F. (2000). Efficiency gains from mergers. *The Research Institute of Industrial Economics*, Stockholm.

Simar, L. and Wilson, P. (1998). Sensitivity analysis of efficiency scores: how to bootstrap in nonparametric frontier models. *Management Science*. **44** (1), 49-61.

Simar, L. and Wilson, P. (2000). Statistical inference in nonparametric frontier models: the state of the art. *Journal of Productivity Analysis*. **13** (1), 49-78

Water Utility Benchmarking for Managerial and Policy Decisions: Lessons from Developing Countries

S. V. Berg[1] *and M. L. Corton*[2]

[1] *Distinguished Service Professor, Economics; sberg@ufl.edu, Public Utility Research Center, University of Florida, Gainesville, FL 32607 Tlf+1 352-392-0132 Fax +1 352-392-7796.*
[2] *PURC Research Associate, University of Florida, mcorton@ufl.edu, Public Utility Research Center, University of Florida, Gainesville, FL 32611, +1 352-392-6148 Fax +1 352-392-7796.*

Keywords: Benchmarking, DEA, SFA, Performance

 Performance Assessment of Urban Infrastrucutre Services
Edited by Enrique Cabrera Jr and Miguel Ángel Pardo. ISBN: 9781843391913.
Published by IWA Publishing, London, UK.

ABSTRACT

Those responsible for utility operations can only manage what they measure, so having information on productivity trends and relative performance enables utility managers to direct attention to shortfalls. Similarly, policy-makers require quantitative analyses in order to identify utilities with strong and weak performance. Examples from Central America are used to illustrate key points.

1 INTRODUCTION

A recent IADB study reports that investments of $40 billion for water assets are needed to meet the United Nation's Millennium Development Goals; wastewater treatment would significantly raise that funding requirement ("Obstacles and Constraints, 2003). A survey of 400 stakeholders included in the study identified inappropriate pricing policy and lack of clarity in regulatory processes as the two major constraints for increasing investment in water and sanitation systems (WSS) in Latin America and the Caribbean. Private sector funding could play a role in expanding or improving urban water systems through either equity investments or the purchase of municipal bonds. However, external financial flows are required to increase significantly absent major improvements in system performance and to develop incentives for better WSS performance – in the sense of increased quality of service provision and cost efficiency of the utility system's operations.

Expecting infrastructure investment to grow in Central America, the Inter-American Development Bank (IADB) funded a PURC study on *Benchmarking Water Utilities* in the region. The aim of IADB was to gauge the impact of loans on network expansion (coverage) and on service quality. In addition, water professionals at international organizations must be able to understand what utilities (and nations) are doing as "best practice" so that incentives can be developed to enhance performance (Baietti, et. al., 2006). The project involved conducting metric benchmarking analysis using data gathered from national regulators and utilities. This work complements and extends the *Association of Water Regulators of the Americas*' (ADERASA) recent empirical study on benchmarking Latin American water service providers.

2. LESSONS

To some extent, academic research tends to be technical—drawing upon advanced statistical and data analysis techniques. Studies useful for decision-making tend to be more intuitive. The latter attempt to translate principles into practice, but excessive emphasis on simplicity can be a problem. Utility managers and regulators in Central America reviewed the PURC research at the recent PURC/IADB Benchmarking Workshop (San Jose, Costa Rica, October 15-16). Participants from six nations identified factors having an impact on data quality, data collection, benchmarking methodologies, and possible policy and regulatory implications of performance rankings. In addition, lessons from academic research underscored the need for sensitivity tests before utilizing "scores" or rankings for setting prices or rewarding managers.

2.1 Lessons from Managers: Keep it Simple

The Workshop provided two opportunities for practitioners to respond to technical studies and to issues raised by regulatory policy-makers. The reflections have been distilled into ten key points, but capturing all the nuances presented by various individuals is beyond the scope of this study.

2.1.1 Data Manager Commitment

The managers responsible for data collection, verification, storage, and processing must be convinced that this activity is vital to the success of the enterprise. Thus, there must be an understanding between the operator and the regulator. The operator should be convinced that the task goes far beyond fulfilling a requirement of a financial statement where numbers are placed in particular columns, but also about how the data is useful for management. Besides serving as a report to the regulator or to any external institution, data must be viewed as important and useful for the company – for strategic, operational, administrative and commercial purposes. Benchmarking is a tool that results in resource savings for the company concerned with performance: the information system pinpoints the areas that should be targeted for future initiatives and allows managers to evaluate the impacts of past interventions.

2.1.2 Data Manager Continuity

There needs to be a high level person responsible for data within the company; however, it is the position (rather than the person) that must have continuity over time within the company. This formalized role is needed to address internal turnover problems which limit data collection, causing gaps in time series and in cross-section observations. Obtaining data series for a number of years also requires the support of funding agencies, including national development banks. The lack of data in systems like IBNET (the World Bank's data library for water utilities) is a sad commentary on past incentives. It suggests that national development agencies and international banks have only been focusing on benchmarking in recent years, perhaps because cutting ribbons for new projects is much more glamorous (and less threatening) than evaluating past decisions.

2.1.3 Centralized vs. Decentralized Systems

Centralization and decentralization are both viable strategies, but the strengths and limitations of each must be recognized as nations and organizations adapt to changing conditions. To some extent, if a system is "working well" there is no need to dramatically change institutional arrangements. Nations and organizations have different historical contexts; the development of new sector policies is highly path-dependent. Thus, within an organization, duplication of data storage files inside the firm and data reports (in specialized formats) to external institutions raises administrative costs. This would suggest centralization—especially given the opportunities associated with new information technologies. On the other hand, centralization without access opens up the possibility for little "Information Empires," where individuals exercise power by withholding data from those who should have access to information. Balancing such considerations leads to different types of policy initiatives. For example, Honduras has gone through a decentralization process (Pearce-Oroz, 2006) while other nations have consolidated their utility suppliers. There is no single recipe for promoting high performance: rather, leaders build upon the experiences and institutions that provide the baseline when making comparisons.

2.1.4 Data Disaggregation

Data disaggregation improves decision-making. Clear customer classification (residential, industrial, and commercial) allows for more accurate information regarding operation and performance of the company. In addition, maintaining data series on particular regions or divisions of a company allows top managers to develop strategies for rewarding strong performance. Disaggregated data allow managers to target areas of sub-standard performance and facilitates quantitative studies of cost and productivity. For example, Mugisha et. al. (2007) show how Uganda's state-owned NWSC utilizes disaggregated benchmarking information to set targets, compare managerial performance, and determine management bonuses (up to 50% of base pay).

2.1.5 Data Definitions and Benchmarking Objectives

Clear variable definitions allow outsiders to interpret information; consistency and clarity are fundamental to the management process. The International Water Association has played a leadership role in this process.

2.1.6 *Operational Data*

Better operational data collection procedures are needed: timely reports that identify patterns mean that network repairs can be addressed in a comprehensive and cost-effective manner.

2.1.7 *Other Data Sources*

Workshop participants urged that data collection not stop at items under the firm's direct control. For example, rural systems must still be monitored—by a government ministry or a regional water utility (not currently serving those potential customers). One representative was adamant that social indicators be included in studies, so that service quality and continuity could be linked to socio-economic characteristics of different regions. Furthermore, factors external to the company affect production costs and opportunities for productivity growth. Distance from water source, geographic characteristics (topology and population density), and a culture of payment (or non-payment) all affect costs. External factors also may have an impact on the collection and storage of data. For example, the existence of records and maps of the city; the frequency of the country census; municipal or city restrictions regarding the network design; number of connections per km, and type of users of the network. Thus, empirical researchers need to take into account factors beyond managers' control just as managers need to draw upon the census and other external data-sets in their own system-planning.

2.1.8 *Information Technology and Management*

Information technology is necessary, but not sufficient, for sound management: information systems should link financial-commercial-operational data. Leaders can only manage what they measure. Aid projects that computerize companies without altering underlying internal incentives cannot have significant impacts on performance. The entire supply chain would be better off if utilities moved more quickly towards best practice: those manufacturing pipes, pumps and other hardware would see a dramatic increase in business in particular countries *if* utilities demonstrated good stewardship.

2.1.9 *Transparency and Public Policy*

The company information needs to be public to promote managerial accountability and citizen confidence in infrastructure services. Some state-owned enterprises become the home of political patronage. When there are poor internal incentives, the performance of an entity is likely to be sub-standard. If customers (and un-served citizens) do not have data on comparable utilities, the citizens are in no position to put pressure on managers to improve performance.

There is evidence that even rough comparisons can put pressure on political leaders to fulfil promises to provide funds for network expansion and on managers to deliver services at least-cost (Rossi and Ruzzier, 2000). However, regulatory commissions and/or water ministries must be committed to institutional reform if they are to risk offending powerful groups (including unions and high income customers now being subsidized).

2.1.10 Methodologies and Capacity-Building

Overall, there is a need within the firm for more information regarding benchmarking methodologies and their application. Larger water utilities have engineers who are familiar with process benchmarking. There is also a need for capacity building in the area of metric benchmarking—starting with trends in Core Indicators, and moving to basic statistical reports and DEA studies. Core indicators include labor productivity, water losses, and trends in average cost. Statistical studies can identify cost drivers and determine how far utilities are from the production frontier.

2.2 Lessons from Academics: Recognize Complexity

The last point from managers provides a natural link to those conducting more technical analyses. Managers are quite aware of the different conditions facing different water utilities and how difficult it is to make valid comparisons across firms and over time. Thus, they may be sceptical of using rankings to reward utilities, set prices, or to reward divisional managers. Nevertheless, information on broad patterns allows political leaders and citizens to "grade" the performance of water utilities.

2.2.1 Sources of Productivity Change

Performance indicators are "partial" in nature: they consider one dimension of performance at a time. By contrast, total factor productivity (TFP) analysis focuses on firms productivity change over time and takes several inputs and outputs into account. Productivity may differ over a period of time for the following reasons:

a. Technological change (frontier shift): a technological change has occurred within the sector producing a variation on firm's productivity.
b. Technical efficiency change (catch-up): A firm has varied its production efficiency over time.
c. Scale efficiency change: A firm change in production efficiency is product of a change in its scale of production.
d. Input mix allocative efficiency: A firm has varied its efficiency by producing outputs using a different mix of inputs, given the input

prices the firm faces;

e. Output mix allocative efficiency: A firm has varied its efficiency by producing a different level of outputs with the same mix of inputs; given the input prices the firm faces.

By understanding the sources of productivity change, managers can focus improvement in areas that seem weak. At the same time, by understanding the sources of productivity change, policy makers, investors and other stakeholders can point to the most productive firms as examples of strong performance—promoting the diffusion of best practice to all firms.

To decompose productivity into the above components requires knowledge of the firm's production technology – the particular way a firm mixes inputs to produce its set of outputs which generally translates into a functional mathematical form for the production process. Quantitative methodologies can be used to isolate the roles of these various components which are generally accomplished by employing econometric procedures to estimate a functional form for the production technology. Statistical tests are performed to check how properly represented is the technology by the selected functional form (garcia and Thomas, 2001).

2.2.2 Strengths and Limitations of Methodologies

A single index of water utility performance has the same problems of any indicator: it will be neither comprehensive nor fully diagnostic. The ratio of water delivered to number of workers does not capture water losses or the sustainability of operations when certain types of expenditures (like maintenance) are deferred. Nevertheless, empirical studies of water utilities within and across countries have yielded insights into the determinants of relative performance by water utilities (Cubbin, 2005).

The development of sound managerial and public policy incentives involves data collection, analysis, and interpretation. Several methodologies should be used to determine the sensitivity of the results (and performance rankings) to model specification and outliers. Data Envelopment Analysis (DEA) and Stochastic Frontier Analysis (SFA) are two approaches to production and cost functions. The limitations to such studies must be balanced against the information asymmetries that arise in the absence of such studies.

The methodologies used by different groups differ in ways that are unsurprising. Practitioners seek decision-relevant studies that will help them document best practice and justify actions (and investments). Policy-makers tend to step back from the actual operations of firms, and seek to document the impacts of policies on sector performance and to identify troublesome patterns. Regulators, in particular, attempt to design incentive plans that utilize relative performance information. Academic researchers, on the other hand, tend to look for sophisticated ways to address more technical issues: the extent of economies of scale, the rate (and direction) of technological change, and sources of inefficiency. Each of these areas of concern justifies some degree of specialization and division of labor. As we shall see, each tends to use different types of benchmarking studies. However, the walls separating practitioners, policy analysts, and academics are higher and thicker than they need to be.

2.2.3 Accessibility of Technical Studies to Practitioners

The reports developed by regulatory commissions and those published by academic researchers have different purposes. Bridging the two types of studies requires one foot in the world of politics and utility operations and one foot in the world of advanced data analysis: few of us have such skill sets. That suggests that teams of analysts from both worlds might have the greatest chance to obtain needed data, ask the right questions, and interpret the results in ways that are intuitively understandable to those without advanced skills. Such studies would include sensitivity tests, so that the results are robust—providing external observers with some confidence in the conclusions. The PURC/IADB Project for Central America (Corton, 2007) represents a starting point for a type of collaboration that shows some promise—as it engaged practitioners in the research process.

3 POLITICAL ECONOMY OF PERFORMANCE INDICATORS

The key lessons regarding collecting and analyzing operating and financial information underscore the practical orientation of decision-makers. Practitioners, policy-makers, and academics all have studied water sector performance. The three groups have tended to operate in relative isolation. Given the important resource allocation issues associated with infrastructure industries in developed and developing nations, greater communication among the three types of researchers is called for. However, such cross-fertilization requires that each group understands the priorities placed on different aspects of inter-firm comparisons. The benchmarking process can be divided into five steps: (1) identify objectives, select methodology and gather data; (2) screen and analyze data; (3) utilize specific analytic techniques; (4) conduct consistency/sensitivity tests; and (5) develop policy implications.

3.1 Steps in Benchmarking

Here, we focus on the first step: objectives and methodologies. Figure 1 depicts the elements and shows four broad methodologies. In addition, several other types of performance indicators must be incorporated into comprehensive benchmarking studies.

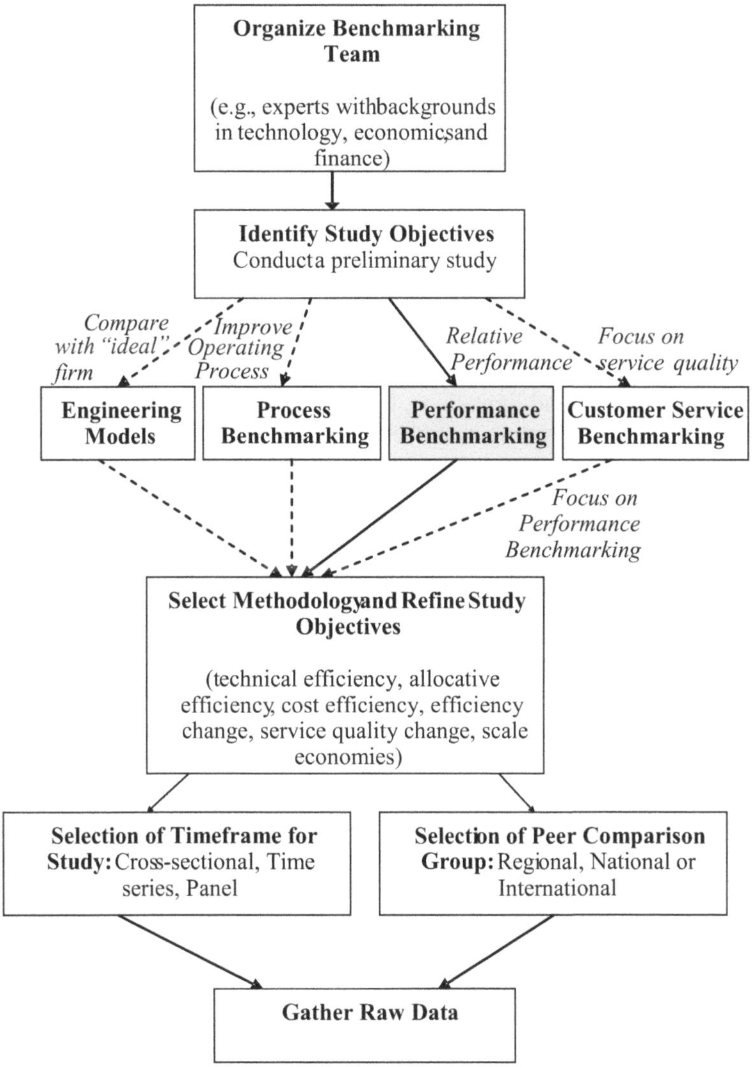

Figure 9. STEP 1: IDENTIFY OBJECTIVES, METHODOLOGY, AND DATA

The first step in the benchmarking process requires the policy-maker or analyst to identify issues to be addressed, the time period to be analyzed, and the types of comparisons to be made. These choices will reflect current analytic capabilities, an initial understanding of data availability, and preliminary methodological choices. The objectives of benchmarking studies tend to depend on emerging policy issues: whether sector consolidation (or disaggregation) might promote greater efficiency, whether privatization is associated with performance improvements (Estache and Rossi, 2002), and which managerial strategies are most effective in containing cost or improving service quality. As Figure 1 indicates, there are a number of alternative approaches to benchmarking, including engineering models, process benchmarking, performance benchmarking, and customer service benchmarking. Regulators utilize all four approaches to different degrees: Chilean regulators use the hypothetical firm (engineering) approach to establish performance objectives, managers tend to focus on specific processes (to identify best practices), regulators everywhere utilize performance indicators to evaluate utilities, and managers conduct customer surveys to determine areas for improvement. The focus here will be on the third category of benchmarking indicators based on performance studies that utilize cost or production functions. Such studies have their own strengths and limitations, but these methodologies provide a valuable framework for investigating a number of questions.

3.2 Four Types of Indicators: Decision-Relevance

3.2.1 Core Overall Performance Indictors

Core Overall Performance Indicators include a number of *Specific Core Indices*, such as volume billed per worker, quality of service (continuity, water quality, complaints), unaccounted for water, coverage, and key financial data (operating expenses relative to total revenues, collections). These partial measures are generally available, and provide the simplest way to perform comparisons: trends direct attention to potential problem areas. Policymakers often combine the specific core indices to create an *Overall Performance Indicator (OPI)*, generally using a weighted average of core indices. Thus, an OPI provides a summary index that can be used to communicate relative performance to a wide audience. Regional associations of water regulators have promoted projects that facilitate comparisons of specific core indicators.

3.2.2 Performance Scores: Production or Cost Estimates

Performance Scores based on Production or Cost Estimates are used to identify the best performers and the weakest performers in a group of utilities. The metric approach (favored by academics) allows quantitative measurement of relative performance (cost efficiency, technical/engineering efficiency, scale efficiency, allocative efficiency, and efficiency change). Performance can be compared with other utilities at a point of time and over time, using statistical and/or nonparametric frontier methods. Thus, performance scores and relative rankings identify under-performing and high-performing utilities. Rankings can be based on production relationships and/or cost structures (Berg, 2007.

3.2.3 Engineering/Model Company Approach

The Engineering/Model Company approach has been used to establish baseline performance. This methodology requires the development of an optimized economic and engineering model: based on creating an idealized benchmark specific to each utility—incorporating the topology, demand patterns, and population density of the service territory. The use of an "artificial" firm that has optimized its network design and minimizes its operating costs can provide insight into what is possible if a firm is starting as a Greenfield Project. As with any methodology, this approach also has its limitations. The engineering models that support it can be very complicated, and the structure of the underlying production relationships can be obscured through a set of assumed coefficients used in the optimization process. Chile and Argentina have used this approach for establishing infrastructure performance targets; however, it is not widely used by policymakers, managers, or economists.

3.2.4 Process Benchmarking

Process Benchmarking focuses on individual production processes in the vertical production chain. Managers actively utilize this technique because it can identify specific stages of the production process that warrant attention. For example, to obtain finished drinking water involves the following steps: pumping up, intake, transport, clarification and filtration of groundwater as well as the purification and treatment of raw surface water. Detailed examination of production facilities and their operations would be the starting point for process benchmarking. Similar studies would be performed for distribution processes (network design, pipeline construction and maintenance), sales processes (including meter reading, data processing, billing, collections, and customer relations), and general processes (like planning, staff recruitment and retention, and public relations). Many water associations focus on process benchmarking as a mechanism for identifying potential benchmarking partners, preparing for and undertaking benchmarking visits, and implementing best practices (Larsson, et. al, (2002)). From the standpoint of public policy, there must clear delineation of utility obligations and regulatory responsibilities so that process benchmarking studies do not lead to undue regulatory interference with managerial decision making.

3.2.5 Customer Survey Benchmarking

Customer Survey Benchmarking focuses on the perceptions of customers as a key element for performance evaluation. Unlike the other approaches, this technique can shed light on consumer concerns, reflected in complaints or captured in customer surveys. One widely-used model identifies five dimensions of service quality as perceived by customers: external characteristics (tidy workplace, employee appearances), reliability (meeting deadlines, consistency in interactions), responsiveness (providing service promptly), consideration (personnel who are courteous, friendly, and helpful), and empathy (giving individual care and attention). Utility managers conduct such surveys to reveal performance trends over time (Parasuraman, Zeithaml and Berry (1985)). Proactive managers who disaggregate complaints by type of customer, location, and type of complaint can identify problem areas.

3.2.6 Other Indicators: Financial and Resource Sustainability

At least two other indicators shed light on other dimensions of sector performance. Financial sustainability is not captured in other indicators. In fact, a utility could defer needed maintenance and reduce operating costs—possibly yielding lower revenue requirements than other firms. Much more complete models would be needed to test for such problems. Similarly, water resource sustainability is not captured in the other indicators, unless a process indicator includes water sources as a factor.

4 CONCLUDING OBSERVATIONS

The specific coefficients of the Central American Water Utility Benchmarking Study are less important than the fact that another region of the world is engaged in a more deliberate process for making performance comparisons. Analysts apply these quantitative techniques to determine relationships among variables: for example, utilities that produce far less output than other utilities (who are using the same input levels) are deemed to be relatively inefficient. Similarly, a utility might have much higher costs than expected (based on observations of others producing the same output level but having lower costs). A finding of excessively high costs would trigger more in-depth studies to determine the source of such poor performance.

This project establishes a strong case for more comprehensive studies in the future—helping to set the stage for creating strong incentives for improved performance. The water regulatory authority can be instrumental in spearheading many of these activities. However, benchmarking will be much more effective if two points are met. First, regulated companies must cooperate with regulators and subscribe to improvement objectives and processes. Second, formal mechanisms need to be established for consumers and other affected parties to raise concerns and suggest modifications to the process. Through incentive regulation and an appropriate price cap formula, regulators can use findings from benchmarking reports to reward high performance companies (Estache and Kouassi, 2002). They can also pressure laggard companies to promote cost-containment and the improvement of service quality. A properly designed benchmarking system should prevent poorly performing companies from increasing prices as much as the "average" water utility to which they have been compared—so long as prices, do indeed, cover cots. If companies operate more efficiently, customers will benefit from lower prices and should continue to expect and receive high quality service. The resulting system is likely to be sustainable—promoting further network expansion and the adoption of best practice by most water utilities.

5 REFERENCES

Baietti, A. W. Kingdom, Meike van Ginneken (2006). "Characteristics of Well-Performing Public Water Utilities," Water Supply & Sanitation Working Notes No. 9, May.

Berg, S. (2007). "Survey of Water Utility Benchmarking Studies," PURC Working Paper.

Corton, M. L. (2007). "Benchmarking Water Utilities in Central America," PURC Working Paper.

Cubbin, John (2005). "Efficiency in the water industry," *Utilities Policy,* 13 (4), 289-293.

Estache, A. and E. Kouassi (2002). "Sector organization, governance and the inefficiency of African water utilities," World Bank Policy Research Working Paper No. 2890, Washington, D.C.

Estache, A. and M. Rossi (2002). "How different is the efficiency of public and private water companies in Asia?" *The World Bank Economic Review* 16(1):139-148.

Garcia, S. and A. Thomas (2001). "The structure of municipal water supply costs: Application to a panel of French local communities," *Journal of Productivity Analysis* 16:5-29.

Larsson, Mats, Renato Parena, Ed Smeets and Ingrid Troquet, (2002) "Process Benchmarking in the Water Industry: Towards a Worldwide Approach", International Water Association.

Mugisha, S., S. Berg, and W. T. Muhairwe (2007), "Using Internal Incentive Contracts to Improve Water Utility Performance: the Case of Uganda's NWSC," *Water Policy*, 271-284

"Obstacles and Constraints for Increasing Investment in the Water and Sanitation Sector in Latin America and the Caribbean: Survey," IADB, Dec 2003, 1-13.

Parasuraman, Zeithaml and Berry (1985) "A Conceptual Model of Service Quality and its implications for future research," *Journal of Marketing* 49 (4), Fall, 41-50.

Pearce-Oroz, Glenn (2006), "The viability of decentralized water and sanitation provision in developing countries: the case of Honduras" ,Water Policy 8 (2006) 31–50.

Rossi, Martin Antonio and Christian Alejandro Ruzzier (2000). "On the regulatory application of efficiency measures," *Utilities Policy,* 9, 81-92.

Service Performance Evaluation within an International Group

P. Jacq

Head of Operation Control, Suez Environnement, 38 rue du Président Wilson, 78230 Le Pecq, France, philippe.jacq@suez-env.com

Keywords: Global and Individual assessment, International Group, Operations

 Performance Assessment of Urban Infrastrucutre Services
Edited by Enrique Cabrera Jr and Miguel Ángel Pardo. ISBN: 9781843391913.
Published by IWA Publishing, London, UK.

ABSTRACT

Since 1998 SUEZ Environment uses a set of Key Performance Indicators to assess the level of services delivered by the different operating companies managed worldwide.

If at the beginning this was only a collection of data and indicators, annual reviews allow now for historical trends. Many improvements were borne to the questionnaires, such as translation in different languages, indication of past years values and use of standard definitions. The assessment is two-fold, one being individual and the second global.

For each company an annual individual report allows for progress evaluation through different approaches, comparing the performance of one operation either to reference values or to group achievements. Reference values are used when some international regulation apply, such as water quality compliance, or when Best Practices define thresholds. Group achievements, ranked by quartiles, are used for indicators linked with topography, such as electricity usage per m3 or length of network per customer, or influenced by local conditions, such as chemical costs per m3.

At a Group level from the presentation of results ranked by value, the use of advanced indicators, it is possible to show geographical or contract type differentiation. If it is well known that water usage is not similar worldwide, relations between other parameters indicate other geographical or contractual differences. This allows for more precise analysis of the results.

In both cases the performance indicators are used as driver for improvements, however they give only a trend that has to be checked by a more detailed approach.

Even if there are many differences between the different operations managed worldwide, in term of contract type and duration, local regulations and economical conditions, performance indicators allow for comparisons that are fruitful.

The process of performance evaluation is at the same time very stable, to show trends, and continuously evolving, to capture new topics.

1 INTRODUCTION

Operating in different countries under different types of contractual arrangement, SUEZ Environment needed to have a common scheme to assess the level of achievements and of service delivered.

The development of a comprehensive benchmarking process took a few years, from the definition of topics to be addressed to the elaboration of a questionnaire, the compilation of the data received and the utilisation of the results.

At the beginning of the process, in 1998, the purpose was only to develop a benchmarking of operations worldwide, to allow for a better management of the whole Group.

Changes in national and international regulations, such as the US Sarbanes-Oxley law, lead to new needs for data acquisition and reports. These new requirements were integrated within the existing questionnaires, that are now the basis for a set of reports on operations and environment. The scope of this document being limited to operational performance.

The whole process is fully auditable, as required by some regulations, and is regularly audited. Since 2005 this process is part to the ISO 9001/2000 Quality Assurance frame developed by SUEZ Environment in the fields of control and support to Operating Companies.

The data collected yearly are in the range of twenty thousand, leading to the development of an important database that allows for new studies on trends, with more and more large scopes.

2 DATA GATHERING

From 1997 a group of managers involved in international operations began to work on the construction of a benchmarking involving all water operating companies within SUEZ Environment. The first tasks were to define the meaningful indicators and the data needed to calculate them.

A detailed questionnaire was then set up, the same format being still used. The questions are arranged by themes: company data, water production and distribution, waste water collection and treatment, customer services and overall business. Even if the questions were drafted by teams of experts, the significantly different answers shown that, in some cases, the understanding was not always similar. The first document was in English, but in order to make it more easy to understand, translation exists in different languages, Czech, German, Hungarian, Spanish and French.

Question		Year	Value	Definition
Alkalmazottak száma		2006		A vízkezelési feladaton dolgozó alkalmazottak teljes száma (nyersvíz szivattyúzás + kezelési folyamat + kimenő víz szivattyúzása)
		2005		
Units	Szám	2004		
		2003		
Population served		2006		The total population, estimated if necessary, receiving the treated water including those through a bulk supply to another distribution company.
		2005		
Units	Number	2004		
		2003		
Number of employees		2006		The total number of employees working in the water treatment function (inlet raw water pumping + treatment chain + drinking water supply pumpage).
		2005		
Units	Number	2004		
		2003		

Figure 10 – EXAMPLES OF QUESTIONS IN ENGLISH AND MAGYAR

The values reported in the previous years are indicated, this allows for a first level of control on data consistence.

In order to allow for further comparisons with other benchmarking schemes, the definition mirror the ones set up by IWA, IBNET and European regulators.

For very detailed matters more support is given to fill in the questionnaire. One example being the Water Balance Table, as defined by IWA, that is attached to the questionnaire, showing the reference to the associated questions.

Sales (QE36)	**Billed Metered Consumption (QB10 first part)**
	Billed Unmetered Consumption
Unbilled Authorised Consumption (QE50)	**Unbilled Metered Consumption (QB10 second part)**
	Unbilled Unmetered Consumption

Figure 2 – IWA WATER BALANCE TABLE (ABSTRACT)

Each year the questionnaires are sent to operating companies in November, answers are expected by January. The data are checked and computed, all anomalies are reviewed with the operators. The review of data includes a control of the comprehensiveness of the answers that should cover all field of activity of the operating company, the verification of each data (comparison with previous years, units,...) and the consolidation by indicators and operating units.

3 REPORTS

Documents are prepared for different clients, from SUEZ Environment General management to Operating Units or expert teams, and from a global approach, such as the Environmental and the Performance Indicators reports, to individual reports.

3.1 Global Approach

From all data gathered more than 250 KPI may be calculated, however such large number is only useful for specialists. An annual "Operational Performance – 100 KPI" is published during the second quarter, showing the achievements of the Operating Unit. This report is designed to be a tool for the management teams of the Operating Units, as the use of Performance Indicators is already the norm in a large number of multinational companies. This technique ensure that the various human, natural and financial resources are efficiently used.

Even though every company is unique, due to its particular social, economical and regulatory context, this report endeavours to make a comparison between SUEZ Environment Group water and waste water contracts:

- Indicators are selected to constitute a common basis for all entities
- Results are presented in the form of ratios, sorted by activity
- Graphs set out results according to performance levels

The results are presented in histograms, showing the apparent better performance to the right. Generally specific comments are attached to graphs, explaining what information it gives and what are the limit values that are regarded as good, medium or poor, according to SUEZ Environment standards.

For the contracts who already participated in the reporting process for two years, the corresponding values are indicated on the graph.

The average has been calculated, and is shown in grey colour.

The colour of each line is defined by the population served, from large – 2 million inhabitants – to medium sized cities – 250 thousand inhabitants.

From few examples it appears that the size of the company does not influence significantly the level of performance. Topographical condition have a strong impact on energy usage for water production.

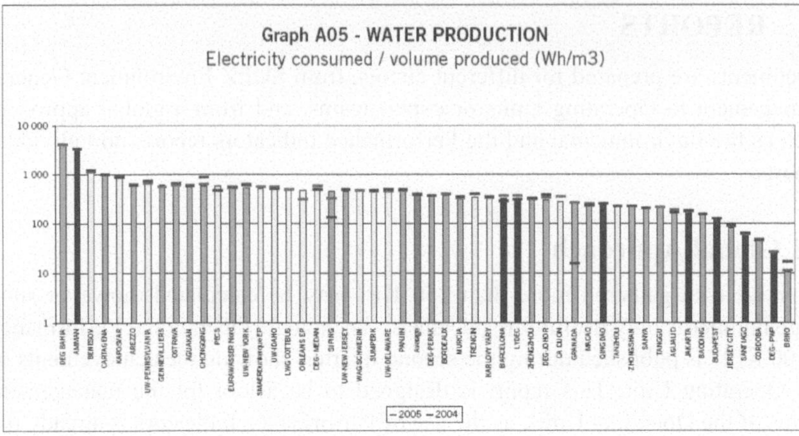

Figure 3 – WATER PRODUCTION – ENERGY CONSUMED PER VOLUME PRODUCED

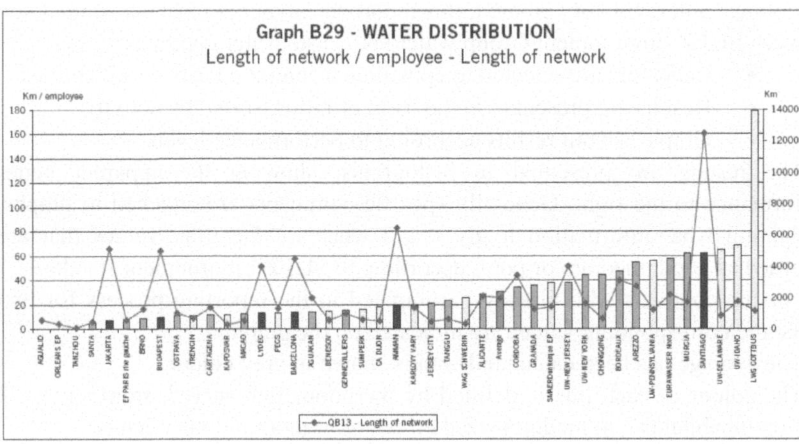

Figure 4 – WATER DISTRIBUTION – LENGTH OF NETWORK/EMPLOYEE

Using the thresholds derived from SUEZ Environment standards, it is possible to prepare a table with only three colours – green, amber and red – that gives a very quick appraisal of the situation. This appraisal being either for one operating company (x-axis) or for one performance indicator (y-axis) as shown hereunder.

Service performance indicators within an international group

	2005	PRODUCTION						DISTRIBUTION										
Country	Population DW or WW (Year 2005)	Total capacity / maximum daily demand	Compliance Rate Water Produced	Bacteriological Compliance Rate	Global analytical Effort after Water Production	Volume produced / employee	Average cost of water produced (OPEX)	Population served / total population	Volume metered / volume delivered	Storage Time	Bursts / length network	Length renewed + rehabilitated / length of network	Lost water / volume input	Lost water / km of network	Compliance Rate Water Distributed	Number of connections / employee	Length of network / employee	Water Distribution OPEX /Length of Water network
USA	785 123																	
France	690 492																	
France	650 317																	
CHINA	630 000																	
MEXICO	625 700																	
China	580 000																	
France	532 588																	
China	527 196																	
SPAIN	445 105																	
SPAIN	435 791																	

Figure 5 – SYNTHETIC REVIEW OF PERFORMANCE INDICATORS

3.2 Individual report

If the headquarter of an international group is mainly interested in global analysis, the manager of an operating company has not the same approach and is expecting an individual report.

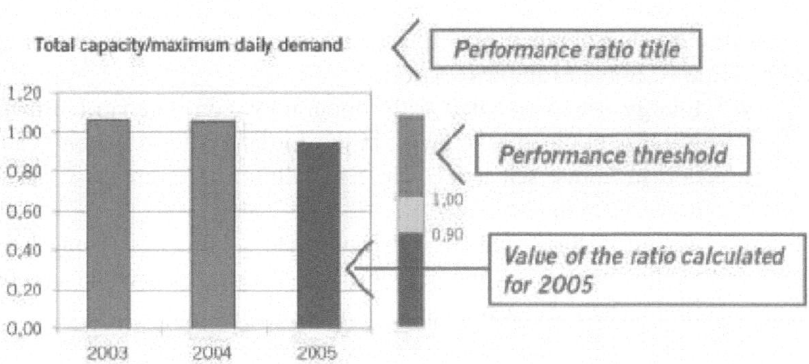

Figure 6 – REFERENCE TO SUEZ ENVIRONMENT STANDARDS

Two different categories of graphs are used, one using SUEZ Environment standards the other ones a comparison to the quartiles of SUEZ Environment Operations.

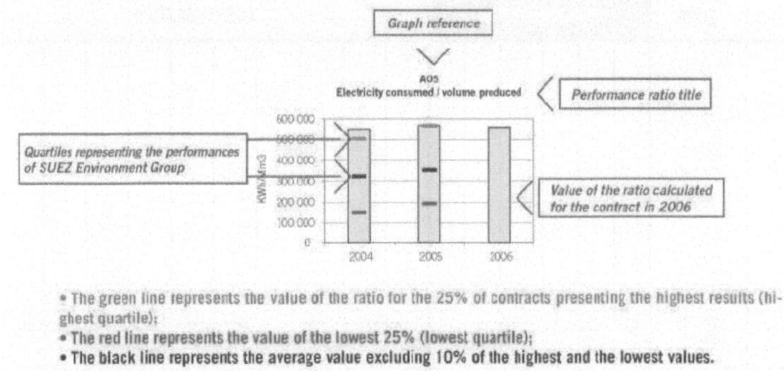

Figure 7 – REFERENCE TO SUEZ ENVIRONMENT QUARTILES

The choice of the type of reference to be used depends on the performance indicator itself.

Standard values are preferred when thresholds are defined, either by regulation or by SUEZ Environment Best Practices:

- Water quality is always defined by national or international standards
- Waste water discharge is generally defined by regulation or contract
- Accepted calls limits are set up through Best Practices on call centers

Quartiles are chosen when performance indicators are strongly influenced by local conditions. Some examples

- Energy use, associated with topography – in Amman (Jordan) raw water is pumped at –250 m, in the Jordan river, to the city centre, 1 000 m above sea level, while in Palmira (Colombia) water flows down from the hills
- chemical costs, linked with raw water – in Pecs (Hungary) high quality water is pumped through deep wells while in Jakarta (Indonesia) raw water has very high organic matter concentration

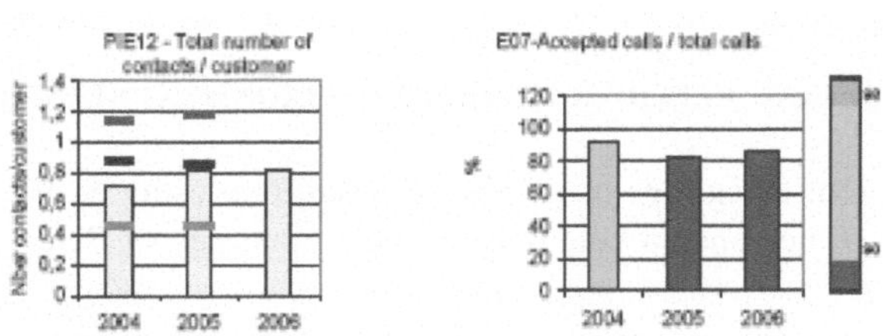

Figure 8 – CUSTOMER SERVICES INDICATORS

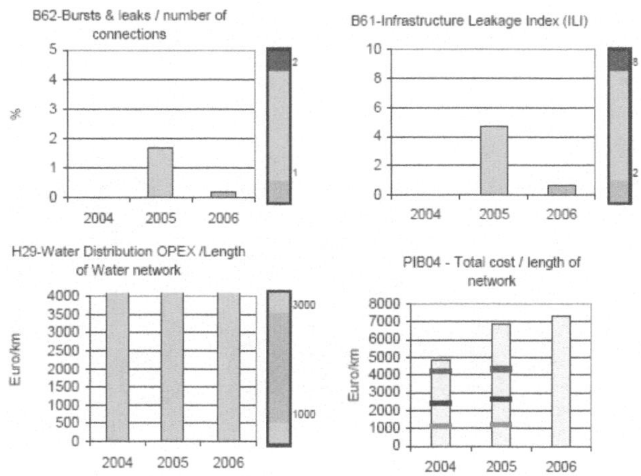

Figure 9 – WATER DISTRIBUTION INDICATORS

These data give a lot of information on the company and its evolution. In the above example, a very good progress is made in ILI value, but with a significant cost increase in the last two years.

Such report is used as a basis for annual reviews of company performances.

4 ADVANCED ANALYSIS

From the data gathered and the performance indicators arrived at, further analysis can be developed. As an example using two indicators as axes gives new sets of information.

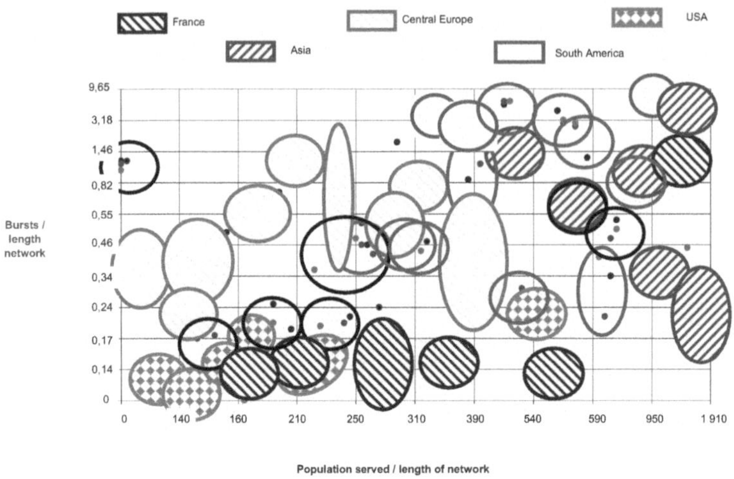

Figure 9 - **POPULATION SERVED AND BURST RATE PAR NETWORK LENGTH**

On this graph, the scales are calculated in such a way that one tenth of the data are in each category. It shows that, for variable population density (x-axis) the burst rate (y-axis) is stable in France and Central Europe, but at different values. US and French operations, but one, present low burst rates, while Asian ones are on the top end on both scales.

5 REFERENCES

Performance Indicators for Water Supply Services – 2006 - IWA Urban Water System Specialist Group – Helena ALEGRE, Jaime Melo BAPTISTA, Enrique CABRERA Jr., Francisco CUBILLO, Patricia DUARTE, Wolfram HIRNER, Wolf MERKEL, Renato PARENA

Process benchmarking as a source for new solutions

Process Indicators for drinking water production – a unified approach and its pilot application in Germany

A. Hein, P. Lévai and W. Merkel

IWW Water Centre, Moritzstraße 26, 45476 Mülheim an der Ruhr, Germany, corresponding author: Andreas Hein, Email: a.hein@iww-online.de, Phone: +49 (0)208 40303-340, Fax: +49 (0)208 40303-82

Keywords: Benchmarking, performance indicators, water supply, process efficiency, sustainability of water production, resources management, water treatment

Edited by Enrique Cabrera Jr and Miguel Ángel Pardo. ISBN: 9781843391913.
Published by IWA Publishing, London, UK.

ABSTRACT

Detailed process analysis for water production has been achieved by a newly developed process model and PI system that corresponds to the existing IWA PI system. Important and new features are different aggregation levels of data analysis and clustering criteria that allow detailed comparisons of performance by analysing efficiency and effectiveness of water production with regard to the process objective, independent of its engineering. The paper covers the methodology of the approach, new developments in terms of cluster criteria and experience from the first validating benchmarking exercise.

1 INTRODUCTION

For overall performance assessment of water supply services, the hierarchically structured IWA PI system (2nd edition) is recognised as a "standard system" in many companies (Alegre et al., 2006). In order to identify specific "best practice" and optimisation measures, a detailed analysis on the process level is favoured by the utilities, for cross-sectional processes like asset management (Foley, 2005) or technical maintenance (Harris et al., 2006) or for specific economic and technical issues (P. Dane, 2007).

The "water production process" summarizing resources management, water catchment and treatment is a key process in the water supply chain. With 12 German water supply companies, a new bottom-up approach was developed and tested, allowing a detailed performance (cost, quality, sustainability) analysis on process level and accounting for differing resources, treatment schemes and assets.

2 METHODOLOGY: PROCESS MODEL, PI-SYSTEM AND CLUSTER CRITERIA

2.1 Process model

Corresponding to the IWA PI system definitions (Merkel/Hirner 2005), the process model is structured in the technical main functions "resources management" (RM), "water catchment" (WC) and "water treatment" (WT), and main tasks "planning/construction/operation/maintenance". The main functions are sub-divided into 14 objects, classified according to their process goal or treatment objective, such as "Fe/Mn removal", "resource protection" or "disinfection". Thus, performance will be analysed in terms of efficiency and effectiveness with regard to a process objective, independent of its engineering.

The main functions or objectives for resources management (RM) and water catchment (WC) are shown completely in Figure. 1. For water treatment (WT), 8 technical objectives were defined: Fe/Mn removal, softening, particle elimination, elimination of organic substances, elimination of inorganic substances, stabilisation, disinfection during water treatment and final disinfection in distribution networks. These technical objectives are completed by the object "buildings" and "energy supply incl. emergency energy supply".

The reference manual of the process model includes detailed process descriptions and a task catalogue document, available for the participants.

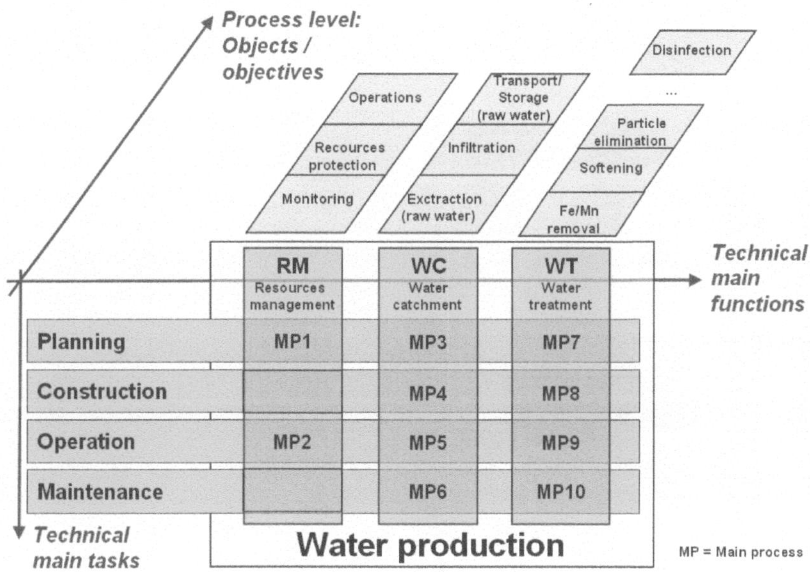

Figure. 1- STRUCTURE OF THE UNIFIED PROCESS MODEL: TASKS, FUNCTIONS AND OBJECTIVES (EXTRACT)

To make water production processes comparable, resource conditions and raw water quality has to be accounted for. Hence, cluster criteria and context information with relevance to the process performance were needed, allowing for proper comparison and detailed performance gap analysis. Essential elements of the process model are different aggregation levels that allow a hierarchical analysis from top-down and bottom-up for the main functions.

2.2 PI-system

Metric benchmarking normally identifies performance gaps and room for improvements on the company level but often fails to clearly indicate measures and actions to be taken. The designed PI-system for water production processes is linked with the IWA PI-system on company level. So, starting up with IWA PIs in a metric benchmarking the process PI-system directly links to it. Because of this strong hierarchy, performance analysis can be systematically expanded to the level of technical processes (**Figure. 2**).

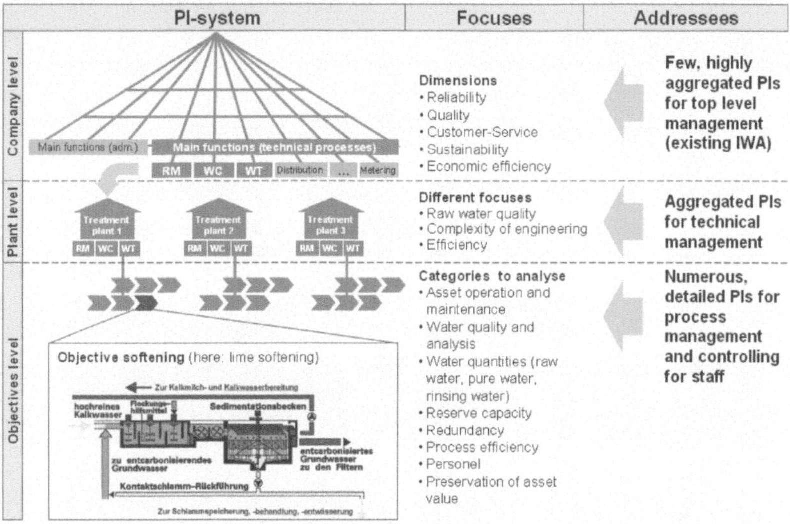

Figure. 2 - HIERARCHY AND CATEGORIES OF THE PROCESS PI-SYSTEM

The adapted PI-system allows a schematic view on different aggregation levels by analysing company level (e.g. data of all treatment plants incl. all personnel, machinery/capacities, raw materials and supplies, buildings, running and capital costs), plant level (one treatment plant incl. catchment, fitting into the company level data set) and the objective level (with a specific objective data set, covering process costs, quality/quantity/dimensioning and sustainability aspects). Likewise, cost and quality of e.g. maintenance processes of water production can be considered (**Table 1**).

Table 1 - STRUCTURE OF KEY PIS AND CONTEXT INFORMATION FOR THE OBJECTIVE FE/MN REMOVAL (EXTRACT)

Code	Name	Unit	Description
Fe/Mn removal: Operation (examples, extract of 12 PIs)			
WA0445	Energy consumption	kWh/m³	Energy consumption / quantity of treated water*)
WA0450	Rinsing Water	‰	Rinsing water use / quantity of treated water*) · 1000
WA0460	Capacity utilisation	%	Maximum quantity of treated water*) per day / rated value of Fe/Mn removal facilities per day · 100
WA0470	Degree of Fe elimination and Degree of Mn elimination	%	Degree of Fe/Mn removal due to German Technical Rules of the German Gas and Water Association (DVGW)
WA0465	Malfunction	%	Number of hours of unscheduled malfunction / Number of hours of planned operating time · 100
WA0441	Personnel for emergency maintenance	P / million m³	Number of staff for emergency maintenance / quantity of treated water*)
Fe/Mn-Removal: Costs (examples, extract of 11 PIs)			
WA0405	Operating costs	€ / t	Operating costs / mass of removed Fe/Mn
WA0417	Costs of sludge disposal	€ / t	Cost of sludge disposal / mass of removed Fe/Mn
WA0430	Maintenance costs	€/1000m³	Maintenance costs / quantity of treated water*)
Fe/Mn-Removal: context information (examples, extract of 21 context information)			
-	Several raw water parameter	-	ph-value, concentration of Fe (average, maximum), Mn (average, maximum), methane, H2S, NH4, DOC
-	Type of process engineering	-	Single-tier/two-tier/other; oxygenation, ozone, chemical dosing, precipitation/flocculation/ sedimentation, filtration, sludge treatment
-	Operation and automation	-	Redundancy, automation, remote operation monitoring
Remark:	*) quantity of treated water means the quantity of treated water in the objective, here: Fe/Mn removal.		

From 2005 to 2008, 12 German water supply companies contributed to the design, practical development and refinement of the PI system, covering 24 treatment plants in different sizes and production conditions with a total of 165 objectives. Therefore, the practical relevance of the key PIs and the feasibility of data collection was being validated in two data collection periods until May 2008.

3 COMPARING A VARIETY OF PRODUCTION SCHEMES

3.1 New explanatory factors

For an accurate comparison of different environmental conditions and water production chains, a "Resource Quality Classification (RQC)" has been developed.

The RQC considers three categories: 1) current resource quality: includes inorganic and organic parameters and current raw water quality impairments, 2) the hazardous potential is assessed in trends of relevant parameters (e.g. nitrate, pesticides, biocides), influences of land utilisation as well as geogenic hazards are important and 3) degree of mitigation strategies and monitoring addresses compensation payments, pre-well monitoring and restricted raw water use. Thus, the RQC contains important, normalised information on the status quo, trends and mitigation measures which can be selected as cluster criteria for RM process performance analysis and as explanatory factors for performance analysis in WC and WT.

The RQC is calculated as weighted index within the interval [0; 9]. So, on the one hand it can be used as explanatory factor for interpretation of PI results and on the other hand RQC shows a resilient cluster criterion.

Likewise a "Quality Safety Index (QSI)" for the produced drinking water has been developed to consider the accordance with the existing standards (which is the case for almost 100 % of the utilities) and to measure additional efforts to aim at higher qualities. The QSI is defined as:

$$QSI = 1 + \frac{c_{\text{limit value}} - c_{\text{measured value}}}{c_{\text{limit value}} - c_{\text{recommended value}}} \qquad (3.1)$$

Legend:

$c_{\text{limit value}}$	= limiting value by German drinking water directive (national quality standard)
$c_{\text{measured value}}$	= average value measured after treatment
$c_{\text{recommended value}}$	= recommended value from technical guidelines (lower than limit value)

High rates are indicating higher water quality after treatment that exceed limit and recommended values. If the QSI is 1 or near 1, the water safety meets ≥100% the quality standards for drinking water concerning the German drinking water directive.

The QSI is interesting in the case for parameters like iron on the one hand, which is irrelevant from the health perspective but lower concentrations protect the distribution system from particles and deposits. On the other hand, many organic trace compounds are extensively removed by expensive treatment due to health-protecting reasons. The additional efforts involved in resource protection and treatment are also clearly identified in the QSI, and help to clarify performance gaps between several production plants.

3.2 Using capable cluster criteria

To form useful comparison groups for process benchmarking, four main criteria have been identified for cluster definition in direct comparisons:

- Raw water type (natural groundwater, enriched groundwater, source water, surface water),
- RQC (applicable for different types of ground water),
- Capacity of the plant/the objective (design per day, output per year),
- Capacity utilisation and type of production plant (base, peak load)

Important steps in the interpretation of benchmarking performance gaps are to formulate hypotheses about potential causes that explain the divergence between the individual suppliers' PI result and the "best practice" value. The process model and PI-system as a toolbox provides these different analysis levels (company, plant and/or objectives level) with all main explanatory factors for the identification of measures for increasing the performance in water production (**Figure. 2**). Keeping the explanatory factors in mind potential deviations from the benchmark can regularly be identified in each cluster group.

For a successful use of cluster criteria, it is strongly recommended to define cluster criteria with significant ranges that produce suitable cluster groups. If the cluster criterion will be defined too narrow, the clustering results will end into too small comparison groups. If the cluster criterion will be defined too wide, the clustering results can deliver implausible correlations between PI results. Even sometimes, clustering is not compulsory necessary. Some experience in benchmarking procedures is needed when cluster limits must accurately be set depending on the compared water supply companies.

Cluster criteria concerning the capacity mostly cannot be used alone. By selecting the cluster criterion "daily design capacity" the focus is on the maximum capacity a treatment plant is designed for. But if the operating capacity of a plant varies strongly over time this cluster criterion is not significant because base load treatment plants are compared with peak load plants. By using the alternative cluster criterion "output per year" the actual quantity of treated water is focused independently of the plant's design.

Depending on the type of plants, clustering with one or the other criteria helps to understand efficiency gaps and to identify optimisation measures.

The criterion "raw water type" is like the RQC very suitable for building cluster groups. Normally it defines the type of plants and processes for water catchment and treatment. Cluster groups can be targeted even more precisely if additionally raw water parameters are considered, e.g. Fe-content, O_2-saturation or turbidity (esp. for surface water).

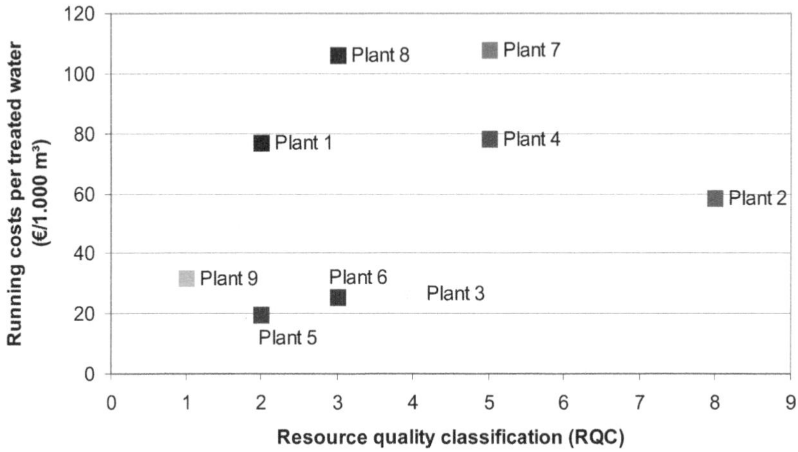

Figure. 3: CORRELATION PLOT: RUNNING COST PER TREATED WATER AND RQC

Figure. 3 shows the correlation between the RQC and the running cost for water treatment. The diagram indicates the expected general tendency: The higher the RQC the higher are also the treatment running cost. There are three variations from this trend. The treatment plants 7 and 8 (at the top of the diagram) cause both comparatively high running costs. This can be explained in each case by a very complex treatment main function which contains a huge number of objectives. Treatment plant 1 has a relatively low RQC but above average costs, resulting from higher maintenance costs.

3.3 Data evaluation and process benchmarking

The process model has been tested in two subsequent data collection periods and has been revised based on the experience of the first trial. Data collection and PI calculation was performed via an Internet-based platform, allowing online quality management, data comparison and individual PI analysis by each participant during the project term. Deviations from benchmarks were discussed in benchmarking workshops, following the scheme of gap analysis.

To start the benchmarking discussions, the PI-results are painted in scatter plots by using the cluster criteria; mostly the result is clearly coherent by the major explanatory factor – if not it needs to be explained. Some further findings resulted from benchmarking discussions:

- Comparing treatment efficiency indicators the role of local operating conditions often strongly influences the process performance and process costs on the objectives level. Sometimes, these external conditions cannot be influenced by the water supplier.
- The knowledge about the specific energy consumption per treatment step and personnel capacity per main function and per objective are important information to evaluate realistic improvement gaps on treatment processes.
- The participants achieved transparency about their own costs (running costs, capital costs) assigned to a particular technical process embedded in the technical structure of the operating system.
- The knowledge about the ratio of costs and types of costs per objective in comparison to other treatment plants is important to start cause analysis and to define necessary measures for improving economic efficiency in treatment processes.
- Comparing different asset management strategies helps to keep the infrastructure on the same level by supporting the decision making for maintenance versus investment measures.

Some final examples with anonymous data are derived from the final report in early 2008.

4 CONCLUSIONS

The application of the methodology has provided important insights for the suppliers in terms of best practice of e.g. operation and maintenance, task fulfilment and efficient treatment processes. Steps for further development are i) adaptation for small water supply companies, ii) enlargement of the data base. The comprehensive system allows also for selective process benchmarking, focussing on specific issues such as comparison of softening treatment plants, cost-effectiveness of resource protection measures or optimised maintenance strategies in water production.

5 REFERENCES

Alegre H., Hirner W., Baptista J.M., Parena R., Cubillo F., Cabrera E., Merkel W., Duarte P. (2006). Performance indicators for water supply services. IWA manual 2nd edition. IWA publishing.

Dane, P. (2007). Benchmarking water supply, waste water management and water control: conference summary paper. Water Practice and Technology 22 (2007), IWA publishing.

Foley A. (2005). Benchmarking asset management. Water Asset Management International 1.1 (2005). IWA Publishing.

Harris P, Saker T, Vincent D (2006). Benchmarking mechanical and electrical maintenance: Key perspectives and emerging challenges. Water Asset Management International 3.2 (2007). IWA publishing.

Merkel, W. und Wichmann, K., (2006). Optimierungspotenziale in der Wasserwirtschaft, Wassergewinnung und Wasseraufbereitung, GWF Wasser Abwasser 147 (2006) Nr. 7-8, 522-527.

Benchmarking the Processes of Customer Meter Reading and Customer Meter Replacement

J. Kölbl[1], H. Theuretzbacher-Fritz[1], R. Neunteufel[2], E. Mayr[2] and R. Perfler[2]

[1] *Graz University of Technology, Institute of Urban Water Management and Water Landscape Engineering. koelbl@sww.tugraz.at, Stremayrgasse 10/I, A-8010 Graz, Austria, Tlf: +43 316 873-8885 Fax: +43 316 873-8376*
[2] *University of Natural Resources and Applied Life Sciences Vienna, Institute of Sanitary Engineering and Water Pollution Control. roman.neunteufel@boku.ac.at, Muthgasse 18, A-1190 Vienna, Austria, Tlf: +43 1 36006-5805 Fax: +43 1 3689949*

Keywords: Austria, customer meter reading, customer meter replacement, process benchmarking

Edited by Enrique Cabrera Jr and Miguel Ángel Pardo. ISBN: 9781843391913.
Published by IWA Publishing, London, UK.

ABSTRACT

After two project runs of metric benchmarking OVGW (the Austrian Association for Gas and Water) started an initiative on process benchmarking in the Austrian water supply sector. In contrast to the IWA approach on process benchmarking the Austrian system is not holistic but analyses various selected processes out of the whole field of activities of a water supply utility. The procedure of benchmarking the different processes consists of four main parts: calculation of process performance indicators (PIs), internal analyses of results, exchange of experience e.g. in form of workshops, realisation of measures. This paper describes the OVGW approach and experiences on benchmarking the processes of customer meter reading and customer meter replacement.

1 INTRODUCTION

Over the last 5 years OVGW implemented a voluntary and anonymous metric benchmarking, which is based on the IWA PI system (Alegre et al., 2000, 2006), in the Austrian water supply sector. More than 70 water supply utilities representing about 50 % of supplied water in Austria participated in the 2005 project (Theuretzbacher-Fritz et al., 2006). One result of this project was the cognition that metric benchmarking is a useful tool to make optimisation potentials evident but the derivation of specific measures on basis of the results of metric benchmarking is often difficult. This was the motivation to start a process benchmarking initiative beginning of 2007 (Figure 1).

2 THE AUSTRIAN PROCESS BENCHMARKING APPROACH

Larsson et al. (2002) describe a holistic approach of process benchmarking like it is practised e.g. in the Netherlands or Scandinavia. The idea of that system is to analyse all the duties of a water supply utility, beginning with the extraction and ending at the sales to customers. The so called processes are highly aggregated sequences of single tasks dealing with e.g. the process of water treatment. However this holistic IWA approach results only in a very coarse division. Another approach can be seen in the selective strategies of process benchmarking which are used in various countries e.g. Australia (Piccinin, 2006) or Bavaria in Germany (Kiesl and Schielein, 2005). For the Austrian requirements the selective strategy seems to be more practicable as the Austrian process benchmarking shall be seen as a complement to the metric benchmarking activities.

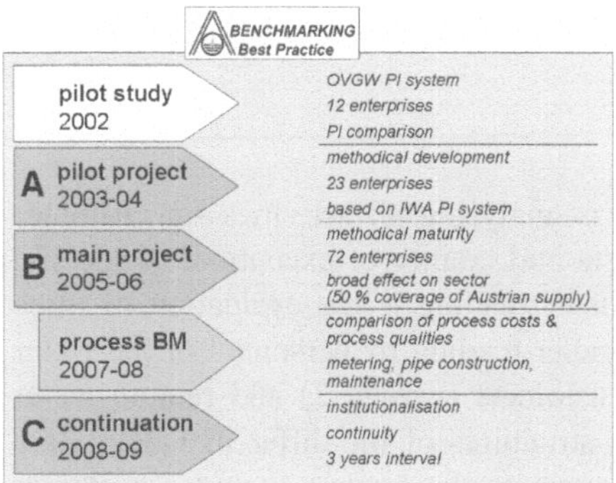

Figure 1 - OVGW BENCHMARKING STRATEGY (THEURETZBACHER-FRITZ ET AL., 2005 AMENDED)

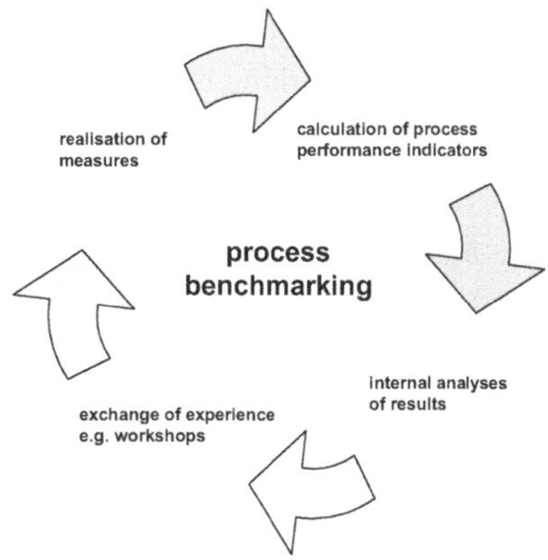

Figure 2 - PROCEDURE OF PROCESS BENCHMARKING

Within the 2007 activities, eight single processes are analysed (Table 1). They are assigned to three subjects (sales process, construction of pipes, mains network operation and maintenance).

Table 1 - PROCESSES OF OVGW PROCESS BENCHMARKING 2008

Sales Process	Customer meter reading
	Customer meter replacement
Construction of pipes	Construction of new mains
	Construction of new service connections
	Rehabilitation of mains
	Rehabilitation of service connections
Mains network operation and maintenance	Water Loss Management
	Network inspection

A precondition for a high-quality process benchmarking is a diligent process structure. Hence, the input and output factors need to be specified for each process and its sub processes to clearly confine the whole process as well as the sub processes. In order to benchmark both technical and economical aspects, quality criteria for each process step have to be determined. A lot of background information about the individual conditions (e.g. structure of water supply system) and about the differences in the operations are necessary (Kölbl et al., submitted). A central point in process benchmarking is the exchange of experiences between the participants. Therefore a one-day workshop on cause analysis and deriving of measures is held for each process. The procedure of process benchmarking is shown in Figure 2.

2.1 Customer meter reading

The aim of benchmarking this process is analysing all activities for customer meter reading and billing regarding costs and quality of execution. A specific challenge in benchmarking customer meter reading is the evaluation of three possible variants in process operation: meter reading by personnel of the water supply utility (variant 1), reading by customers (variant 2) and remote meter reading (RMR, variant 3). The process structures of the different variants are shown in Figure 3. Depending on the variants the process includes customer information about the meter reading (fixing a date), the meter reading itself (by personnel of the water undertaking or outsourced personnel, by customer or with remote meter reading) but also additional meter readings and processing of meter readings by customers (postcards, online data management). Furthermore, accounting and debtor management are analysed. The aim is the comparison of different meter reading and accounting methods in order to find an optimal (efficient) approach for each company (depending on the frame conditions).

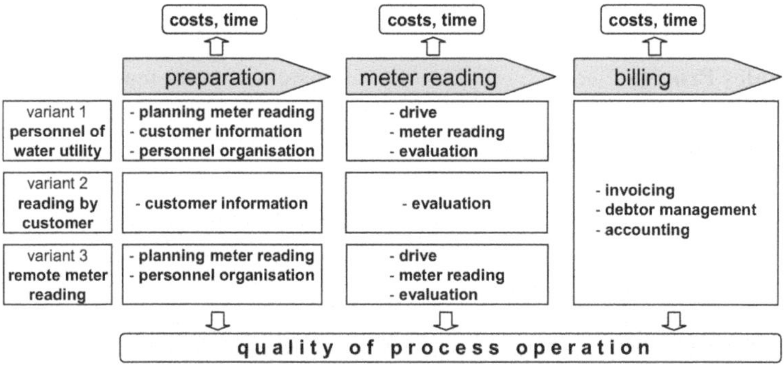

Figure 3 - PROCESS STRUCTURE OF CUSTOMER METER READING

To evaluate the process quality a quality index was developed. This index considers qualification of staff, inspection of service connections (e.g. visual check for false connections), leak detection but also aspects of customer service (e.g. customer information, number of invoices per year) and aspects of data quality (plausibility check) and quality management (work instructions).

Results of benchmarking the process of customer meter reading

Beside evaluations of each variant and to each sub process also the overall performance is analysed. Therefore the costs per meter as well as the effort in working time per meter are calculated.

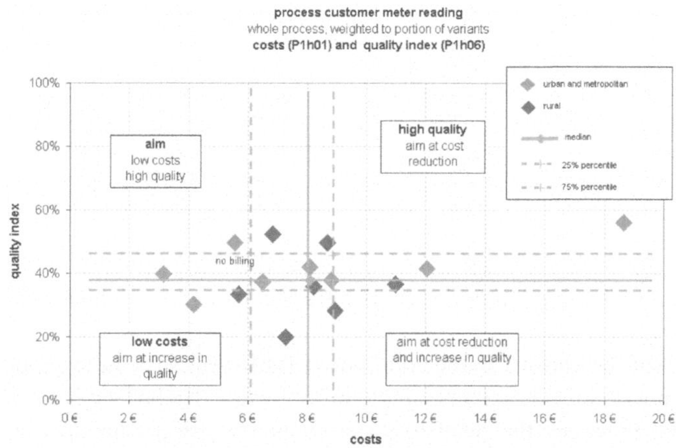

Figure 4 - COST-QUALITY-MAP OF PROCESS OF CUSTOMER METER READING

Figure 4 shows the cost-quality map for the whole process. If more than one variant is practised the weighted costs for each variant are considered. Within the 15 participants the range of costs is very wide and reaches from about 3 € per meter reading (incl. billing) up to more than 18 € per meter reading at a utility with a special situation (actually implementing a new variant and therefore higher efforts). The median value is about 8 € per meter reading. The utility with the lowest costs is in the lucky situation that the reply-quota for meter readings by customers is very high and therefore the effort for additional meter readings is very low.

The average time needed for operating the whole process is between 12 and 17 minutes (Figure 5). To find saving potentials it is often better to analyse the working time instead of the costs because the costs also reflect variations in personnel cost between different utilities which can not be changed in short time. Therefore measures for improving the efficiency have to be focused on a reduction of working time.

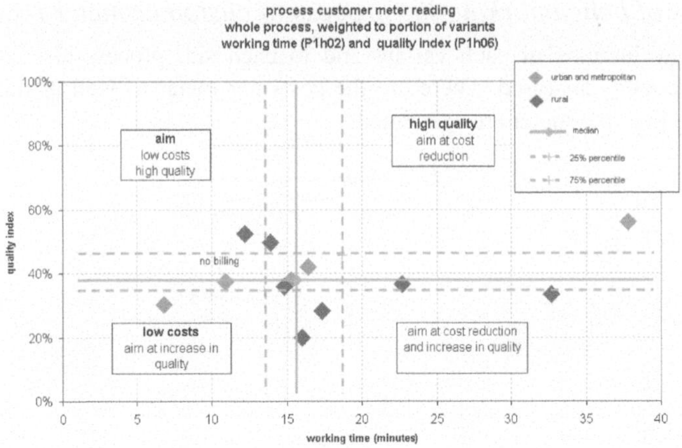

Figure 5 - **TIME-QUALITY-MAP OF PROCESS OF CUSTOMER METER READING**

In fact of the various frame conditions of different structured supply systems it is not possible to define one single best practice. Rather we have to speak of some best practices that lead to an efficient and successful operation of this process. It can be shown that also a mix of two or three of the possible variants can be advantageous. Nevertheless with experienced staff the use of only one variant can result in high efficiency as well. Following statements are an extract of recommendable procedures which were worked out in the workshop on cause analysis by the participating utilities. Of course the individual situations of each utility have to be considered before setting measures.

Variant 1: Meter reading by personnel of water supply utility

- Each reader should read the same meters each year (this will increase the efficiency because of the knowledge where the meter is situated, when the customers are at home etc.).
- It is important that the customer meter data base contains a detailed description where the meter is situated.
- The reader should know the last year consumption of each customer to be able to carry out plausibility checks on the scene.
- It is recommendable to combine the meter reading with a leak detection of the service connection (e.g. with sounding stick).
- Well qualified and experienced personnel are an important factor for the efficiency and the quality of the process operation (this is also valid for the other variants).

- The use of PDAs (Personal Digital Assistant) instead of a written documentation is recommendable for utilities with more than 1000 service connections. A mixed use of PDAs and written documentation decreases the efficiency significantly.

Variant 2: Meter reading by customer

- The customers not replying with the meter reading value are often the same each year. In this case the efficiency can be increased with a selective mix of variants. "Reliable" customers are requested to read their meters on their own and others are read by personnel of the water supply utility.
- Reply cards should be stored for several years (as evidence).

Variant 3: Remote Meter Reading (RMR)

- As a result of the technological developments of the last years the recent types of remote meters should be used.
- As in variant 1 also the readers of the remote meters should read the same meters each year (knowledge of local characteristics).
- An automatic plausibility check (comparison with last year consumption) should be integrated as a leakage-warning.
- The great advantage of RMR is the possibility of a fixed date of meter readings. This enables better water balancing and thus more reliable water loss calculations.

Sub process billing (same for all three variants)

- An invoicing together with other branches e.g. waste water, gas or electricity (e.g. possible in multi-utility companies) saves costs.
- Another cost saving potential is to force debit orders because of the reduced effort in accounting in comparison to payment with paying-in form or standing order.

2.2 Customer meter replacement

To benchmark this process, all activities for routine meter replacements (in accordance to Austrian law every 5 years) are analysed.

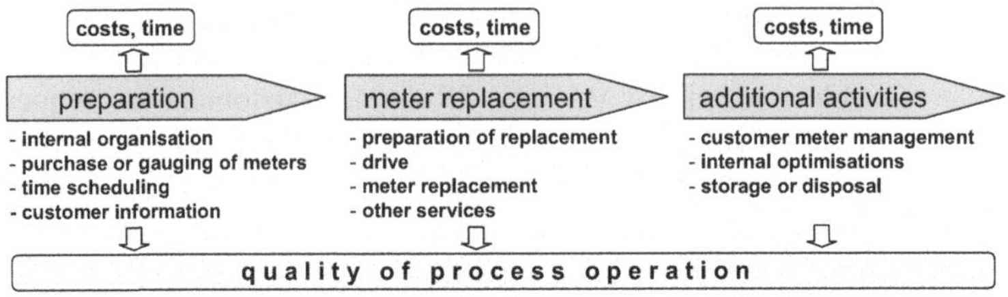

Figure 6 - PROCESS STRUCTURE OF CUSTOMER METER REPLACEMENT

The first step is customer information and fixing a date (postcard or telephone call). Changing meters can be done by personnel of the water utility or by external companies. But also the management of customer meters and storage or disposals of used meters are part of this process (see Figure 6). To avoid a comparison of apples with oranges three different variants have to be considered: replacement of conventional meters, replacement of remote meters and installation of new remote meters.

Results of benchmarking the process of customer meter replacement

As in the process of customer meter reading also for this process the costs per meter as well as the effort in working time per meter are calculated for each variant and each sub process.

Figure 7 shows the cost-quality map for the whole process. If more than one variant is practised the weighted costs for each variant are considered. The median value for average costs of meter replacement in the field of 16 participants is about 50 € per meter (including meter purchase). The results show that a high quality process operation is possible at this cost level. Another finding of the analysed sample is that outsourcing of (at least) parts of the process does not reduce the average costs per meter.

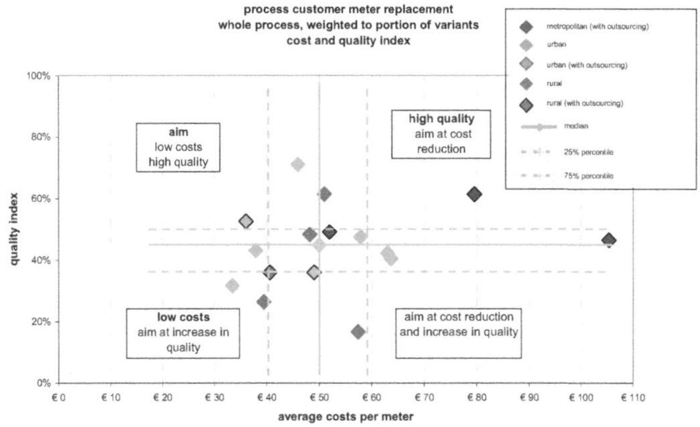

Figure 7. COST-QUALITY-MAP OF PROCESS OF CUSTOMER METER REPLACEMENT

The median average working time per meter is about 1 hour per meter, whereby most of the values are between 45 and 75 minutes.

Similar to the process of meter reading the definition of one single best practice is not possible because of the various frame conditions. The following statements are an extract of recommendable procedures which were worked out in the workshop on cause analysis by the participating utilities.

Sub process preparation

- A good route planning saves time and money.
- Harmonising the replacement interval for whole zones (e.g. replacements of all meters of a DMA in the same year) enables optimal time scheduling.
- In general customer information is not necessary but selective customer information for sensitive customers is recommendable.

Sub process meter replacement

- Target numbers of meters to replace per person and day are useful to increase the efficiency of the staff. Depending on the frame conditions 15 to 20 meters per person and day are realistic.
- Contact to customers is important for all participants. If the meter replacement is done by external companies (outsourcing) the contact to the customers is lost and additional tasks like inspection of service connections are not possible in combination with the meter replacement.
- Experienced personnel are more efficient in process operation (knowledge of local frame conditions and experience in work-flow).

- Leak detection (e.g. sounding stick) at service connections in combination with the meter replacement does not increase the working time significantly, but gives useful additional information.

Sub process additional activities

- Customer meter management is very important for scheduling and optimising the work-flow.

2.3 Methodical results

The experiences and also the feedback of the participants are predominantly positive for the first project run of the two processes. Nevertheless there are some aspects that should be improved for the next project run.

The exchange of experiences in the workshops is a central part in process benchmarking. As a basis for the discussions in the workshops performance indicators are absolutely necessary for the overall process but also for sub processes and different variants (e.g. costs per meter or working time per meter) to find out optimisation potentials and to derive measures for improvement.

One problem is the small number of participants which was 15 and 16 in these processes. Experiences from the metric benchmarking project show, that a higher number of participants allow a more detailed evaluation for different peer groups and a better consideration of frame conditions. In both of the described processes three different variants are considered and therefore setting up peer groups was not always possible.

Another finding is that the motivation to participate is obviously higher for utilities which actually have or anticipate problems in operating these processes. So the “real” best practice utilities probably did not participate.

Another aspect is how to consider the higher investment costs of remote meters in the process of customer meter reading. In principle investment costs are considered in the process of customer meter replacement but some utilities only participate in process of customer meter reading, so there is a need to consider these costs for decision support between different variants. In general the higher investment costs for remote meters are about 5 € to 7 € per meter per year. Remote meters have advantages especially for meters in shafts or meters at inaccessible properties.

3 CONCLUSIONS

Complementary to metric benchmarking OVGW initiated a project on process benchmarking. Within this project selected processes are analysed. First experiences are predominantly positive. A flexible process data system enables the comparison of different variants in process operation for the processes of customer meter reading and customer meter replacement.

On basis of various process PIs and the workshop experiences optimisation potentials can be detected and measures can be set.

Another aspect is the use of the performance comparison as a decision support to find the optimal strategy for each utility (meter reading by own personal or outsourcing, meter reading by customer or remote meter reading). On basis of these process benchmarking data advantages and disadvantages in process quality and also in efficiency of each variant can be assessed.

Many participants recommend a selective strategy for the use of remote meters. Especially for meters in shafts or meters at inaccessible properties remote meters are efficient solutions. The usage of remote meters for whole systems does not yet seem to be efficient in many cases even if there are examples with good experiences. Maybe solutions together with other branches (e.g. gas, electricity) can increase the efficiency of remote meters in the future.

The aim for following project runs is to verify the current findings by an increasing number of participants allowing for an improved evaluation of different peer groups and a better consideration of frame conditions.

3 REFERENCES

Alegre H., Baptista J.M., Cabrera E., Cubillo F., Duarte P., Hirner W., Merkel W. and Parena R. (2006). Performance Indicators for Water Supply Services. - Second Edition, Manual of Best Practice, IWA Publishing, London, UK. ISBN 1843390515.

Alegre H., Hirner W., Baptista J.M. and Parena R. (2000). Performance Indicators for Water Supply Services. - Manual of Best Practice, IWA Publishing, London, UK. ISBN: 1900222272.

Kiesl H. and Schielein J. (2005). Effizienz- und Qualitätsuntersuchung der kommunalen Wasserversorgung in Bayern (EffWB) 2004. – Final Public Report on Bavarian project 2004/05). Nuremberg, Germany.

Kölbl J., Liemberger R., Theuretzbacher-Fritz H., Neunteufel R., Mayr E. and Perfler R. (submitted). Mapping the process of water loss management. – Paper submitted for IWA World Water Congress 2008, Vienna, Austria.

Larsson M., Parena R., Smeets E. and Troquet I. (2002). Process benchmarking in the water industry. - IWA-Publishing, London, ISBN 1-84339-010-8.

Piccinin C. (2006). From metric to process benchmarking. – Presentation at IWA Conference Benchmarking water supply, wastewater management and water control, Amsterdam, The Netherlands, July 2006. http://www.iwahq.org (accessed 10 September 2007)

Theuretzbacher-Fritz H., Neunteufel R., Kölbl J., Perfler R., Unterwainig M. and Krendelsberger R. (2006). Benchmarking und Best Practices in der österreichischen Wasserversorgung – Stufe B - Final Public Report on Austrian project 2005/06, Graz / Vienna / Wr. Neustadt, Austria.

Theuretzbacher-Fritz H., Schielein J., Kiesl H., Kölbl J., Neunteufel R. and Perfler R. (2005). Trans-National Water Supply Benchmarking: The Cross-Border Co-operation of The Bavarian EffWB Project and The Austrian OVGW Project. - Water Science & Technology, Water Supply **6** (2005), p. 273-280, London, UK.

Benchmarking Laboratory Processes for Drinking-Water Quality Monitoring

D. Klein[1], D. Petersohn[2] and B. Paetsch[3]

[1]Hessenwasser GmbH. detlef.klein@hessenwasser.de, Taunusstr. 100, D-64521 Gross-Gerau, Tlf +49 (0) 69 25490 2102 Fax: +49 (0) 69 25490 4109
[2]Berliner Wasserbetriebe. dietmar.petersohn@bwb.de, Neue Jüdenstrasse 1, D-10179 Berlin, Tlf +49 (0) 30 8644 3288 Fax: +49 (0) 30 8644 3339
[3]Westf. Wasser- u. Umweltanalytik GmbH. barbara.paetsch@wwu-labor.de, Willy-Brandt-Allee 26, D-45891 Gelsenkirchen, Tlf +49 (0) 209 708 373 Fax: +49 (0) 209 708 666

Keywords: drinking-water supply, laboratory quality control, process benchmarking

Edited by Enrique Cabrera Jr and Miguel Ángel Pardo. ISBN: 9781843391913.
Published by IWA Publishing, London, UK.

ABSTRACT

Six major German water laboratories have developed a process benchmarking system specific to the quality monitoring tasks of drinking-water supply companies. Based on an analysis of laboratory characteristics, relevant tasks, and processes have been defined. Relevant data can be classified into data on service output, laboratory infrastructure, personnel, finance and structure. Improving comparability of laboratory services with respect to the indicators provided by existing benchmarking systems was a major concern. Project partners set up standardised service specifications, distinguishing between tests that are required by applicable standards / regulatory legislation, or other objectives. Compilation of financial data had to be adjusted, because cost allocation varies significantly when the various individual benchmarking objects such as laboratory space, vehicle fleet, IT-services and overhead-costs are considered.

After testing a range of performance indicators for water-quality monitoring, a total of 46 indicators have been retained covering five dimensions. Network delivery rate and the scale of abstraction and treatment plants have been identified as main explanatory factors for specific service output variations. The proportion of organic testing also accounts for significant differences. Due to factor cost variation most financial indicators appear not to be useful for evaluation of process efficiency.

Participating laboratories involved with monitoring drinking-water supplies obtain valuable insights for optimised capacity utilisation in different aspects. The regional and strategic setting of water undertakings and their associated infrastructures have an impact on the highly integrated performance indicators such as water-quality monitoring costs per cubic metre accounting for variations which are more than ten times higher than variations in process efficiency.

1 INTRODUCTION

Water suppliers depend greatly on the support of competent, effective and reliable laboratories to monitor the quality of their drinking water during its production and processing. In order to meet legal requirements, it is not enough for them to simply guarantee that the design, construction and running of their waterworks are done according to accepted technical standards. On the contrary, the processes surrounding the production and distribution of drinking water are only complete when it has been analytically proven that the drinking water conforms to the necessary, high standards of quality. This means that laboratories perform a critical end function within the process chain.

Over the last few years, many drinking water suppliers in Germany have participated in benchmarking projects as part of a national (and European) modernisation strategy. Wide-spread benchmarking is considered as the most appropriate way to further develop the German drinking water sector as it is mainly composed of municipal undertakings or mixed-ownership undertakings of which the majority of shareholders and operators are public. Useful background information regarding benchmarking the performance of water undertakings exists in Germany (DVGW 2005) and internationally (Alegre *et al.* 2006). However, the practical application of these guidelines is limited when it comes down to benchmarking the typical services of laboratories that test the supply of public drinking water.

Therefore, six laboratories belonging to or working on behalf of drinking water suppliers (associated lab facilities), voluntarily involved in a pilot benchmarking project (Huschens *et al.* 2007). The project implementation included a code of conduct allowing for non-anonymous handling of data and results. The assessment period used for pilot-testing was calendar-year 2005.

2. WATER QUALITY CONTROL (WQC)

Internal and integrated quality control of products is very common within the market-based food and drinks industry. Water suppliers provide one of the most important food products; therefore the industrial approach of internal WQC already represents an initial benchmark. WQC involves much more than carrying out and presenting the results of analyses. Successful WQC requires the following basic conditions:

- The direct and active involvement of laboratories and the regular participation of their managerial staff in company meetings, guarantees the required level of expert knowledge for all areas of WQC as well as ensuring the safe supply of drinking water (analysis of weak points, intervention, crisis guidance).
- Sampling plans/ schedules and the monitoring of input, general operations and the finally delivered product all need to be properly coordinated. It should be possible to adjust operations on short notice, if required.
- Knowledge of the various localities and systems in use as well as unhindered access simplify the planning and implementation of quality control procedures. This is also true for determining the cause of system failures and/ or drops in water quality.

•Any immediate measures that turn out to be necessary as a result of analysis and observation can be discussed directly with the responsible operators and the effectiveness of any countermeasures can be monitored and documented.

•Reporting and archiving of results are not only carried out according to legal and regulatory requirements, but are also made permanently available for extensive, comparative evaluation.

•The direct availability of experts from lab facilities within company communication and customer service departments helps to gain and maintain the customer's confidence, thereby adding extra growth potential to water supply companies.

3 SETTING UP THE PROCESS MODEL

Any valid process model must include the thorough benchmarking of the entire process chain, from the design phase right through to data analysis as well as customer service and archiving of data etc. 'Methodical development techniques' were used as a basis for designing the surveying processes. The services offered by laboratories are not limited to testing drinking water according to the legal requirements. They are in fact a great deal more extensive, see Fig.1.

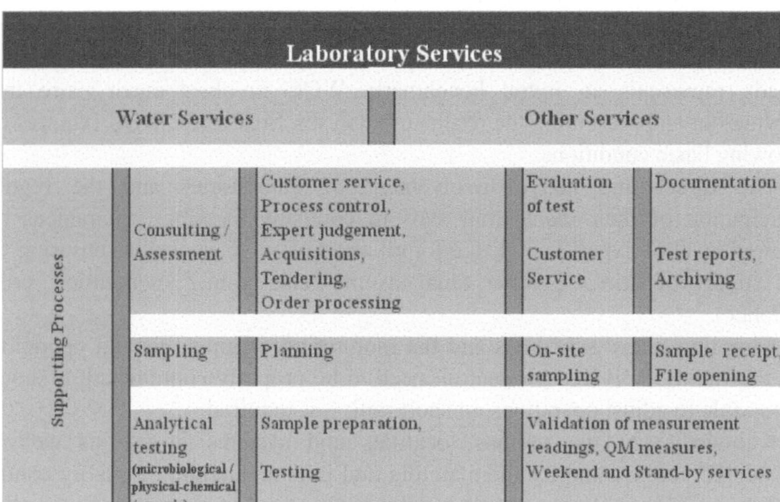

Figure 1 - **OVERVIEW OF LABORATORY SERVICES**

The main process (MP) is divided into two sub-processes (SP); Water Services and Other Services (research, process engineering, online monitoring, etc.). The benchmarking model is not limited to the analytical services that are provided, but includes all relevant activities (planning and carrying out sampling, evaluation of results, presenting reports). Both the sub-processes can be further divided into subroutines:

•Consultation/ Assessment includes activities such as customer service, coordination of project goals and any possible outsourcing of testing (not considered in this study) as well as the acquisition of internal and external customers, providing quotes/ cost planning, i.e. using LIMS (Laboratory Information Management System) and/ or SAP, order processing (again with the support of LIMS) and sampling information.

•Analysis is divided according to its different specialist areas.

•Sampling covers planning, on-site sampling including documentation, sample receipt (reception control, if necessary pre-treatment of samples) and file opening (LIMS, container labelling etc.).

Supporting processes include tasks, which indirectly affect the three subroutines (consulting/ evaluation, analysis, sampling) in various ways, i.e. personnel management, quality management, secretary/ office tasks, accounting, equipment servicing, IT coordination, laboratory controls, vehicle management, cleaning of laboratory glassware and the lab itself.

4 PROCESS INDICATORS

4.1 Definition and selection

To improve comparisons between the various laboratory services, a uniform specifications sheet was created. Standardised unit prices were determined using averaging. A new unit based on the Euro (L-EUR or L-TEUR), was applied to the resulting service output thus distinguishing it from the usual cost and returns items.

To allow for the significance of personnel data (in man-years, MY) to get reliable results, a much more extensive definition was developed for the existing benchmarking systems. Working time regulations, longer periods of downtime and the effects of partial retirements all need to be individually analysed in order to accurately calculate the man-years.

The entire laboratory expenses were then analysed according to the different cost types to obtain a set of reliable financial data. Cost assignment policy turned out to vary significantly considering such items like premises, car pool, IT, social insurance costs, management, funding. To help with this, the so-called 'comparative total expenses' were worked out as an intersection quantity. Due to differences in the supply structure, an auxiliary variable 'managed water quantity' was added to the structural data. Five groups of indicators were tested in an iterative process. 75 indicators were pre-selected in total; 46 indicators were retained for evaluation (Table 1).

Table 1: DISTRIBUTION OF INDICATORS ACCORDING TO TYP

Set of indicators	Number of indicators
Supply / Service structure	16
Productivity of (labour)	10
Financial efficiency	12
Laboratory infrastructure	5
Personnel structure	3

Ranges of indicators were determined for evaluation in order to fully accommodate the undertaking's structural conditions whilst assigning well defined benchmarks. The ranges of indicators are calculated using a robust assessment method that supplies accurate results with both right-skewed and normal distribution models. The Q method is used to determine the deviation from the standard of comparison. The average was determined using the Huber-Schätzer statistical approach. Both methods conform to applicable quality assurance standards (AQS 2001).

4.2 Evaluation: Structural indicators

Structural indicators are data sets largely determined by external factors and can generally only be influenced over the medium to long-term. Achieving fundamental structural similarities was one of the goals when deciding on the initial composition of the pilot benchmarking group:

- The function of a central drinking water testing laboratory as part of a water supply company
- Organisation as a legal and/ or administrative, independent unit
- Comparably high drinking water supply volumes of the suppliers involved:
  - Between 80 and 250m. cubic metres/ year
  - In total approx. 23,150 kilometres of drinking water mains and 850m. cubic metres of supplied drinking water (around 18 % of German drinking water supply)

However, large structural discrepancies remain. For Water Services alone, the calculated service output varies between 1.7 and 4.9m. L-EUR. Further differences are represented by various additional services and secondary business. They represent a share ranging from 10 to 50%.

There are fundamental differences in specific service output (L-EUR per cubic metre) which go up to 270 % within the pilot group. These are largely explained by the supply structure of the various companies. Two main structural indicators are shown in Fig. 2.

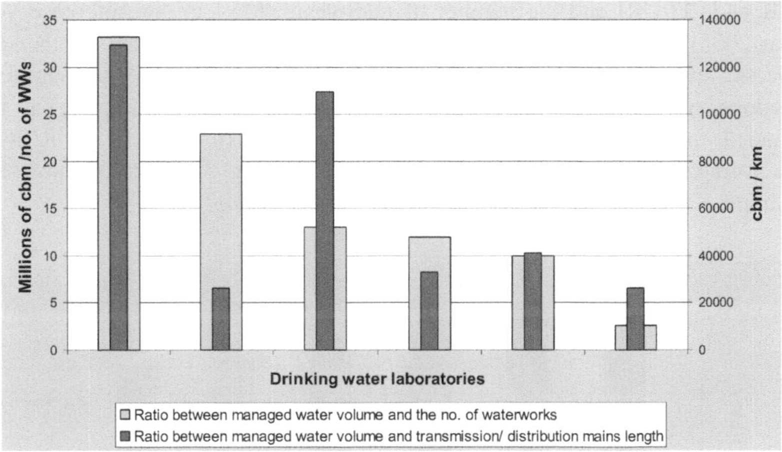

Figure. 2: FOCUS OF SUPPLY: WATERWORKS VS. DISTRIBUTION

The service output has been evaluated according to defined sub-processes and specialist areas. The importance of the sub-processes may vary. For instance, values for Consulting and Assessment lie at 15% for suppliers that employ distribution to households; this figure is only up to 10% for bulk water suppliers. In each case, these engineering services represent a significant share of the overall services. Analytical services make up 60 to 70% of the services. Noticeable however, are the differing weights of the analytical specialist area organic testing, which range from less than 20% to over 50%. The importance of Sampling within the whole remit of services provided varies considerably, ranging anywhere from around 15% to as much as 30%.

The share of legally required testing hovers between 69 and 87% whilst the percentage of testing required by operation fluctuates between 13 and 31%. Looking at the different specialist areas individually, testing required by regulation presents the following shares:

- Microbiology 93 – 98 %
  Inorganic testing 90 – 96 %
- Organic testing 52 – 91 %

The analytical differences can be explained by the range of services within organic analysis. Amongst other things, the main explanatory factors are the local level of environmental impact and the type of water resources.

4.3 Evaluation: Indicators for productivity

Indicators for productivity result from the ratio between the service output (in L-EUR or L-TEUR) and the number of man-years (MY) in one reference year (year 2005). A range of 116 to 144 L-TEUR/ MY was calculated for the sub-process level. If structurally influenced sampling (distance to sampling locations, proportion of relatively time-consuming resource sampling) is not included, then the productivity corridor narrows to 116 - 136 L-TEUR/ MY. The average in each case is 130 L-TEUR/ MY (see Fig. 3). Therefore, potential for improvement appears to be in the range of 10 to 15%.

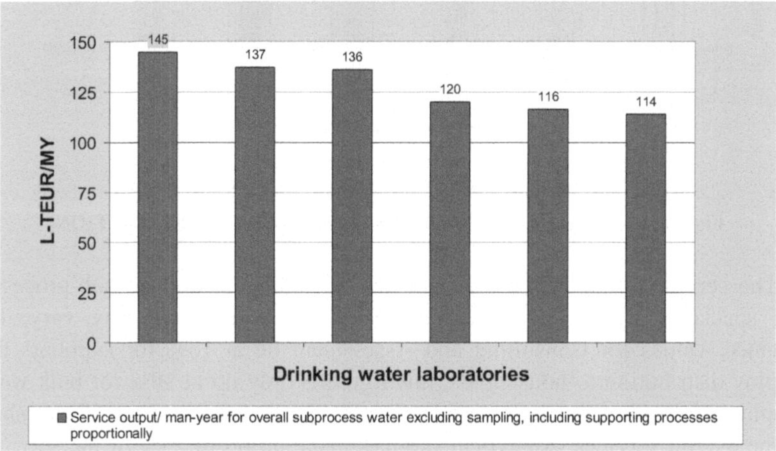

Figure. 3: PRODUCTIVITY (OF LABOR)

Appropriate allocation of personnel within supporting processes is around 1MY to 4 or 5MY of the production processes.

Dividing analysis once more into its various specialist areas results in the following (L-TEUR/ MY, averages in brackets) productivity indicators:

- Microbiology 152 – 212 (182)
- Inorganic testing 155 – 195 (175)
- Organic testing 102 – 180 (141)

4.4 Evaluation: Financial indicators

The personnel make up the largest share of the total expenses. However, for the specific labour costs the differences can be as much as 42%. Between the participants, cost allocation for the use of buildings and space is handled quite differently. In the assignment of cost types, building/ space expenses appear either as depreciation or miscellaneous operational costs. When accounted for specifically (EURO/ square metres), huge differences result from the various accounting practices (rent, amortisation or 'additional expenses').

Personnel expenses and any expenses allocated for building/ space make up around 80% of the total comparable expenditure. Regional differences in income (in particular east-west differences) and varying accounting practices for space impact heavily upon the individual position of participants when comparing total costs. However, as service provision is linked to the supply area and the location of abstraction and treatment plants the level of both these cost factors needs to be individually evaluated, taking account of the regional economical context as well as company policy and its standards for safe drinking water supply.

4.5 Interpretation within the context of various indicator sets

The service output (L-EUR) per cubic metre of managed water volume varies by a factor of 2.7. There is a clear relationship between the services required by applicable standards / legislation and the structural indicators network delivery rate (cubic metre of water per kilometre of mains) on the one hand and the scale of abstraction (measured by the average output of waterworks) on the other. If interaction is positive with the specific cost differences outlined in chapter 4.4, then the differences in specific costs of water quality control per cubic meter supplied may increase to a factor of 3.5.

Benchmarking of lab equipment has shown that outdated equipment can have a negative effect on maintenance expenses and productivity (i.e. equipment failure/ downtime). An appropriate annual investment volume of 4 to 5% has been estimated preliminarily.

The personnel structure was shown to significantly influence productivity. Differences in the number of days absent (with an overall average of 50 days per MY) due to illness or varying holiday entitlements, can cause productivity to fluctuate by as much as 3%.

5 CONCLUSIONS

The results of the benchmarking project have outlined areas of potential process improvements for the various drinking water laboratories within the following areas:

- The degree of efficiency in selected subroutines
- Deployment concepts for logistics and personnel, with special focus on sampling
- Measures to reduce employee absenteeism
- Automation of the process chain
- Profitability of extraordinary testing exceeding routine analysis, in particular in the field of organic testing
- Medium to long-term investment strategy for equipment

In previous benchmarking projects (including this one), extreme differences of over 300% have been shown by using the comparatively highly aggregated indicator 'water quality monitoring costs per cubic metre of water'. This represents the theoretical potential for improvement. As far as the project's participants are concerned, these differences relate mainly to the supply structure and/ or relevant factor costs. Around 10 to 20% can be assigned to internal potential for improvement of laboratories' process efficiency.

6 REFERENCES

Alegre, H., Baptista, J. M., Cabrera Jr., E., Cubillo, F., Duarte, P., Hirner, W., Parena, R. (2006) Performance indicators for water supply services. Second Edition, 290 p., IWA, London.

AQS (2001) Information about Suggested Statistical Methods. AQS 10. Lfg. III/0I, Appendix A-3, January 2001.

DVGW (2005) Water Information No. 68. DWA Themes: Benchmarking Compendium for Water Suppliers and Wastewater Disposal Companies. November 2005.

Huschens, 0., Karrenbrock, F., Klein, D., Krueger, M., Paetsch, B., Petersohn, D. and Volquardson, T. (2007) Benchmarking drinking water laboratories. Final project report, June 2007.

Case Studies

WaBe project - Performance indicators of water utilities in Czech Republic

L. Tuhovcak and T. Kucera

Brno University of Technology, Faculty of Civil Engineering, Institute of Municipal Water Management, Zizkova 17, Brno, Czech Republic, e-mail: tuhovcak.l@fce.vutbr.cz, kucera.t@fce.vutbr.cz

ABSTRACT

The performance benchmarking analysis of water utilities is currently becoming topical also in the Czech Republic. The possibility of basic benchmarking of these utilities based on the IWA methodology is offered by the WaBe project

Edited by Enrique Cabrera Jr and Miguel Ángel Pardo. ISBN: 9781843391913.
Published by IWA Publishing, London, UK.

(Water Benchmarking). The WaBe project is coordinated with the European project COST Action C18: Performance assessment of urban infrastructure services: the case of water supply, wastewater and solid waste.

The objective of the WaBe project is to implement a system of IWA performance indicators and to develop a web application, which will be able to present and compare water utilities from Czech Republic.

The basic 24 performance indicators have been selected in the first stage of the project. The individual water utility can be evaluated on the basis of these indicators. The values of indicators are presented in a graphic and table form. It is also possible to draw tables and charts for individual utility and to monitor the development of selected indicators of this utility in time or to benchmark the selected indicator of each utility over chosen years amongst themselves and to make comparisons with the average, minimum and maximum achieved values of the selected indicator. The other objective of the project is to estimate recommended values for each indicator.

Presently, a scheme of the WaBe tool has been developed for the benchmarking and structure as an open web application. The water utility will be able to perform benchmarking on the basis of entered data there. The part of the WaBe application (www.wabe.cz) will be public; the other part will be accessible only for water utilities. The first results and examples from WaBe application are presented here in the conference.

1 INTRODUCTION

The "Wabe" project is implemented as a research project „Performance assessment of public water supply systems and water utilities" within the national programme COST of the Ministry of Education, Youth and Sports of the CR. The duration of the project is 3 years.

The preparation and the proposal of the aforementioned research project has been motivated by several facts. Water companies are generally a natural monopoly. In the Czech Republic the prices of their services are controlled according to the national legislation. The eligibility of the costs is inspected by the Ministry of Finance. However, this is purely a price limitation. The regulator in the proper sense of the word, i.e. the one who would assess the costs in terms of their effective use and the price of drinking water, has not been established in the Czech Republic.

Many from more than a hundred water companies in the Czech Republic use systems evaluating the performance and efficiency of operation, however, the applied methods differ in various companies. Such an evaluation may be very subjective, above all with respect to mutual comparison between the companies.

The basic objective of the project is therefore to propose a set of basic performance indicators (PI´s) for the analysis, evaluation, assessment and benchmarking of public systems of drinking water supplies and water companies which ensure the operation of water pipe networks. The project includes a uniform methodology for the determination, calculation, and evaluation of each of the proposed indicator. The project outputs include the proposed methodology as well as the Internet application „WaBe“, which will make it possible to administer data, perform and present results of the basic benchmarking analyses.

2 PERFORMANCE INDICATORS

One of the many definitions that characterises the performance indicators reads as follows: "the performance indicators are measures the use of which makes it possible to map development and drawbacks in the company."

2.1 Selected performance indicators

The aim of the project is to introduce a uniform system at the national – Czech level with a coordinated procedure recommended by IWA. When selecting the basic performance indicators for the application as part of the Wabe project we therefore used the system of indicators proposed by IWA (Alegre 2006) as the basic supporting document. At the same time, some indicators applicable in the ČR were introduced. For the Wabe project, the following 24 basic performance indicators were selected:

- Water produced for distribution per employee [$m^3.year^{-1}.empl^{-1}$]
- Water produced for distribution per employee (drinking water section) [$m^3.year^{-1}.empl^{-1}$]
- Water produced for distribution per service pipe [$m^3.year^{-1}.serv^{-1}$]
- Water produced for distribution per km of water network [$m^3.year^{-1}.km^{-1}$]
- Non-revenue water [%]
- Non-revenue water per employee [$m^3.year^{-1}.empl^{-1}$]
- Non-revenue water per service pipe [$m^3.year^{-1}.serv^{-1}$]
- Non-revenue water per km of water network [$m^3.year^{-1}.km^{-1}$]
- Non-revenue water per supplied inhabitant [$l.pers^{-1}.d^{-1}$]
- Unit leakage – non-revenue water per km of network per day [$m^3.km^{-1}.d^{-1}$]
- Unit leakage – water losses [$m^3.km^{-1}.d^{-1}$]

- Specific water consumption produced for distribution per inhabitant [l.pers.$^{-1}$.d^{-1}]
- Specific water consumption invoiced total per inhabitant [l.pers.$^{-1}$.d^{-1}]
- Specific water consumption invoiced to households per inhabitant [l.pers^{-1}.d^{-1}]
- Specific water consumption invoiced to other consumers per inhabitant [l.pers.$^{-1}$.d^{-1}]
- Total storage capacity of water reservoirs per average daily consumption of water produced for distribution [%]
- Total capacity of water reservoirs per inhabitant [m^3.pers^{-1}]
- Total capacity of water reservoirs per km of network [m^3.km^{-1}]
- Density of service pipes [serv.km^{-1}]
- Number of connected inhabitants per employee of the company [pers.empl^{-1}]
- Number of connected inhabitants per employee of the company – drinking water section [pers.empl^{-1}]
- Number of employees working in the field of drinking water [%]
- Number of employees per service pipe [empl.serv^{-1}]
- Number of employees per km of network [empl.km^{-1}]

For the next project stage we plan to introduce other indicators, for example:

- Water network burst rate [no.100km^{-1}.year^{-1}]
- Water network burst rate per service pipe [no.1000serv^{-1}.year^{-1}]
- Number of justified complaints per supplied inhabitants [no.year $^{-1}$]
- Inhabitants supplied with drinking water (coverage) [%]
- Average age of water network [years]
- Number of water supply systems [no.]
- Water losses per service pipe [m^3.serv^{-1}]

Each performance indicator along with the method of its determining and the necessary input data is described in the proposed methodology. To determine the proposed basic 24 indicators a total of 32 input data is necessary.

3 WABE TOOL

As part of this project, the WaBe tool has been since the beginning developed as a web application with various access levels from the Internet for various types of users. At present, a structure and functions of the application are designed, programming will start in the near future. The tool will be able to administer data, calculate the performance indicators and provide a complete presentation of outputs in the table and graphic formats.

3.1 Data structure

The application administers two basic types of data: basic (entered) and calculated. The basic data are understood as general data of the water companies (name, contact, shareholders, ...), the input data for the calculation of PI´s and the national statistics (for monitored indicators).

The calculated data is the PI´s and the statistics of involved parties. Marginal extreme values (min. or max.) or recommended optimum values are proposed for each indicator

3.2 Comparative analysis

The main purpose of developing the WaBe tool is to conduct basic comparative analyses of water companies and their operations. Water companies can perform several types of comparisons. Where desirable, a comparison with the national average will be enabled with the best or recommended value.

As the data is sensitive from the point of view of the water companies, the principle of confidentiality will be followed, i.e. the anonymity of someone else's data will be maintained when comparing concrete companies.

The first type of the data analysis is the „development". Here it is possible to monitor the year-to year development of one or several indicators of one operator. A sample of the output of this type is shown in fig. 2.1.

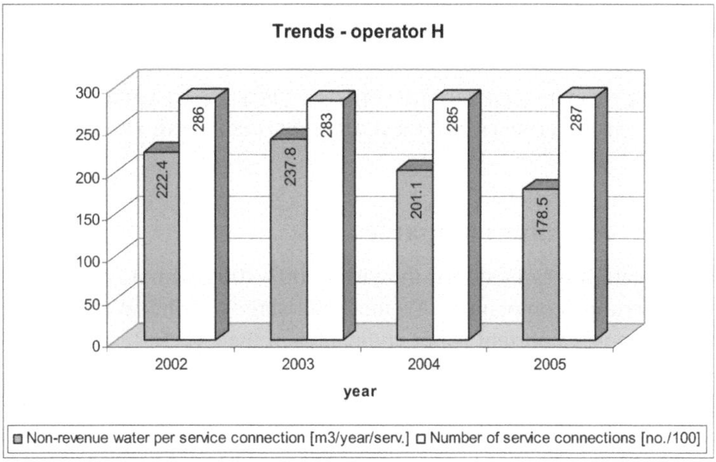

Figure 2.1 EXAMPLE OF THE GRAPHIC OUTPUT: TRENDS IN SELECTED INDICATORS OF OPERATOR H

Another type of available analyses is „comparing". This is a comparison of one up to three indicators for a maximum of five companies in a year. As an option it is possible to show the average achieved in the participating companies (WaBe average) and the average in the whole ČR (CZ average). Fig. 2.2 shows the output of the comparison of the indicators of „unit leakage – water losses (U11)" and the „density of service pipes (U19)" for 4 companies in 2005. The graph also sets the maximum acceptable value for the indicator (U11).

The presented graphs show realistic data of the companies involved in the WaBe project.

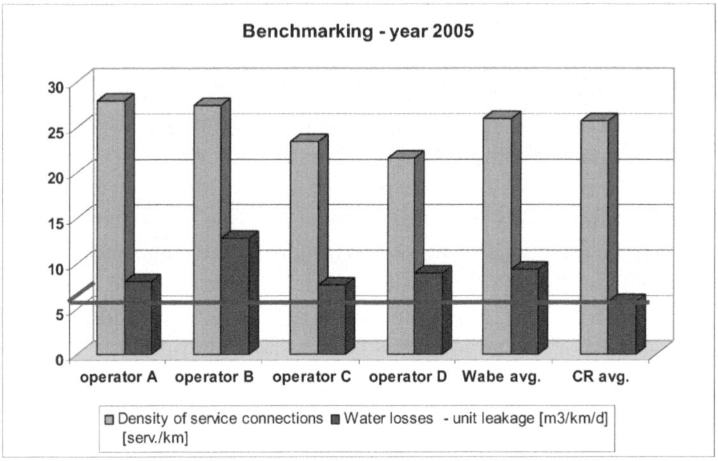

Figure 2.2 EXAMPLE OF THE GRAPHIC OUTPUT: COMPARISON OF THE INDICATORS OF SEVERAL OPERATORS IN ONE YEAR

3.3 Involved water companies

The data collection was made for the years 2002-2005 with a smaller or nigger success at various companies. At the first stage of the Wabe project we approached 9 most important water companies in the Czech Republic, the list of which, along with the service areas is shown in Tab. 2.1. It also presents the share of obtained (availability) and the percentual share of the companies in the water supplies in the Czech Republic.

Table 2.1. OVERVIEW OF INVOLVED WATER COMPANIES

Water company	area	Data availability [%]	Share in water supplies to inhabitants [%]
1. JVS, a.s.	Southern and western Bohemia	98	2,2
Brněnské vodárny a kanalizace, a.s.	Brno and surroundings	47	4,2
Ostravské vodárny a kanalizace a. s.	Ostrava	81	3,4
Severočeské vodovody a kanalizace, a.s.	Regions of Ústí and Liberec	61	11,8
Severomoravské vodovody a kanalizace Ostrava a.s.	Northern Moravia and Silesia	55	7,8
Středomoravská vodárenská, a.s.	Olomouc, central Moravia	94	1,5
Vodárenská akciová společnost, a.s.	South Moravian region and Vysočina	59	5,6
Vodovody a kanalizace Hradec Králové, a.s.	Hradec Králové	51	1,6
Vodovody a kanalizace Jižní Čechy, a.s.	České Budějovice and surroundings	38	3,6

On the basis of the provided data we quantified the aforementioned 24 performance indicators. The average values for the monitored companies in 2002-2005 are presented in tab. 1.2. (two of these indicators are not presented due to the absence of some of the input data).

Table 2.2. AVERAGE VALUES OF DETERMINED INDICATORS

performance indicator	units	year			
		2002	2003	2004	2005
Water produced for distribution per employee	[m^3.year^{-1}.empl.$^{-1}$]	20 362	30 513	26 580	26 611
Water produced for distribution per service pipe	[m^3.year^{-1}.serv.$^{-1}$]	408	436	432	401
Water produced for distribution per km of water network	[m^3.year^{-1}.km^{-1}]	10 938	11 433	10 661	9 940
Non-revenue water	[%]	15,5	17,9	17,2	16,4
Non-revenue water per employee	[m3.year-1.empl-1]	4 494	7 513	6 833	6 462
Non-revenue water per service connection	[m3.year-1.serv-1]	99	103	89	78
Non-revenue water per km of water network	[m3.year-1.km-1]	2 378	2 475	2 205	1 974
Non-revenue water per supplied inhabitant	[l.pers.-1.d-1]	34	39	35	33
Non-revenue water -unit leakage per km of network per day	[m3.km-1.d-1]	6,5	6,8	6,0	5,4
Water losses - unit leakage	[m3.km-1.d-1]	5,5	4,9	4,6	4,1
Specific water consumption produced for distribution per inhabitant	[l.pers.-1.d-1]	166	187	179	173
Specific water consumption invoiced total per inhabitant	[l.pers.-1.d-1]	160	159	154	151
Specific water consumption invoiced to households per inhabitant	[l.pers.-1.d-1]	101	101	88	85
Specific water consumption invoiced to other consumers per inhabitant	[l.pers.-1.d-1]	59	58	47	48

Total storage capacity of water tanks per average daily consumption of water produced for distribution	[%]	36	36	38	40
Total capacity of water storage tanks per inhabitant	[m^3.pers.$^{-1}$]	0,191	0,238	0,319	0,327
Total capacity of water storage tanks per km of network	[m^3.km.$^{-1}$]	35,0	39,3	46,1	45,2
Density of service connections	[serv.km^{-1}]	21	23	26	26
Number of connected inhabitants per employee of the company	[pers.empl^{-1}]	330	451	417	435
Number of employees working in the field of drinking water	[%]	3,5	3,3	6,7	6,9
Number of employees per service pipe	[empl.serv.$^{-1}$]	1,1	1,1	1,1	1,1
Number of employees per km of network	[empl.km^{-1}]	3,4	3,5	3,4	3,1

4 CONCLUSIONS

The WaBe application is at the testing stage now. The first achieved results will be the intervals of the achieved values of the individual performance indicators. The WaBe application will be publicly accessible and will allow for comparing the individual water companies. The question is still the interest of water companies in the participation in this project and its potential extension by other performance indicators.

5 REFERENCES

Alegre, et al. (2006) Performance Indicators for Water Supply Services. IWA Publishing. Second Edition, ISBN: 1843390515.

Tuhovcak L., Chalupnikova P. (2000): *Benchmarking and Evaluation Process of Water Utilities Using Performance Indicators – the Czech Experience*, IWA specialised conference „Water Management for the 21st Century", Berlin, p.19-26

Tuhovcak, L. (2002): *Transformation of the water supply sector in the countries of Central Europe*, NATO Advanced Research Workshop, Borovetz, p.109-117

Water Utility Sector Performance Monitoring and Benchmarking in Slovak Republic

S. Stanko[1] , *K. Tothova* [2] *and J. Kleinertova* [3]

[1] *University of Technology Bratislava Faculty of Civil Engineering, Radlinskeho 11, 813 68 Bratislava, Slovak Republic stefan.stanko@stuba.sk*
[2] *University of Technology Bratislava Faculty of Civil Engineering, Radlinskeho 11, 813 68 Bratislava, Slovak Republic katarina.tothova@stuba.sk*
[3] *LARIVE Slovakia s.r.o. , Heydukova 5, 811 08 Bratislava , Slovak Republic, janka.kleinertova@larive.sk*

Edited by Enrique Cabrera Jr and Miguel Ángel Pardo. ISBN: 9781843391913.
Published by IWA Publishing, London, UK.

Keywords: Performance monitoring, water management, investment needs.

ABSTRACT

The paper describes the existing framework of water supply and sewerages in Slovak Republic. Next emphases the paper suggests to sustainable development of public water supply and sewerage system and its investments necessity. Requirenment is to build new and infrastructure asset and renovation of present state. This is very difficult to compensate general lack of operational and investment financial sources from the last era of central direction now. Cost requirements are higher than owns sources and higher than population possibilities. Strict water price regulation in Slovakia is involved by consumer protection. However, where is the limit of "appropriate" price?

Next paper shows the role of the Slovakia benchmarking initiator which represents the Regulatory Office for Network Industries (URSO). The regulatory office has executed its activities in water sector since 2003 and it has performed pilot benchmarking project. The project was evaluated as metric benchmarking for years 2003-2005. The quantitative evaluation of different companies were make by using the performance indicators, base on elementary technical, technological and financial data. The same project has exposed many discrepancies in common terms and definitions of performance data. The conclusions of realized benchmarking consist of various recommendations for performance monitoring in Slovak water sector.

The scope of the paper is to show the questions of the water supply and waste water system performance monitoring in Slovakia.

1 INTRODUCTION

Considering the huge backlog due to former inadequate maintenance of water supply and wastewater systems as well as the huge investments in this sector that are necessary to fulfil the acquis communautaire, the public authorities are increasingly forced to attract private capital in the form of private-public-partnership. Consequently, in the future, the authorities' responsibility will be limited to defining public objectives (in line with the aquis communautaire) and to monitoring, regulating and providing public funding.

To accurate level of water supply and dry weather flow dewatering safety needs the significant volume of financials. Governmental paper "The Plan of Water Supply and Waste Water Dewatering" doesn't solve the financial investments covering, but only shows the trends of water supply and waste water dewatering.

The solution for water companies (WC) is to use the bank loans, or public investments, and own sources (accumulated sources). The public sources are cofinancing from EU and governmental funds. This type of financial is very problematic in last time. There is real threat that the part of these non-self investments must be refunded to EU. The own sources of WC are limited by controlled benefit. On the other side strict water price regulation in Slovakia involves by Slovak consumer protection, the reason is low purchase power of inhabitants compared to EU countries.

Assurance of development and necessary investments is covered by performance monitoring of water companies, which is created together with every investment application for non-self financing. On the other side, the goal of regulation office is the evaluation of operation and operation costs effective. The regulation office was initiator of water companies benchmarking. The results show strong and soft aspects of performance indicators. The project determined the others ways of benchmarking exploitation concerning water companies benefit.

2 WATER SUPPLY AND WASTE WATER SEWERAGE – PRESENT STATE

2.1 Water supply

Recently, the public water systems supply 85 % of population in Slovakia. (it represents 4.54 million population). From the 2,883 municipalities in Slovakia about 2,170 of them are connected to public water supply systems; in 110 municipalities is water supply system under construction; without the public water supply system are 710 municipalities. The coverage of the public supply by the public water systems differs per region, and moves from 76.2 to 95.5 %. The total length of water pipes does approx. 25,700 km.

Despite the rising tendency in the number of supplied inhabitants, decrease in drinking water consumption and household specific water demand remained. This phenomenon is caused due to increasing drinking water tariffs. The consumption of drinking water is falling deep below the average (in households dropped from the level of 195 l/capita/day to 100 l/capita/day), leaving people dangerously close to approaching the minimal hygienic standards. This trend also applies to other water use categories (industrial and other users).

Non-uniform occurrence of water in time and space in the different regions of the country has led in the past to development of large-scale water supply systems. Due to there are 47 regional water supply networks currently supplying some 70 % of the inhabitants with good quality drinking water.

2.2 Waste water Collection and Treatment

The total number of inhabitants connected to public waste water sewerage in SR is 57.6 %; the annual increase is about 0.5%. From the 2,883 municipalities in Slovakia only 594 of them are connected to public sewerage systems. The number of municipalities connected to public sewage systems with following treatment in waste water treatment plans (WWTPs) is 492. There are 49 public sewerage systems without WWTP. The public sewerage system coverage s is very different in Slovak regions. Some aldermanries are only 40% and others over 90 %. The total length of waste water pipes does approx. 7,550 km.

2.3 Legislative and Regulatory Framework

Substantive legal rules of Slovak Republic declare that groundwater and water sources are within the ownership of the state. Responsibility to provide water supply and sewerage were delegated to municipalities through actual law. Municipalities became owners of infrastructure assets of water and waste water systems. In 2003 the transformation process of former governmental water companies to join-stock companies was finished. The authority to operate the public water supply or public sewage systems is the individual or legal entity which was granted a license for operating public water supply systems or public sewage systems. The Ministry of Environment of the Slovak Republic administers central evidence of licenses issued.

The water price is regulated by independent Regulatory body - Regulatory Office for Network Industries (URSO), in Slovakia, since 2003. The regulatory body gradually eliminated cross subsidy between inhabitants and corporate entity, which meant artificially low price for households. After 1989 in generally the water price has markedly increased and it is expected despite the strict price regulation to increases moreover because of financial requirement to build-up, renewal and reconstruction of abandoned infrastructure asset.

The range and structure of lawful charges, process of appropriate profit estimation and details for price submission are publish in law document URSO, which institutes the range of price regulation for water production, drinking water distribution and waste water collection.

The water price in Slovakia were establish by uniform methodology for all consumers under the same water company in 2007, Average water price is 0.7 EUR per cubic meter, perform 4.2 percentage decreasing against 2006. The water price depend on water company is from 0.56 to 0.89 EUR/m3 without VAT. Household water and waste water price together perform average 1.3 EUR/m3

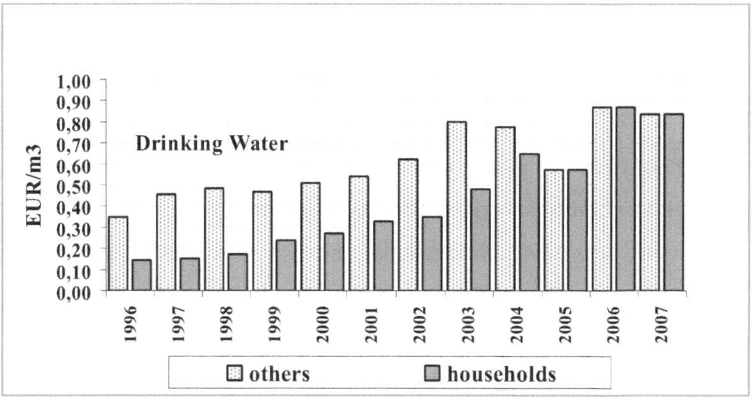

Figure 2 DRINKING WATER PRICE IN SR

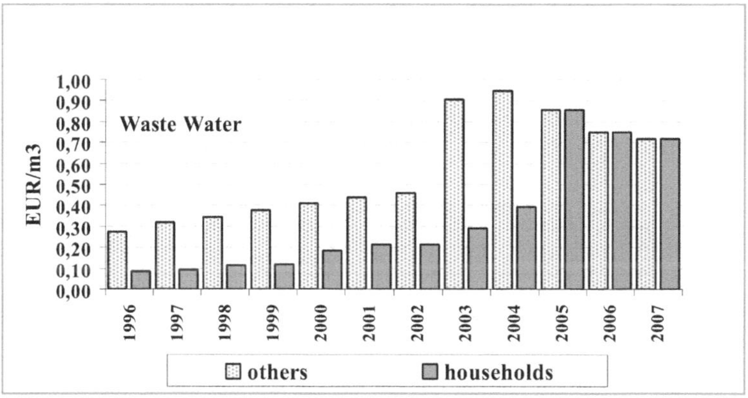

Figure 3 WASTE WATER SERVICES PRICE IN SR

3 WATER UTILITY SECTOR PERFORMANCE MONITORING

3.1 Performance Monitoring from the point of arrangement of investments requested by legislation

In the following year, the high legislative pressure will attack water companies in Slovakia. Due to EU accession requirements, Slovak water companies have time only until 2010 or 2015 to solve many problems especially in sewerage and waste water treatment.

The development of public water supply and waste water treatment is defined in several strategic documents. It is mainly "The Concept of water sector SR policy till 2015 year" and "The Plan of development of public water supply and public sewerages until 2015 year" (hereinafter called "Plan of development"). The Plan of development takes into account all legislative requirements until 2015, based on the creation of water rural agglomerations.

The Plan of development expects the need of financial means for investment according to table 1:

Table 1 EXPECTED INVESTMENT

Investment	**mil €**
water supply	1,564
sewerage&WWTPs	2 390
Total	**3 954**

For calculation in EUR there was used exchange rate EUR:SKK 1:33. The value of investment is in prices of 2004.

We can state that investment will be made in very short time because water companies are bound by legal conditions and therefore, to survive their business, they have to invest. Of course, there are prepared instruments to softer conditions for water companies from financial point of view. European Commission and the Slovak Republic prepared financial means within the Operational Program of Environment:

Table 2. EXPECTED INVESTMENT

EU funds	State budget of SR	Total
915 mil €	cca 158 mil €	1,073. mil €

The rest of missing financial means is on shoulders of water companies and/or municipalities. It is important the arrangement of all needed financial means to fulfill legislation requested investment. Without such arrangement we can in the future expect that performance monitoring of water utility sector will state only problems in connection with requested technical conditions.

In relation with these facts there are several questions. In following lines, we will try to answer them.

Question 1st – Public sources versus value of investment needs

The financial means arranged from **public sources** within Operational Program of Environment for period 2007-2013 **represent 27% of total needs** of requested investments.

Still there are two years till 2015 for which additional financial means could be allocated. However, we have to keep in mind that decision process including financial arrangement will have to be done by 2013, latest.

Sub question 1.1 What will be the percentage of public sources participation after taking into account the inflation in contracting sector?

In last 3 years the inflation in construction sector is about 6% per annum. In Slovakia in last years is high boom in construction of infrastructure. The demand for capacities is high. Thus we can estimate the inflation in this sector at least 2.5% p.a. (calculated from 2008). If we will theoretically assume to fix prices for all construction contracts in 2009, then public sources will represent only **21% of investments in current prices**.

Question 2nd – What are other public financial sources?

Envirofond

It is the public fund used by municipalities for smaller projects. If the budget will be kept on the similar level as it was in last years then this financial source represents only of € 0.36 mil. Unfortunately the source is too small to significantly increase potential of financial sources.

Question 3rd – What are the own sources of water companies?

Deprecations of existing equity

Cash generating from depreciations of existing equity could be used for time being as the source for new investments. For sure, it could not be used in full amount for new investments, because then the problem with reproduction of existing equity will outstand. If we will assume to use 50% of cash generated by depreciations within next 8 years then water companies have financial means in amount of € 418 mil. This source represents in current prices of 2009 about 10% of calculated needs.

Profit after tax of water companies

In 2004 the cross subsidy in prices between households and other clients of water companies was eliminated. The prices started to reflect better the costs of services and water companies started to create the profit.

As it is visible form graph on the figure 4, water companies generate profit, but it is not so high. The regulation is still rather strong. The concept of regulation is based on "costs model" and thus there is almost no place to take into account new – required investments.

If the regulation will continue in the same way then the maximum of profit that could be used for new investments for next 8 years will be at level of € 163 mil. It is 4% of required needs.

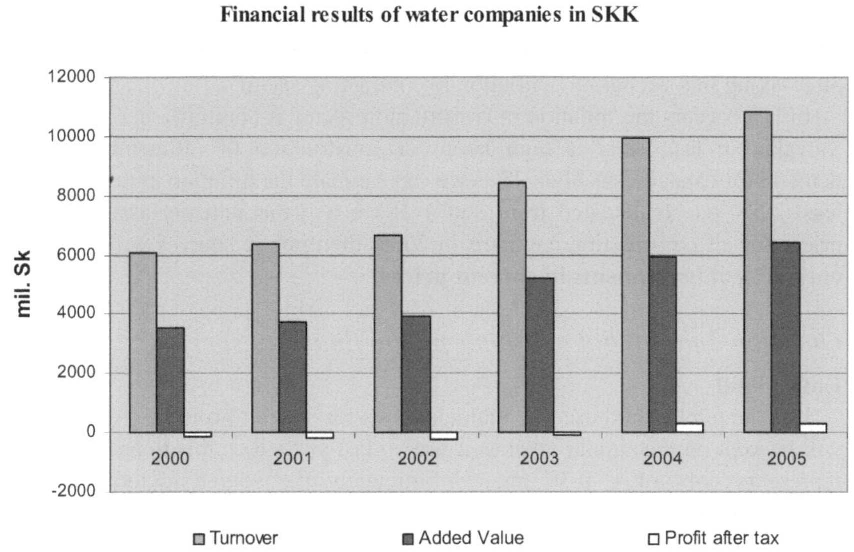

Figure 4 . FINANCIAL RESULTS OF WATER COMPANIES IN SKK.

Question 4th – What are the foreign sources?

Bank loans

Bank loans could use water companies for investments, but there is the barrier of absorption made by banks. The practice, up till now the project's financing was limited by value of collaterals. In practice it was finally on level 15% from investment costs. Using the same calculation then the source of bank loans could represents € 662 mil, what is 17% of investment needs.

Other foreign sources

For investment could be used also other financial products like leasing, back-leasing, but these products have again the barrier of collateral.

PPP projects in Slovakia are not so popular up till now.

4 BENCHMARKING OF WATER UTILITIES

In 2005, initialised by the Regulatory body URSO a methodology for water losses evaluation in public water supply systems was developed, based on the IWA recommendations. The first benchmarking was connected with PI system only considering PIs connected to water loss assessment. The 9 largest water utilities in the Slovak Republic participated in this project, alltogether supply 93% of the population.

The URSO decree 2006 was the start for benchmarking study, which included a wider range of PIs (especially economic indicators). There were used 3 years period 2003, 2004, 2005 for evaluation. The range concerned 14 operational water companies. The goal of the pilot project was to evaluate a create the recommendations the selected performance indicators (PI). Project focused the metric not process benchmarking. PI were derived from basic data, consist from 3 groups:

- water supply services data (operating and balance data)
- waste water services data (operating and balance data)
- financial data

Derived performance indicators were divided into next sections:

- operational covering
- specific technical data for water supply system and waste water system
- payments and revenues
- costs and employees

The realised benchmarking shows the lacks of water company data basement in Slovakia. The performance indicators are marked by inaccurate data definition and annual data demand changes. From this the evaluation need to extend with the missing data of specific year and definition of exact data.

The 3 year period data is non adequate for benchmarking evaluation. The performance indicators values have sensible dispersion, with impossibility of causer determination. The result of PI evaluation were some recommendations concerning mainly uniform technical-balance data declaration and recommendations goes from present state of financial undersizing of new investments and reconstruction of water supply and waste water supply systems.

The necessity is to find the ways, how to invest the infrastructure assets. The ultimate of water price raise is determined by consumption basket of goods and services. The present state of water price loading is 1,4% in 2006 of Slovakia consumption basket.

The highly-developed EU countries the water price loading is 2%. This is the real water price rise expectation, with condition of benefit reinvestment. Profit ratio appraisal must make provision for non only profit ration, but take consider on population impact including population loading. Appraisal the influence impact of population needs to make separately for each region.

The detailed resulting data are blinded, and the reports of this benchmarking are not made public. The benchmarking data and the results of benchmarking will be used for future regulatory purposes and for periodic revue of water companies in Slovak Republic.

5 CONCLUSIONS

Development of public water supply and sewerage systems means then new infrastructure assets building and restoration of existing. The expected investment until 2015 is really significant. Potential financial sources that could be used represent 58% of total needs defined in strategic documents of SR. Evident created deficit in financial sources produces the problem of difficulties in fulfilling the required legislative obligations. In real terms, it will cause Slovakia a lot of problems in water utility sector performance in general.

Market economy brings the necessity of new approaches and innovations in financial gaining. The new approach – public-private sector, capital market, or various obligations offer the chances of investments, except additional supporting, own sources and loans.

These new chances and questions of required water supply and waste water sewerage assurance are discussed in project Expert Assistance for Slovak Regional Water Companies covered by Ministry of Environmental of Slovak Republic. The project is financing by EU funds and performance technical and expert assistance in level increasing for water supply, waste water sewerage and waste water treatment.

The scope of this paper is showing the focused problems in water supply and waste water sewerage and waste water treatment in Slovakia.

This work was supported by Science and Technology Assistance Agency under the contract no. APVT-20-031804.

6 REFERENCES

Act No.442/2002 Dig. On public water and sewerage systems

Act No. 276/2002 Dig. On regulation in network industries

Janka Kleinertova. *Water price regulation and investment activity.* Vodárenské pohl'ady 4/2006

Tóthová K., Dubová V, Barloková D,(2005) : *Benchmarking of Water Supply – Water Losses Assessment.* In.: NATO Advance Researche – Workshop Integrated Urban Water Resource Management, Proceedings, Senec Slovakia, oktober 2005, p.129-136., ISBN 80-86020-47-9

http://www.statistics.sk

A PI system for drinking water treatment plants – framework and case study application

P. Vieira[1], *C. Silva*[1], *M. J. Rosa*[1], *H. Alegre*[1], *H. Lucas*[2], *R. Sancho*[2] *and P. Ramalho*[1]

[1] *Laboratório Nacional de Engenharia Civil. pvieira@lnec.pt, csilva@lnec.pt, mjrosa@lnec.pt, halegre@lnec.pt, pramalho@lnec.pt, Av. do Brasil 101, 1700-066 Lisboa, Portugal, Tlf: +35121844384.*
[2] *Águas do Algarve. h.lucas@aguasdoalgarve.pt, r.sancho@aguasdoalgarve.pt, R. do Repouso 10, 8000-302 Faro, Portugal., Tlf: +351289899070.*

Keywords: Drinking water, case study, performance indicator, treatment plant

 Performance Assessment of Urban Infrastructure Services
Edited by Enrique Cabrera Jr and Miguel Ángel Pardo. ISBN: 9781843391913.
Published by IWA Publishing, London, UK.

ABSTRACT

The traditional approach of providing safe drinking water to consumers is being enhanced by the water utility concern of doing it in a cost-effective and sustainable way. Consequently, some research projects have developed performance assessment systems (PAS) mainly consisting of performance indicators for water supply systems. However, none of those PAS addressed the water treatment plants in a specific way. This paper presents a proposal for a PI system that applies to drinking water treatment facilities and describes the results of its application to a full scale case study.

1 INTRODUCTION

The traditional approach of providing safe drinking water to consumers is being enhanced by the water utility concern of doing it in a cost-effective and sustainable way. The focus on design and construction of new water supply systems has changed to the efficient and effective operation of the existing ones. The need to assess the condition and behaviour and to compare different systems has revealed the lack of performance assessment methodologies in the water sector. Consequently, some research projects developed performance assessment systems (PAS) mainly consisting of performance indicators (PI).

The most relevant PI systems, developed for supply systems, were those promoted by the International Water Association (IWA) (Alegre *et al.*, 2000; Alegre *et al.*, 2006) and by the World Bank (Worldbank, 1999; Worldbank, 2006), but several others can be named such as the "Six Scandinavian Cities Group" (Stahre and Adamsson, 2002), the program QualServe from AWWA (Crotty, 2003), the UK Ofwat scheme (Ofwat, 2004) and a series of national projects that were modified versions of the IWA system. IWA concepts and methodology have also been incorporated in recently approved international standards ISO 24500 on *Service activities relating to drinking water supply systems and wastewater systems* (ISO/DIS24510, 2006; ISO/DIS24512, 2006).

However, none of these PAS addresses the water treatment plants (WTP) in a specific way. European and national environmental policies, the strong need for treatment plants rehabilitation and the effective implementation of Quality Assurance Systems (*e.g.* ISO 9000) are also driving PAS for WTP.

This paper presents a proposal for a PI system that applies to drinking water treatment facilities and describes the results of its application to a full scale case study – a WTP from Águas do Algarve, S.A. located in the south of Portugal. The system developed aims at being as universal as possible but is anchored to the Portuguese situation.

2 THE PI SYSTEM FOR DRINKING WATER TREATMENT PLANTS

The PI system developed fits the principles established in ISO 24500, as well as in the IWA PI system for water supply systems. It has been developed in close cooperation with a water supplier (Águas do Algarve S.A., Portugal) to take into account the end-user point of view and data availability in routine operation of a treatment plant.

In total, 77 PI have been defined and classified in the following seven evaluation domains: *Treated water quality*, *Plant reliability*, *Use of natural resources and raw materials*, *By-products management*, *Safety*, *Human resources*, and *Economical and financial resources*. The structure of the PI system is presented in **Table 1** but detailed characterization of each PI (objective, definition, processing rule, units of measurement, data required, results analysis, etc.) is made elsewhere (Vieira *et al.*, 2007).

3 CASE STUDY APPLICATION

3.1 Description of the case study – Alcantarilha WTP

To test the applicability and feasibility of the WTP PI system, a set of selected indicators has been tested in a full scale case study that is a drinking water treatment plant from a Portuguese water utility – Águas do Algarve, S.A. This WTP located in Alcantarilha is responsible for providing a reliable supply of safe drinking water to *ca.* half million people in southern Portugal (Algarve) since 2000. It was designed to treat up to 3 m^3/s (*ca.* 1 million people, year 2020) of surface water from Funcho Dam reservoir (*ca.* 200 km^2 and 43.4 hm^3) by conventional treatment. The treatment sequence consists of pre-ozonation, adsorption with powdered activated carbon, coagulation (using an aluminium polyhydroxychlorosulphate of high basicity), flocculation (with flocculant addition, whenever needed), sedimentation, rapid sand filtration and chlorination. Sludge from settlers is mechanically dewatered in a filter press, after addition of lime and a cationic polymer.

The flexibility needed by Alcantarilha WTP to face strong seasonal variations in raw water quality together with seasonal water demand is provided by three treatment lines in parallel (1 m^3/s each).

Águas do Algarve is a bulk supplier and the water produced at Alcantarilha WTP is delivered to service reservoirs from several municipalities. Raw water arrives at the plant by gravity and treated water is pumped to those reservoirs.

Table 1. PI SYSTEM FOR DRINKING WATER TREATMENT PLANTS

Domain of evaluation	N. of PI	PI code	Objective
Treated water quality	8	WQ	To evaluate performance in terms of compliance with criteria established by the water utility for the water at the exit of the WTP and in terms of the WTP ability to produce water that has an adequate quality at the consumption point
Plant reliability	35	ER	To evaluate performance in terms of WTP technological robustness and in terms of flexibility of the WTP to respond to variations that occur in raw water
Use of natural resources and raw materials	7	RU	To evaluate performance in terms of efficiency of the use of WTP inputs: water, energy, chemicals and filter media.
By-products management	7	BP	To evaluate performance in terms of the amount of process treatment by-products and the adequacy of corresponding management practices
Safety	3	S	To evaluate the performance in terms of plant safety and environmental safety
Human resources	8	Pe	To evaluate the performance in terms of the availability of WTP personnel, the adequacy of personnel qualification and training, absenteeism and overtime work
Economical and financial resources	9	Fi	To evaluate the performance in terms of running costs and capital costs

3.2 Application of selected PI to Alcantarilha WTP

Although 47 PI (*ca.* 60% of total proposed PI) have been determined for Alcantarilha WTP, in the present paper, PI results are only illustrated by some examples. Depending on the PI, calculation has been made for the years 2003 to 2006. Six evaluation domains were studied.

The assessment period for the majority of the PI was one year, but in cases where availability of data allowed, the PI calculation was additionally made for periods of highest water production and for periods of lowest water production to understand the seasonal changes on both the raw water quality/quantity and on the drinking water demand found in the Algarve region.

3.3 PI results and discussion

3.3.1 Evaluation domain "Treated water quality"

Results from PI in the domain *Treated water quality* are shown in Figure 1.

WQ1–Compliance of water analysis assesses the performance of the WTP in terms of compliance of treated water quality with predefined criteria. In this case, the Portuguese drinking water quality legislation was adopted as criteria. As shown in Figure 1, this indicator reveals high levels of compliance ranging from 98.5% to 99.9% of the total analyses performed. The lowest value was observed for 2005 when, due to a severe drought felt in Algarve, Alcantarilha WTP had to treat groundwater or a mixture of groundwater and surface water which negatively impacted the water quality, particularly in what concerns manganese, aluminium, coliform bacteria and bromate parameters.

In 2003 not all the parameters requested by law were analysed in treated water (WQ3–*Parameters analyzed*). However, in subsequent years, the concern with water quality control has sharply increased as shown by *WQ3* values always higher than 100% (145-167%) indicating that, in recent years, the water utility has a monitoring plan far stricter than legislation.

- **WQ1 – Compliance of water analysis** [%, ♦] =
  (no. of treated water analysis complying with criteria defined by water supplier/no. of analysis performed)*100
- **WQ3 – Parameters analyzed** [%, □] =
  (no. of parameters analysed/no. of parameters legislated)*100

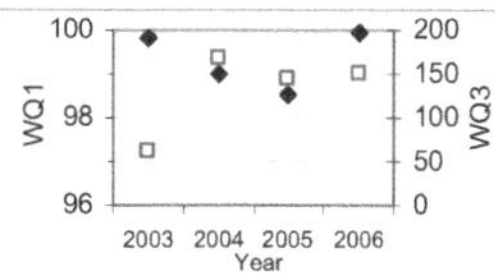

- **WQ4 – WTP storage tanks cleaning** [%/year] =
  ((volume of treated water storage tank cells that have been cleaned*365/assessment period)/volume of treated water storage tank cells)*100
- **WQ5/6 – Water quality at distribution system consumption points that have low/high chlorine concentration** [mg/L] =
  Average of the 10% lowest/highest values of free chlorine recorded in the distribution network
- **WQ7 – Water quality at distribution system consumption points that have high THM concentration** [μg/L] =
  Average of the 10% highest values of THM recorded in the distribution network

	2005	2006
WQ4	100%	100%
WQ5[1]	0.41 mg/L	0.39 mg/L
WQ6[1]	1.13 mg/L	0.98 mg/L
WQ7[1]	49 μg/L	51 μg/L

[1]PI determined only for service reservoirs fed exclusively by the WTP

Figure 1 . PI RESULTS FROM DOMAIN "TREATED WATER QUALITY".

The extent to which treated water storage tanks at the plant are cleaned assesses the existence of conditions that minimize the contamination and degradation risk for the treated water after leaving the WTP. This aspect is measured by *WQ4–WTP storage tanks cleaning*. Besides WQ4, the ability of the WTP to produce water that has an adequate quality at the consumption point is also assessed by *WQ5 – Water quality at distribution system consumption points that have low chlorine concentration*, *WQ6 – Water quality at distribution system consumption points that have high chlorine concentration* and *WQ7 – Water quality at distribution system consumption points that have high THM concentration*. Results (Figure 1) show that all the tanks are cleaned once a year (WQ4=100%), THM levels in service reservoirs are well below the legislated value (100 μg/L) and that chlorine concentration is adequate for bulk distribution.

3.3.2 Evaluation domain "Plant reliability"

Alcantarilha WTP was designed to be supplied by surface water from Funcho dam and by Odelouca dam which is currently under construction. The adequacy of Funcho water source utilisation was assessed by *ER1 – Water source utilisation*. Figure 2 shows that in 2003 (ER1=127%) the WTP was using more water than the total annual abstraction allowance of the source (in this case, considered to be the contractual volume), this situation being caused by the delay in the construction of the second surface water source that, when in operation, will bring to an end Funcho overexploitation. Due to the severe drought felt in 2004 and 2005, ER1 has lowered to 86% and 11% as there was not enough water in the source to treat (during this period the WTP used other sources – groundwater – in order to compensate the lack of surface water).

ER2 – Raw water storage capacity, *ER4 – WTP utilisation*, *ER5 – Treated water storage capacity* and *ER7 – Pumping capacity downstream the WTP* measure the reliability of the WTP in terms of existing infrastructure capacity.

The lack of raw water storage capacity at Alcantarilha WTP (ER2 = 0) is a weakness of this system because i) in case of source contamination, the plant has to stop production and ii) in case of sudden changes in source water quality (*e.g.*, turbidity increase) there is no buffer capacity that allows to change treatment conditions in accordance. The treated water storage capacity (ER5, Figure 2) allows to continue supplying treated water during 0.18-0.34 days (*ca.*4-8h) in case of interruption of the treatment. These values are even lower if only the period of maximum production is considered (0.12-0.26 days; summer time). However, as the treated water is pumped to service reservoirs, it is more important to have a storage capacity already in service reservoirs and not inside the plant, which is the case of this distribution system that has a total storage capacity for 1-2 days.

ER4 (Figure 2) shows that, during the period analysed, Alcantarilha WTP was oversized, since values were lower than 26% in average, in the range 26-54% considering the maximum production day, and lower than 10% considering the minimum production day. The variation in plant utilisation between maximum and minimum production situations shows the high seasonality associated with this system. The relatively low values at the maximum production day reflect the water scarcity in Funcho dam (ER1). The variation observed from 2003 to 2006 is due to the drought in 2004-2005 (see ER1). As expected, this variation is not felt in the case of minimum production.

As the previous four PI, ER7 (Figure 2) also shows that there is an adequate capacity at Alcantarilha, as on the day of maximum pumping (that occurred in 2003) only 56% of the installed pumping capacity was used. The variation over the studied period (decrease from 56% in 2003 to 26% in 2005) followed the decrease in WTP utilisation due to water scarcity. Comparison of ER7 and ER4 calculated for maximum production day (ER4 slightly lower than ER7) indicates that the existing pumping capacity would not be enough to cope with the treated water produced in a situation where full treatment plant capacity is used. In this situation, the water would have to be stored in the treated water reservoir inside the plant.

Concerning continuity of operation of the WTP, *ER11 – Daily period of WTP operation* shows that the plant operated during 24 h/day, as recommended in order not to cause disruptions in the treatment process.

A lack of a regular procedure for the optimization of chemicals dosage by jar tests may negatively impact *ER12 – Adjustment of chemicals dosage to raw water quality variation* (values between 0 and 1, Figure 2) which assesses the plant reliability to cope with raw water quality variations.

In terms of WTP automation degree, all filters (*ER15 – Filters control* = 100%) and sludge discharges (*ER16 – Sludge discharge from settlers* = 100%) are fully automated, and water quality monitoring is partially made online (*ER13 – Online water quality monitoring*, Figure 2).

An adequate process monitoring is implemented in Alcantarilha WTP as demonstrated by the high values of *ER17 – Water quality monitoring* (Figure 2) and *ER18 – Process supervision* (100%). Monitoring points exist in raw water, between each treatment step and in treated water.

As the reliability of a treatment plant is higher if the possibility of dosing alternative chemicals is planed (*e.g.*, powdered activated carbon, if necessary, during some periods of the year), this performance aspect was evaluated through *ER19 – Possibility of dosing alternative chemicals* (0.4-0.5, Figure 2). This PI showed that Alcantarilha may use a number of alternative chemicals (*i.e.*, used less than 50% of the time) that is half of the chemicals used in regular operation (*i.e.*, used more than 50% of the time).

The accurate measure of the dosages of all these chemicals is feasible as all dosing systems have changeable set points (*ER20 – Chemicals dosing systems with changeable set point* = 100%) and their calibration procedures have improved since 2003 to a frequency of *c.a.* once per year per dosing system (*ER21 – Chemicals dosing systems calibration*, Figure 2).

Formal procedures for equipment inspection and calibration exist at Alcantarilha WTP as shown by *ER21 – Chemicals dosing systems calibration, ER22 – Pump inspection, ER23 – Emergency equipment inspection, ER24 – Signal transmission equipment inspection, ER25 – Electrical switchgear equipment inspection, ER26 – Filter media inspection, ER27 – Flow meters calibration, ER28 – Water level meters calibration, ER29 – Pressure meters calibration* (Figure 2). The high values of *ER30 – Online water quality meters calibration* (Figure 2) express the concern with the accurate control of water quality, also reflected by WQ3. Between 2003 and 2006 there was a clear increase in the plant performance concerning this aspect. In 2006, all pumps, signal transmission and online water quality meters were inspected/calibrated more than once. However, there is still a need for an improvement in electrical switchgear equipment, filter media, flow meters, water level meters, pressure meters and, particularly, emergency equipment calibration/inspection as not all the equipments are being inspected/calibrated once a year (related PI are less than the guideline value 1/equipment/year).

In average, 0.8 failures/pump/year have occurred (*ER32 – Pump failures*, Figure 2). Other type of failures has caused the unplanned interruption of the WTP operation during 13.2-16.1 hours/year (*ER33 – Interruptions of WTP operation*, Figure 2). Although the reliability of the electrical supply system increased from 2005 to 2006, energy failure has still been the cause for 60% of the total interruption time in 2006 (*ER34 – Interruptions of WTP operation due to energy failure*, Figure 2).

- **ER1 – Water source utilisation** [%/year, ♦] =
  ((volume of water at the entrance of the WTP*365/ assessment period)/total annual abstraction capacity of the source)*100
- **ER7 – Pumping capacity downstream the WTP** [%, □] =
  (volume of treated water pumped on the day of highest volume of water pumped/maximum treated water pumping capacity in one day)*100

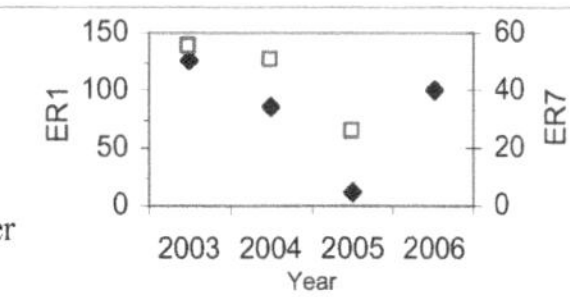

- **ER4 – WTP utilisation** [%,♦] =
  volume of treated water/maximum WTP capacity

  □: maximum production; ○: minimum production

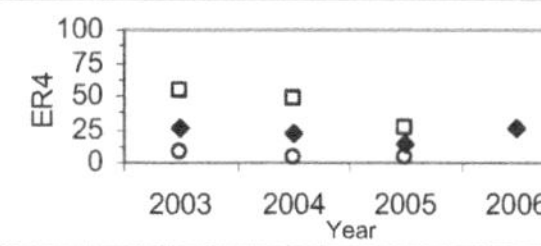

- **ER5 – Treated water storage capacity** [day] =
  (volume of WTP treated water storage tanks*assessment period)/volume of treated water

  □: maximum production

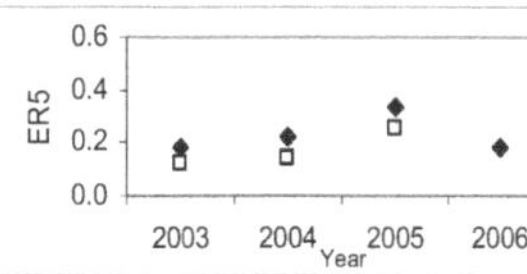

- **ER12 – Adjustment of chemicals dosage to raw water quality variation** [no./year, ♦] =
  no. of jar tests performed*365/assessment period
- **ER19 – Possibility of dosing alternative chemicals** [-,□] =
  no. of dosing systems for chemicals not used in a regular basis/no. of dosing systems for chemicals used regularly during WTP operation

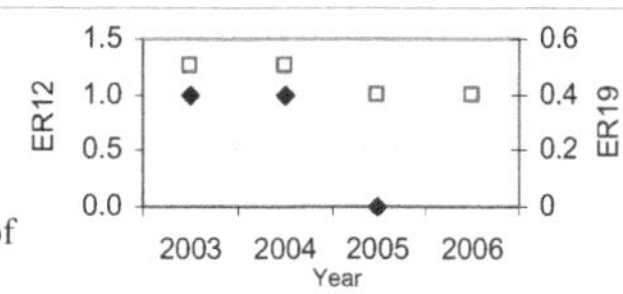

- **ER13 – Online water quality monitoring** [no./UPO] =
  no. of sampling points with online water quality monitoring/ (no. of UPO of the WTP+1)
- **ER17 – Water quality monitoring** [no./UPO] =
  no. of sampling points for water quality monitoring/(no. of UPO of the WTP+1)

	2003-2006
ER13	0.86
ER17	0.86

(UPO: Unit process/operation)

- **ER21 – Chemicals dosing systems calibration** [no./dosing system/year] =
  (no. of chemical dosing systems calibrations*365/assessment period)/no. of chemical dosing systems

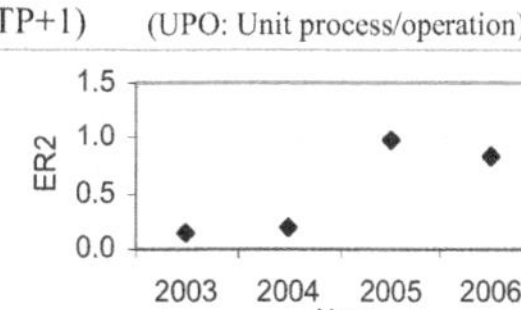

Figure 2 . PI RESULTS FROM DOMAIN "PLANT RELIABILITY".

- **ER# – Equipment inspection** [no./equipment /year] =
  (no. of equipment inspections*365/ assessment period)/no. of equipments

Where #: 22 for pump inspection (□), 24 for signal transmission equipment inspection (♦), 25 for electrical switchgear equipment inspection (○)

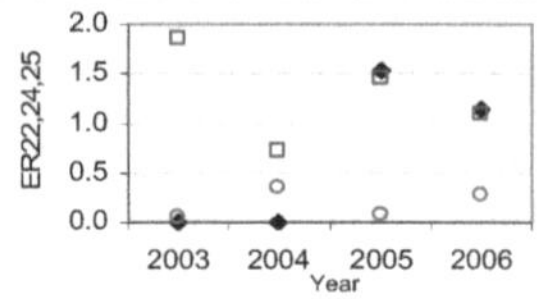

- **ER23 – Emergency equipment inspection** [%/year, ♦] =
  ((sum of the nominal power of the emergency equipment subjected to inspection x365/assessment period)/total nominal power of the emergency equipment)*100
- **ER26 – Filter media inspection** [no./filter/year, □] =
  (no. of filter inspections*365/assessment period)/no. of filters)*100

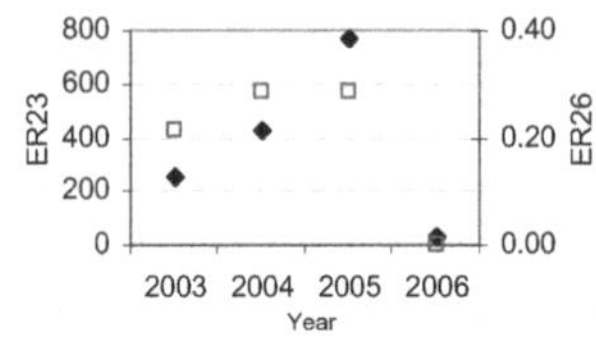

- **ER# – Equipment calibration** [no./meter/year] =
  (no. of meter calibrations*365/ assessment period)/no. of meters

Where #: 27 for flow meters calibration (♦), 28 for water level meters calibration (□), 29 for pressure meters calibration (○), 30 for online water quality meters calibration (×)

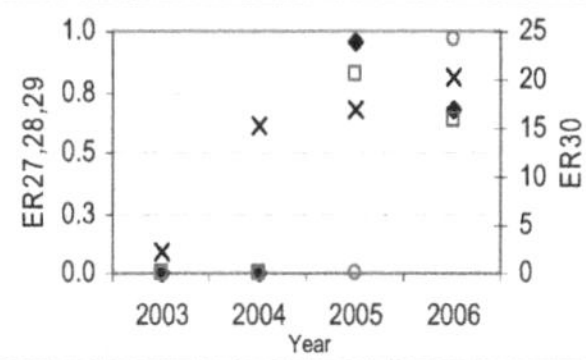

- **ER32 – Pump failures** [no./pump/year] =
  (no. of pump failures*365/assessment period)/no. of pumps

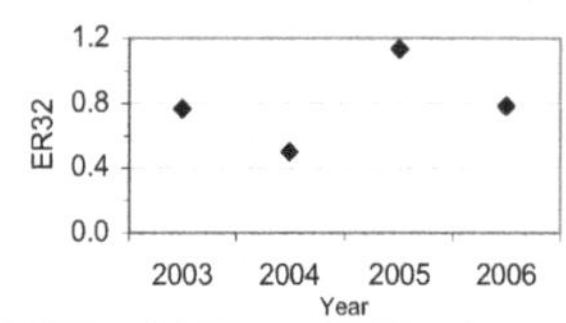

- **ER33 – Interruptions of WTP operation** [h/year] =
  (no. of hours that the WTP was out of service*365)/assessment period
- **ER34 – Interruptions of WTP operation due to energy failure** [h/year] =
  (no. of hours that the WTP was out of service or was operated with emergency energy supply, due to failure in the electricity supply *365)/assessment period

	2005	**2006**
ER33	13.2	16.1
ER34	13.2	9.6

Figure 2 . PI RESULTS FROM DOMAIN "PLANT RELIABILITY". (cont.)

3.3.3 Evaluation domain "Use of natural resources and raw materials"

Efficiency in water at the WTP was assessed by *RU1 – Efficiency of water use at the WTP* (Figure 3) that was always above 98%, revealing a high performance. Values slightly above 100% (2005, 2006) were due to flow measuring errors, which can be related to the already mentioned need for improving equipment calibration (ER27).

3.3.4 Evaluation domain "By-products management"

The amount of waste production has been evaluated only for sludge. *BP1 – Quantity of sludge generated* was in the range 3.6-19.9 g/NTU/m^3 (higher values occurred in 2004-2005 when the WTP was treating groundwater with very low turbidity and hence more difficult to remove) and the sludge had 33-44% of dry weight (*BP2 – Quality of sludge*, Figure 4). In 2003-2004 none or very few sludge valorisation was made (0-41%) and storage was made at the WTP. In 2005, all the previously stored sludge was sent to valorisation in cement industry (*BP4 – Final disposal of sludge* = 301%, Figure 4).

During the period analysed, the plant has never discharged to the environment process wastewater (*e.g.*, from filter backwash, sludge dewatering) that could cause negative environmental impacts (*BP7 – Process wastewater without treatment that is discharged in the environment* = 0).

- **RU1 – Efficiency of water use at the WTP** [%] =
  (volume of treated water/volume of water at the entrance of the WTP)*100

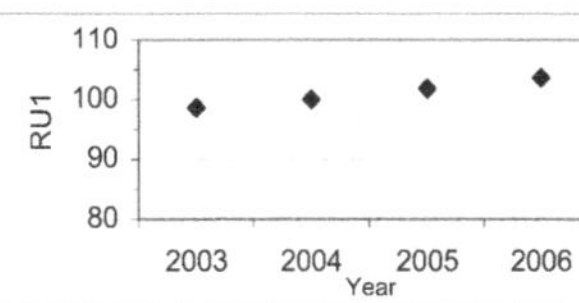

Figure 3 . PI RESULTS FROM DOMAIN "USE OF NATURAL RESOURCES AND RAW MATERIALS".

- **BP1 – Quantity of sludge generated** [kg/(m^3.NTU), ♦] =
  Quantity of sludge generated/(volume of water at the entrance of the WTP *raw water turbidity)
- **BP4 – Final disposal of sludge** [%, □] =
  (quantity of sludge subjected to valorisation/quantity of sludge generated)*100

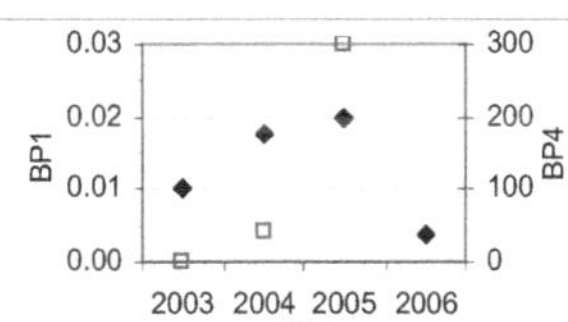

- **BP2 – Quality of sludge** [% w/w] = Dry weight of sludge

	2004	2005	2006
BP2	44.0	32.7	38.0

Figure 4 . PI RESULTS FROM DOMAIN "BY-PRODUCTS MANAGEMENT".

3.3.5 Evaluation domain "Safety"

In terms of environmental safety performance, there is no record of accidental chemicals spillage or leakage at Alcantarilha (*S1 – Chemicals spillage* = 0). Plant safety performance, assessed by *S2 – Working accidents*, showed that only two working accidents requiring medical attention (S2=0.039) took place at this facility in 2006 (Figure 5).

3.3.6 *Evaluation domain "Human resources"*

Personnel indicators address the efficiency of human resources.

In total, Alcantarilha WTP has 2.08 full time employees/$10^6 m^3$ of treated water (*Pe1 – WTP personnel*, Figure 6). Outsourcing personnel is not considered in this PI. A more in-depth analysis of how the staff is distributed according to qualification (*Pe2 – Personnel with a university degree*, *Pe3 – Personnel with basic education*, Figure 6) shows that the human resources structure seems adequate, the major component being personnel with basic education (56%) and almost one third personnel with a university degree.

Personnel training is mainly (99% of total training time, 66.9 h/employee/year) done externally to the company (*Pe4 – Total training time*, *Pe5 – Internal training time*, Figure 6). *Pe7 – Absenteeism due to work accidents or illness at work* assesses absenteeism due to work accidents or illness (Figure 6).

- **S1 – Chemicals spillage** [kg/m^3] =
  Quantity of chemicals spilled/volume of treated water
- **S2 – Working accidents** [no./employee/year] =
  (no. of working accidents*365/assessment period)/no. of full time equivalent employees of the WTP

	2004	2005	2006
S1	0	0	0
S2	0	-	0.039

Figure 5 . PI RESULTS FROM DOMAIN "SAFETY".

- **Pe1 – WTP personnel** [no./$10^6 m^3$] =
  no. of full time equivalent employees of the WTP/volume of treated water
- **Pe7 – Absenteeism due to work accidents or illness at work** [day/employee/year]=
  (no. of days of absenteeism due to work accidents or illness at work of WTP employees*365/assessment period)/ no. of full time equivalent employees of the WTP
- **Pe8 – Overtime work** [%] =
  (hours of overtime work of the WTP employees/total hours of regular work of the WTP employees)*100

	2006
Pe1	2.08
Pe7	2.5
Pe8	4.0

- **Pe2 – Personnel with a university degree** [%] =
  (no. of WTP employees with a university degree/no. of full time equivalent employees of the WTP)*100
- **Pe3 – Personnel with basic education** [%] =
  (no. of WTP employees with basic education/ no. of full time equivalent employees of the WTP)*100

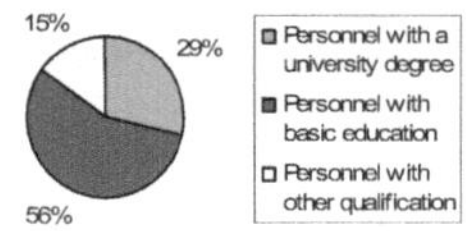

- **Pe4 – Total training time** [h/employee/year] =
  (no. of WTP employees training hours*365/assessment period)/no. of full time equivalent employees of the WTP
- **Pe5 – Internal training time** [h/employee/year] =
  (no. of WTP employees internal training hours*365/assessment period)/no. of full time equivalent employees of the WTP)

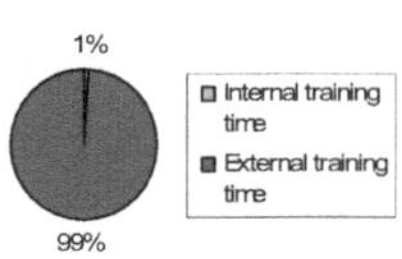

Figure 6 . PI RESULTS FROM DOMAIN "HUMAN RESOURCES".

4 FINAL REMARKS

In this paper, a PI system for drinking water treatment plants and its application to a full scale case study are presented.

The study evidenced that the proposed PI system is able to assess the WTP performance and it is also sensitive to changes in this performance. In addition, its application clearly indicates the major reasons for these changes and their impacts on WTP performance, as was the case of Alcantarilha WTP with the severe drought in 2004-2005. Nevertheless, the plant was always able to overcome these difficulties and fulfil the objectives set by the water company during this period: to produce high quality water regardless the water source characteristics (groundwater and/or surface water).

The outcome from this field test will serve as a basis for the refinement of the PI system. The final PAS for WTP will integrate the refined PI system with a set of measures for assessing operational performance (covering the aspects of overall and unit processes/operations efficiencies and their relation with operating conditions), which is being developed by the authors.

5 ACKNOWLEDGEMENTS

This project is partially funded by the Portuguese Foundation for Science and Technology and by Águas do Algarve, S.A. The authors acknowledge Eng. Marisa Viriato and Eng. Nuno Martins from Águas do Algarve.

6 REFERENCES

Alegre, H., Hirner, W., Melo Baptista, J., Parena, R. (2000) *Performance Indicators for Water Supply Services*. IWA Publishing. London.

Alegre, H., Melo Baptista, J., Cabrera Jr, E., Cubillo, F., Duarte, P., Hirner, W., Merkel, W., Parena, R. (2006) *Performance Indicators for Water Supply Services*. IWA Publishing. London.

Crotty, P. (2003) *Selection and definition of performance indicators for water and wastewater utilities*. American Water Works Association Research Foundation. USA.

ISO/DIS24510 (2006) *Service activities relating to drinking water and wastewater – Guidelines for the improvement and for the assessment of the service to users.*

ISO/DIS24512 (2006) *Service activities relating to drinking water and wastewater - Guidelines for the management of drinking water utilities and for the assessment of drinking water services.*

Ofwat (2004) *Updating the overall performance assessment (OPA) - Conclusions and methodology for 2004-05 onwards.* Office of Water Services. UK.

Stahre, P., Adamsson, J. (2002) Performance benchmarking. A powerful management tool for water and wastewater utilities. *Watermarque* **3** (5).

Vieira, P., Alegre, H., Rosa, M. J. (2007) *Avaliação de desempenho de Estações de tratamento de água. Revisão do estado da arte e proposta de Sistema de Avaliação de Desempenho.* Report 215/07-NES. LNEC. Lisboa. (in Portuguese)

Worldbank (1999) *Benchmarking water and sanitation utilities: a start-up kit.* World Bank. Washington.

Worldbank (2006) *IBNET indicator definitions*. World Bank. Washington.

Assessing performance: The case of Jakarta

M. Iwanami[1]

[1]*Department of International Development, School of Public Policy, The University of Birmingham, Edgbaston, Birmingham B15 2TT.*
Email: mxi497@bham.ac.uk ; iwanami772@hotmail.com , Tel: +44 (0)121 414 2297; Fax: +44 (0) 121 414 5032

Keywords: Customer perception, performance, private sector participation, urban water supply, stakeholders

Edited by Enrique Cabrera Jr and Miguel Ángel Pardo. ISBN: 9781843391913.
Published by IWA Publishing, London, UK.

ABSTRACT

Water supply services in Jakarta are privately operated with TPJ (Thames PAM Jaya) operating eastern Jakarta and PALYJA (PAM Lyonnaise Jaya) operating western Jakarta. Each company has signed a 25 year concession contract with PAM Jaya (Jakarta City Water Company), effective from February 1998 but was renegotiated in October 2001. The contract includes 10 performance indicators that TPJ and PALYJA should improve, which are monitored by PAM Jaya and JWSRB (Jakarta Water Supply Regulatory Body). Since the commencement of the contract, there have been improvements in the 10 performance indicators, although they do not necessarily reach the targets set in the contract. Such improvements are, however, not necessarily felt and shared by the key stakeholders (i.e. customers, PAM Jaya, JWSRB and the companies). In light of this fact, this work finds out why so by identifying the aspects of water supply services that are important to the key stakeholders. In doing so, this work reveals that NRW (Non Revenue Water) is not necessarily an appropriate indicator to be monitored in assessing performance of water supply services in the case of Jakarta.

1 INTRODUCTION

Water supply services in Jakarta are privately operated with TPJ (Thames PAM Jaya) operating eastern Jakarta and PALYJA (PAM Lyonnaise Jaya) operating western Jakarta. Each company has signed a 25 year concession contract with PAM Jaya (Jakarta City Water Company), effective from February 1998. The service areas are divided by the Ciliwung River, as can be seen from Figure 1 as below. Moreover, each service area has three service zones with TPJ having service zone 2, 3 and 6 whilst PALYJA having 1, 4 and 5.

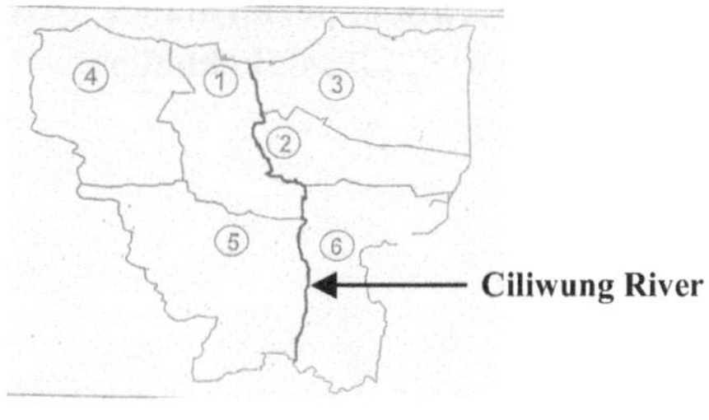

Figure 7. THE SERVICE AREA

Source: PT Kekarpola Airindo (1995)

Since the commencement of the contract, due to the East Asian Crisis, this contract was renegotiated in October 2001 and an independent regulatory body was established: JWSRB (Jakarta Water Supply Regulatory Body).

Private sector participation (PSP) in water in Jakarta was motivated by its poor performance of water services such as low service coverage, low quality of water, high losses of water and intermittent water supply as well as the government's financial constrain to improve the network (Lanti 2006). The Indonesian government thus decided to transfer Jakarta's water supply services to two foreign private companies one from UK (TPJ) and the other from France (PALYJA) to improve its services.

There are 10 performance indicators (5 technical targets and 5 service standards) that the companies (TPJ and PALYJA) should improve and are monitored by PAM Jaya and JWSRB as shown in Table 1.

Table 1. THE PERFORMANCE INDICATORS

Technical Targets	Service Standards
Volume of water billed (sold)	Water quality[1]
Water production from water treatment plant	Water pressure at the consumer's tap
NRW (Non revenue water)	Customer care[2]
Number of connection	Routine interruption in distribution network[3]
Service coverage ratio	Time for new connection

Source: Cooperation Agreement Amended and Restated (RCA) 2001

After the commencement of the contract, although they do not necessarily reach the targets set in the contract, there have been improvements in them. However, previous research tells that such improvements are not necessarily felt by the key stakeholders: the customers, PAM Jaya, JWSRB, TPJ and PALYJA.

[1] Needs to be compliance with the Clean water standards and Portable water standards.

[2] For example, response time to routine telephone call.

[3] This means the response time when there is a interruption tertiary, secondary and primary pipes.

Table 2. TIMELINE OF SELECTED PERFORMANCE INDICATORS OF WATER IN JAKARTA

	98	02	04	05
Population in JKT (m)	9.38	9.99	9.99	9.99
Population served (m)	4.04	5.23	5.72	5.73
Number of connections	487,978	649,429	705,890	708,919
Service coverage ratio (%)	43	52	57	61
Production from Water Treatment Plant (m3)	396	407	430	449
NRW (%)	56	48	47	50

Source: JWSRB (2006)

Deriving from this fact, the aim of this work is to reveal why there is a disparity in recognising improved performances. This is revealed by asking the stakeholders "Which aspect of water supply services to you think is important?" since it can be assumed that different stakeholders use different criteria in assessing changes in performance. The views of the stakeholders were attained by Focus Group Discussions (FGDs) for collecting the views of the customers and interviews for collecting the views of PAM Jaya, JWSRB and the companies.

2 BACKGROUND OF WATER IN JAKARTA

2.1 Raw water supply

The main source of raw water for Jakarta is the Jatiluhur Dam. This is located east of Jakarta and approximately 80% of the raw water comes from this dam. The remaining 20% comes from rivers such as Cisadane, Krukut and Pesanggrahar which are located on the west of Jakarta.

Jatiluhur Dam is a multi purpose dam of which 75% of its water is for irrigation, 15% for electricity and industrial usages and 10% for drinking water. Many rivers join the Western Tarum Canal (where the water of the Jatiluhur Dam runs though to reach the city of Jakarta). The fact that many rivers join the West Tarum Canal is posing a problem to the water resource for Jakarta since such rivers contain industrial waste that firms dispose at the upstream of the rivers. This results in worsening the water quality of the raw water raising health concerns and also implies increase purification cost to the water companies.

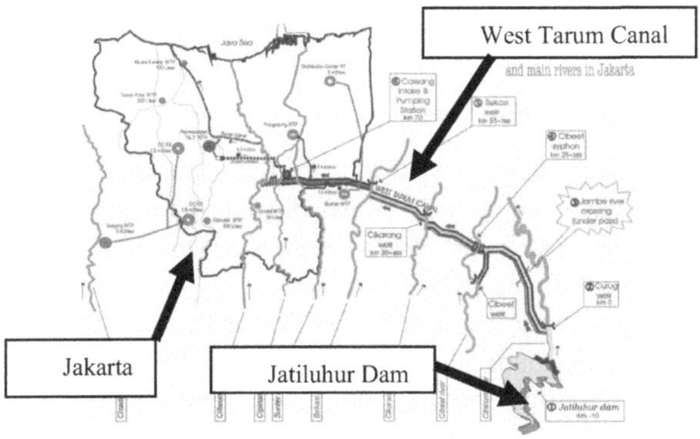

Figure 8. SOURCES OF RAW WATER FOR JAKARTA

Source: PALYJA presentation (2005)

2.2 Type of customers and water tariff

In Jakarta, customers are categorised into seven types and they are K1, K2, K3A, K3B, K4A, K4B and K5 and the customers under each type is as shown in the table below. Household users are under from category K1 to K4B and they are differentiated depending on the size of the house (m2).

Table 3. WATER TARIFF BY CUSTOMER CATEGORY

Customer Category	Tariff (IDR/m3)		
	0-10 m3	11-20 m3	>20 m3
K1	950	950	950
K2	950	950	1,425
K3A	3,260	4,280	4,990
K3B	4,465	5,475	6,775
K4A	6,200	7,400	8,850
K4B	11,325	11,325	11,325
K5	13,200	13,200	13,200

Source: PALYJA/TPJ water tariff brochure (2006)

3 PERCEPTIONS OF THE STAKEHOLDERS

3.1 Perceptions of the customers

The respondents were housewives in the K2 customer band since it is housewives that engage in various household chores using water. Six focus group discussions were carried out (one for each service zone) and there were 4 to 8 participants in each focus group.

Water continuity was the most common aspect that customers' value most. Five out of six Focus Groups identified this aspect. Customers say that water continuity is most important because they face intermittent water supply and thus cannot use water when they want to use it. The strategy to deal with intermittent water supply was to store water when it is available, use underground water and buy water from water vendors. The aspect that the other Focus Group identified was the price of water.

Participants were not satisfied with the piped water service not only due to the intermittent water supply but also due to the low quality of water (such as strong smell of chorine and yellowish water from the tap). However, customers say that quality and price does not matter as long as the water continues from the tap.

3.2 Perceptions of PAM Jaya

All respondents (seven respondents) had a negative impression of water supply in Jakarta due to a) high NRW and that companies are not achieving their NRW targets, b) limited raw water supply and c) not 24 hours water supply.

Most of the respondents think that NRW is the most important aspect. The reason why people at PAM Jaya think that NRW is important is because they believe that a decrease in NRW would result in an increase of water available to customers since there will be more water available to distribute through the network. They also say that it also means that there would be a reduction in losses of revenue (i.e. there would be an increase of revenue for the companies since the volume of water sold should increase). Moreover, some respondents were saying that the reduction of NRW means that the raw water would not be wasted since Jakarta is facing a shortage of raw water supply.

3.3 Perceptions of Jakarta Water Supply Regulatory Body

JWSRB (six people were interviewed) also had a negative impression about the water supply services in Jakarta. Their reasoning for their negative response were because of a) not 24 hours water supply, b) the high level of NRW and c) low water pressure.

As for which aspect of water supply services is important, the answers were a) water continuity (24 hours water supply), b) NRW and c) water pressure.

The reason why high NRW is not good is because, respondents explain, was that this means that at the end, it is the customers that have to burden for the company's losses via increase of the water tariff.

3.4 Perceptions of the companies

Respondents for both companies (6 people each for TPJ and PALYJA) mentioned a variety of aspects that they think is important. They were such as customer satisfaction, water availability (which implies the continuity of water) and sales. The reasoning for this is because improvements in such aspects have a link to better services to the customers, which means an increase of the water available to the customers.

Respondents for both companies perceive water supply services in Jakarta is improving due to increase in the service coverage ratio, water quality is compliant with the Ministry of Health standards, decrease in NRW and increase water in the distribution network.

Some respondents perceive water in Jakarta is bad since the NRW is at a high rate of 50%. Some explain the reason why the NRW is high is because of condition of the distribution network is not good due to old pipes and bad instalment of pipes. Other reasons why respondents say water in Jakarta is bad are because customers face intermittent water supply and low water pressure.

The respondents at both companies explained the reason why NRW remains high is because first, the condition of the distribution network is not good due to old pipes and bad instalment of pipes. Respondents at PALYJA said that although they know that they need to rehabilitate the network, the bottleneck in doing so is that companies have limited capital expenditure to spend and that investment plans needs to be agreed with PAM Jaya. They also explained that it is difficult to regulate illegal connections since they are in illegal settlements. Since companies do not have the right to make connections at illegal settlements, they cannot regulate such illegal usages of water. Respondents at TPJ explained that (in addition to the reasons that PALYJA explained) the process for making new connections takes time.

4 FINDINGS

Interviews and FGDs revealed that there are similarities and differences of which aspect of water supply services is important amongst the various stakeholders.

As for the customers, the most common aspect that they value was the continuity of water. Their reasoning for this can be categories into two reasons. First is because water is not available when they need it and second is because they have to pay for the piped water although it is not available when they need it. The next important aspect seems to be the price of water since customers were often saying that they do not mind paying more for water as long as they get it. Their common complaints were "Why do we have to pay of what we don't get?".

As for PAM Jaya, the most important aspect was the NRW. The reasoning for this was because the private operators were making losses (i.e. not making revenues), wasting raw water and a reduction in NRW will automatically increase the volume of water delivered through the network.

The respondents of the JWSRB said that the most important aspect was the continuity of water, which is the same as what the customers think are important. Their reasoning for selecting that aspect was, so that customers can use water, because wastewater may intrude to pipe of not 24 hours water supply and also the NRW should be very low.

Respondents for the companies had a variety of aspects such as customer satisfaction which can be attributed to water pressure, continuity of water and good quality of water (no colour and no smell). Other aspects were water availability and sales.

Therefore, comparing such aspects with the monitored performance indicators stated in the contract, NRW which PAM Jaya think is important is indeed one of the indicators monitored. However, water continuity which customers and the JWSRB think is important is an aspect which is not monitored. This means that although water continuity was the most common aspect that customers and JWSRB value, this is not one of the aspects monitored in the contract. TPJ and PALYJA had a mixture of what they think is important that is either monitored or not monitored.

5 DISCUSSIONS

As PAM Jaya explains, indeed, a decrease in NRW would theoretically mean that there will be an increase in the water available to customers since there would be less water lost from the distribution network. Therefore, this should result in an increase in water pressure, which also implies that more water would be delivered to the periphery of the network. Thus, customers should see improvements (i.e. increase in water pressure and water continuity) of their water supply services and therefore, NRW should be a suitable indicator in understanding performance of water supply services. This explanation of how a reduction in NRW leads to an increase of water available to the customers seems logical. However, this mechanism seemed to have not worked for Jakarta although the NRW has decreased from 58.5% (in 1998) to 50% (in 2005) since all areas where the focus group discussions where held were facing water continuity problems.

In Jakarta's case an increase of the water pressure would not necessarily result in an increase of water available to the customers. This is because as companies have explained, the network of Jakarta is said to be badly installed and old. The fact that participants of the focus groups saw yellow water is an indication of such network condition. Pipes are broken and thus soil and wastewater intrude the pipes. Thus an increase in pressure will make additional leakages at the further end of the network since the conditions of the pipes are poor: the pipes will not be able to endure the additional pressure. Also an increase in pressure would also increase in NRW because more water would be leaking from the old/badly installed pipes since the water pressure will also increase within the distribution network.

However, NRW consists of technical losses and commercial losses. The above discussion would become more valid if the NRW was only due to technical losses. Within the 50% of NRW (as of 2005) in Jakarta, half of the losses is said to be due to technical losses and the remaining half is due to commercial losses (illegal connection and illegal consumption). However, it is still not clear up to date, the exact numbers of each type of losses (TPJ/PALYJA personal communication 2006).

Therefore, in the case of Jakarta a decrease of NRW would not necessarily mean that there would be an increase of water available to the customers. One of the companies was expressing that, under the current distribution network of Jakarta, if the priority were to increase the water available to the customers by increasing the water delivered from the water treatment plant, the NRW would increase. However, again, NRW consists of not only physical losses but also commercial losses. Thus, in the case of Jakarta, to reduce the NRW would be by decreasing the water losses by rehabilitating the existing network and tackling with illegal connections and consumptions.

6 CONCLUSION

In conclusion, it can be said that the reason why customers do not necessarily perceive improvements despite improvements in other performance indicators is because customers receive intermittent water supply and water continuity, what customers value most, is not monitored. Moreover, the reason why improvements in performance are not felt amongst the stakeholders is because they have different criteria in assessing performance which are not necessarily monitored. It also can be concluded that NRW is not an appropriate indicator to be monitored in understanding performance of water supply services in Jakarta. It would be more appropriate to monitor specifically which type of water losses (i.e. physical and/or technical) that companies should reduce.

7 REFERENCES

Cooperation Agreement Amended and Restated (RCA) as of 22 October 2001 Concerning Clean Water Supply and Service Improvement for the Eastern Part of Jakarta (Schedules). Jakarta.

Cooperation Agreement Amended and Restated (RCA) as of 22 October 2001 Concerning Clean Water Supply and Service Improvement for the Western Part of Jakarta (Contract). Jakarta.

Asian Development Bank (ADB). (1993). Water Utilities Data Book: Asia and Pacific Region. Manila: ADB.

Asian Development Bank (ADB). (1997). Second Water Utilities Data Book: Asia and Pacific Region. Manila: ADB

Behn., R., D., (2003). Why Measure Performance? Different Purposes Require Different Measures. Public Administration Review, Vol. 63, No. 5, pp586-606.

Brown, S.W. and Swartz, T.A. (1989). A gap analysis of professional service quality. Journal of Marketing. Vol. 53 No.2 pp92-98

International Water Association (2000). Losses from Water Supply Systems: Standard Terminology and Recommended Performance Measures. London: IWA.

Johnstone, N., Wood, L. (2001). Private Firms and Public Water. Realising Social and Environmental Objectives in Developing Countries. Cheltenham: Edward Elgar Publishing.

JWSRB (Jakarta Water Supply Regulatory Body). (2006). Jakarta Water Supply Report 1998-Apr 2006. Obtained through personal communication.

Lanti, A. (2006). 'A Regulatory Approach to the Jakarta Water Supply Concession Contracts', in Cecilia Tprtajada, Olli Varis, Jan Lundqvist and Asit. K. Biswas (ed.), Water Management for Large Cities. (2006). Oxon: Routledge, pp. 72-93.

McIntosh, A. (2003). Asian Water Suppliers: Reaching the Urban Poor. Manila: Asian Development Bank.

Nickson, A., and Franceys, R. (2003). Tapping the market. The Challenge of Institutional Reform in the Urban Water Sector. New York: Palgrave Macmillan.

PALYJA (2005). Presentation: Water for all 2005. Obtained through personal communication.

PALYJA/TPJ (2006). Water tariff brochure.

PT. Kekarpola Airindo (1995). Perusahaan Daerah Air Minum. Daerah Khusus Ibukota Jakarta. Unpublished.

Uitto, Juha I., Biswas, Asit, K. (2000). Water for Urban Areas. Challenges and Perspectives. Tokyo: United Nations University Press.

World Heath Organisation: WHO. (2000). Tools for assessing the O&M status of water supply and sanitation in developing countries. Geneva. WHO.

World Bank. (1994). World Development Report 1994: Infrastructure for Development. New York: Oxford University Press.

WB (1997). Toolkits for Private Participation in Water and Sanitation. Washington D.C.: The World Bank

WB (2003). Efficient, Sustainable Services for All? –An OED Review of the World Bank's Assistance to Water Supply and Sanitation. Washington D.C.: The World Bank.

A Tale of Two Benchmarking Initiatives: A South African Case Study

N. Moodley[1]

[1] *PDG. nishendra@pdg.co.za, P.O. Box 53123, Kenilworth, Cape Town, South Africa, 7745. Telephone: ++27 21 7973660. Fax: ++27217973671*

Keywords: Comparison of Sectoral & Municipal Benchmarking Initiatives in South Africa

Edited by Enrique Cabrera Jr and Miguel Ángel Pardo. ISBN: 9781843391913.
Published by IWA Publishing, London, UK.

ABSTRACT

In the last four years South Africa has seen two significant national benchmarking processes in the public sector. The first is the Vuna Awards for Municipal Performance Excellence conducted since 2003. The second is the more recent National Benchmarking Initiative for Water & Sanitation conducted since 2005.

This paper aims to compare, in the same developing country context, these very different initiatives and explore potential for linkages and alignment between municipal-wide benchmarking initiatives and sectoral benchmarking initiatives.

While both these benchmarking initiatives differ significantly in style, scope and depth, there are important lessons that each have for the other and their relative strengths and weaknesses are complementary. However, a common weakness is that both these processes need to expand their successes in metric benchmarking to make greater progress on process benchmarking.

There are also important linkages that can be drawn between these two initiatives and other sectoral initiatives that may be developed for solid waste, roads and other public infrastructure services.

1 INTRODUCTION

Performance measurement and benchmarking is of particular importance to public institutions in South Africa. Because of the history of apartheid and the deliberate lack of accountability of public institutions to majority of South Africa's people, public services such water and wastewater services, refuse removal and other municipal services were either non-existent or, where they existed, highly insulated from the needs and oversight of the public.

Because of this insularity, municipal services have been traditionally found to be ineffective in providing services to all households requiring services, inefficient in the use of resources, unresponsive to the changing needs of consumers and unaccountable and opaque with regard to oversight.

Through public policy and legislation government has introduced improved mechanisms of public accountability and transparency, which has required the introduction of performance management systems and the use of performance indicators in municipalities, since 2000.

As a consequence over the last four years, South Africa has seen the emergence of two significant benchmarking initiatives focusing on municipalities and municipal services.

The first is the Vuna Awards for Municipal Performance Excellence conducted since 2003. The second is the more recent National Benchmarking Initiative for Water & Sanitation conducted since 2005. The former is driven by a partnership of the Department of Provincial & Local Government (DPLG), the Development Bank of Southern Africa (DBSA), the South African Local Government Association (SALGA) and the National Productivity Institute (NPI). The latter is driven by a partnership between the Department of Water Affairs & Forestry (DWAF), SALGA & the Water Research Commission (WRC).

This paper aims to compare, in the same developing country context, these very different initiatives and explore potential for linkages and alignment between municipal-wide benchmarking initiatives and sectoral benchmarking initiatives.

2 DEVELOPING COUNTRY CONTEXT

2.1 Restructuring of Municipalities

The introduction of democracy in South Africa saw approximately 7000 municipalities rationalised into 284 municipalities by 2000. Recent demarcations have reduced this to 283 municipalities in 2005:

- 6 Metropolitan Municipalities (Category A)
- 231 Local Municipalities (Category B)
- 46 District Municipalities (Category C)

All local municipalities can be found within a district. This is not the case for metropolitan municipalities, which are single-tier. A district municipality generally has about five local municipalities within its boundaries.

This has represented a massive restructuring exercise in local government which needed to be dealt with while extending basic services to unserviced households.

In the division of powers and functions for public services the role of water service authority (which includes water & wastewater services) is always assigned to the Metropolitan Municipality where it exists, and is either assigned to the local or district municipality, more often the local municipality. However 21 of the 46 districts are water service authorities. All local and metropolitan municipalities are authorities for electricity and solid waste. There is still a lack of clear division of responsibility between districts and local municipalities with regard to the roads function.

2.2 Access to Services

As a result of Apartheid policies the majority of South Africa's citizens were left either unserviced or inadequately serviced. While significant progress has been made in rolling out new infrastructure, the backlog, particularly in urban areas has, with rapid urbanisation, grown considerably.

The Department of Water Affairs (DWAF) keeps regular updates of the situation with regard to backlogs with the current situation being:

Water supply:

- 0.65 million households (2.9 million people or 6%) do not currently have access to infrastructure.
- 1.65 million households (7.0 million people or 14%) do not currently have access to an adequate water supply (minimum standard which implies good quality water from a reticulated supply within 200 metres of the dwelling).

Sanitation

- 3.53 million households (14.3 million people or 29%) do not have access to adequate sanitation (minimum standard which implies a Ventilated Improved Pit latrine or equivalent).

Providing an effective service to all households requiring a service is still the most important priority nationally.

2.3 Poor Data

Both as a result of the amalgamation of municipal entities and a historical lack of development of good information systems, there is a very weak information base from which to draw good data for benchmarking. This is however better for most services regarded as engineering services and at early stages of development for other services.

2.4 Low Technical Capacity

The building and retention of capacity, particularly technical capacity, has never been higher on the agenda for local government in South Africa. Partly as a result of the tertiary education system not delivering the required technical skills needed and a 'brain drain' of expertise to both the private sector and other countries, municipalities have in general been unable to attract or retain the necessary skilled individuals. This is particularly severe for both engineers and planners.

2.5 Service Delivery AND Transformation

While a balanced approach is necessary in most contexts for benchmarking, the South African approach needs to emphasise together with effectiveness and efficiency considerations, notions of equity and institutional transformation priorities such as racial representivity and skills development. In other words our best performing public institutions should not be seen as those only delivering good services but those delivering good services while also transforming their institutions to deliver services sustainably and enjoy legitimacy in the eyes of those they serve.

3 COMPARING THE TWO INITIATIVES

This section provides a comparative analysis of the Vuna Awards for Municipal Performance Excellence and the National benchmarking Initiative for Water Services.

3.1 Origin

The Vuna Awards were initiated in 2003 as a way to celebrate three years of democratically demarcated and elected local government. In contrast to poor media coverage of local government at the time, the awards were introduced to recognise, award and celebrate successes. Its intention was to conduct these awards on an annual basis and it has been conducted annually since then except for 2005, when partly due to local government elections, it did not happen.

The Vuna Awards relies on voluntary participation of municipalities, who apply by submitting amongst other requirements, performance data required for performance indicators, for the financial year under review.

In its initial year, the process was conducted entirely by the Department of Provincial & Local Government using their internal capacity. While the process required performance data from municipalities the indicators were highly subjective and poorly formulated, undermining the credibility of the assessment process. In subsequent years external capacity has been used to plan for and manage the benchmarking process that culminates in the awards. In progressively moving the emphasis from the awards to strengthening the benchmarking process that underlies the assessment of municipalities, a more rigorous process and set of indicators has been adopted since 2004. The credibility of the awards process has subsequently improved significantly.

In contrast to the 'big bang' approach of the Vuna Awards, the NBI has been more cautiously planned and has evolved progressively, while being conducted annually since 2005. However, its origins lie in an earlier benchmarking initiative for 6 metropolitan municipalities and three of the secondary cities that formed a learning network called the City Water Managers Forum. Having implemented a benchmarking initiative for this comparatively better capacitated group, the approach was incrementally introduced to other municipalities who were authorised to perform the water function.

3.2 Process Followed

The Vuna cycle is ideally represented by the diagram below.

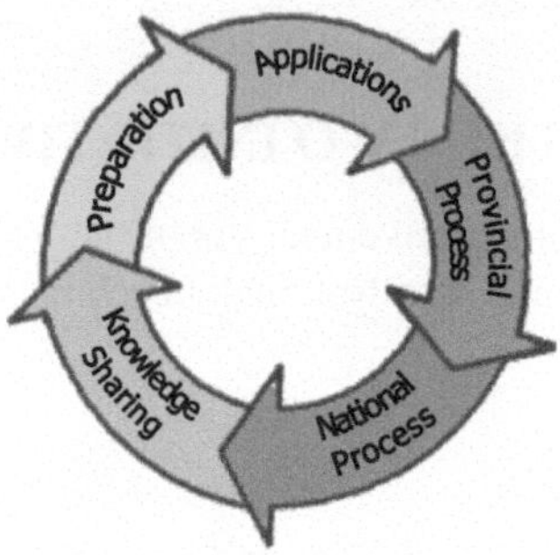

Figure 1: THE PHASES OF THE ANNUAL VUNA CYCLE

The Vuna process is made up of the following stages:

- A preparation stage in which the rules and indicators bare consulted upon and adopted
- An applications stage when municipalities voluntarily apply by submitting comprehensive performance data and other requirements.
- A provincial process by which participating municipalities are benchmarked, evaluated and shortlisted for further verification and interaction, before winning municipalities are adjudicated for provincial awards.
- A national process by which provincial winners and best performers in each key performance area are further evaluated and benchmarked together before a national adjudication for national awards are handed out at an awards ceremony.
- A knowledge sharing process by which innovative and best practices from leading municipalities are profiled. It has been the intention to introduce a conference to share best practice and to generate reportcards, so that individual municipalities receive feedback.

The NBI process is indicated in each of the steps described below undertaken annually:

1. **Confirmation of indicators**: (March and April.)
2. **Circulation of templates**: An induction pack (for new participants), and the data templates (for both existing and new participant) are circulated. (April to June.)
3. **New participant engagement**: SALGA invites new participants. (April to June.)
4. **Data collation and verification**: After the end of the Financial Year (June), municipalities collate and verify all of their performance data (with support from the service provider). (July to November.)
5. **Analysis**. Analysis takes place in the period December and January.
6. **Conference preparation**. Intensive activities related to the conference take place in the two months running up to the conference.
7. **Conference**. It is the intention to hold the conference in mid-February annually. The conference is 2 or 3 days in length depending on the programme agreed by the steering committee comprising DWAF, SALGA and WRC.
8. **Annual benchmarking report**. The report is produced after the conference, taking into account the feedback from the conference. The aim is to have a final draft available at the end of March and for this to be circulated to municipalities as soon after this as practical.
9. **Reportcard**. An individualized reportcard is produced for each municipality indicating how they performed in terms of performance indicators calculated from data provided and how they compare with peers.

3.3 Participation

The graph below indicates participation in the Vuna process from 2004 to 2007[4], by municipal category.

[4] There was no Vuna Awards process in 2005

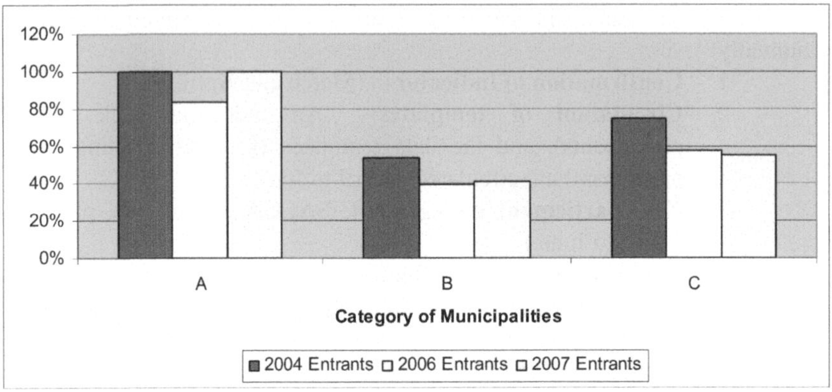

Figure 2: PARTICIPATION RATE OF MUNICIPALITIES BY CATEGORY

The participation rate of cities (category A) is high due to there being a small number (6) and that they have relatively much higher capacity. The decrease, since 2004, in participation is assumed to be largely due to the increased rigor of improved indicators and the lack of good information at many municipalities.

Table 1: ACTUAL PARTICIPANT NUMBERS FOR 2004 TO 200

Category	*2004 Entrants*	*2006 Entrants*	*2007 Entrants*	*Total Eligible*
A	6	5	6	6
B	124	91	94	231
C	35	27	26	47
Total	165	123	126	284

Participation in the NBI process has been incrementally extended since 2005. The South African Local Government Association (SALGA) has the primary responsibility to identify potential participants and to engage with them to obtain their willingness to participate in the benchmarking process.

In the first round (for 2004/05) 40 participants were identified and 22 responded. In the second round (for 2005/06) 70 participants were identified and 48 responded. This represents an increase in the response rate from 55% to 68%.

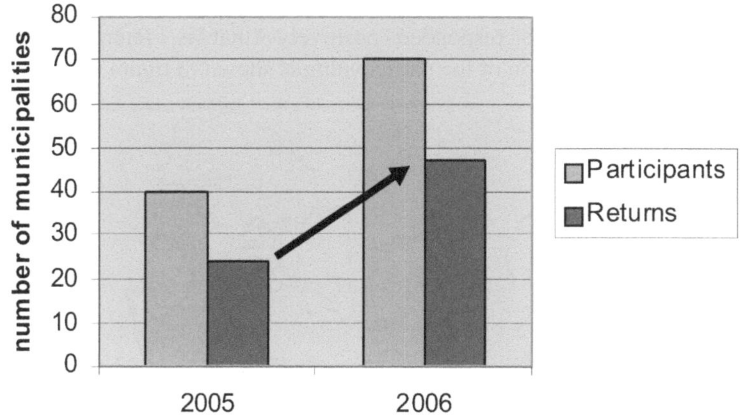

Figure 3: PARTICIPATION FOR EACH ROUND

The participation in 2005/6 for Metros, Districts and Locals is shown in the figure below. The participation rates (identified participants versus those who were able to provide information) for Metros was 100%, for Districts 75% and for Locals 56%. The higher participation rates in the Metros and Districts possibly can be attributed to the active peer networks operating for these water services authorities.

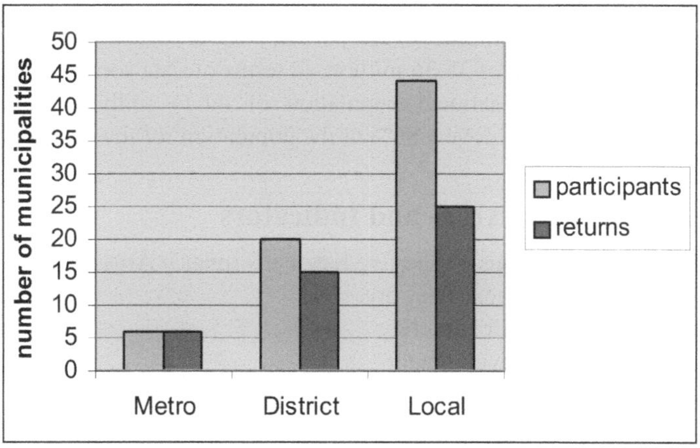

Figure 4: PARTICIPATION RATE FOR METROS, DMS AND LMS

In 2006, Seventy municipalities were identified by SALGA and accepted by the National Benchmarking Initiative steering committee. Of the 70 municipalities identified and approached to participate in the National Benchmarking Study 48 responded positively (that is, returned data). A geographical representation of the participating is shown in figure below.

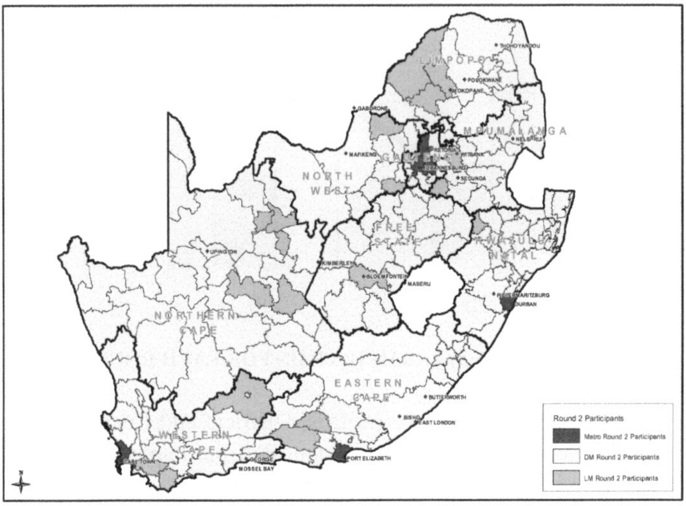

Figure 5: **GEOGRAPHICAL REPRESENTATION OF PARTICIPATION**

The figure above provides a geographical representation of the coverage of this benchmarking exercise.

The water service authorities that participated in 2006 round of the NBI account for a population of 38.36 million, in terms of data sourced from Census 2001. Against a total national population of 44.74 million in 2001, this benchmarking study thus covers 86% of the population[5] of the country.

3.4 Performance Areas and Indicators

The Vuna Awards focus on the five Key Performance Areas (KPAs) for local government in South Africa. They are:

- Basic Services & Infrastructure
- Local Economic Development
- Financial Management & Viability
- Institutional Transformation
- Good Governance

[5] South African Census 2001 Population distribution was used to calculate this.

Over the 5 KPAs, 141 fields of data are required to calculate 109 performance indicators.

Informed by the national Strategic Framework for Water Services (2003), the NBI involves indicators in the following performance areas:

Performance areas used in NBI:

1. Access to basic water supply
2. Access to basic sanitation supply
3. Quality of services: Potable water quality
4. Quality of services: Continuity of supply
5. Access to free basic services (water)
6. Access to free basic services (sanitation)
7. Financial performance: Affordability and debtor management
8. Asset management: Metering coverage and unaccounted-for water
9. Protection of the environment: Effluent discharge quality

It can be observed that the Vuna indicators are more balanced in their focus on effectiveness and efficiency of services, as well as paying attention to good financial management, transforming municipal institutions and engaging in good governance activities. The NBI indicators are purely focused on effective and efficient service delivery, with some attention to financial performance. The one staff related indicator which is 'staff per 1000 connections' is purely an efficiency indicator. The NBI indicators can be significantly challenged to, in future, include more indicators reflective of the transformation priorities of government balanced with the emphasis of efficient and effective service delivery.

Given that Vuna focuses institutionally on the whole of a municipality and reflects national priorities with regards to municipalities, and that NBI is dealing with the water sector only, it is no surprise that their indicators will differ in terms of scope and depth. The municipal-wide focus of Vuna means that only a few indicators can be selected from each sector, based on the strategic importance to overall municipal performance and national priorities. NBI indicators that relatively narrowly focus on the water sector only can ensure significant operational depth.

The following schematic tries to capture this. If the circle below is viewed as a cone from above then the national municipal-wide indicators are the tip of the iceberg. The detail lies in sectoral benchmarking initiatives in each sector.

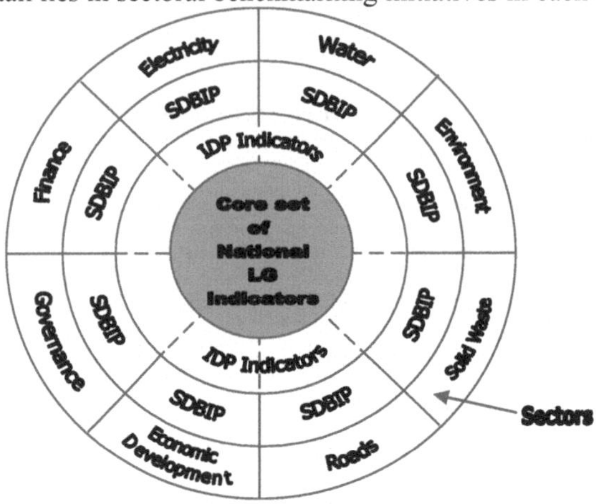

Figure 6: **SCHEMATIC OF RELATIONSHIP BETWEEN MUNICIPAL-WIDE AND SECTORAL INDICATORS**

IDP indicators refer to the municipality's Integrated Development Plan, required by legislation, as the strategic plan of the municipality across sectors. This reflects the broader set of municipal priorities. SDBIP are Service Delivery Business Improvement Plans that are also required by legislation for each municipal department often related to a sector. The outer circle reflects additional indicators valued by sectoral benchmarking initiatives.

3.5 Data collection

In both NBI and Vuna most of the required data is collected from the municipalities concerned. While Vuna relies on voluntary application or submission of information, supported by briefings held in each province, NBI relies on a similar process but provides significant capacity (between 2-3 days of contact time per municipality by the service providers conducting the benchmarking exercise) in order to support municipalities to understand the data requirements, definitions and process.

3.6 Data Quality

The Vuna process does not have significant quality control, except that data submitted must be signed off by the municipal manager who is the chief accounting officer of the municipality. Most information obtained is generally expected to be information audited together with the municipality's financial statements and annual performance report, it is legally required to publish. The provincial process is used to encourage provincial officials to query and verify data either randomly or by close inspection.

In reporting data into the NBI process municipalities are expected to qualify each field of data provided with its data quality. Each performance indicator (or grouping of like indicators) was coupled to one of the following "level of confidence" indicators:

- Not stated,
- Estimate,
- Reliable information, or
- Audited information.

Using a numerical equivalent for each level (as indicated above), the average level of confidence of the data received for the Metros, Districts and Local Municipalities are calculated. The average data confidence in the 2006 process for Metros, as an example, is shown in the figure below. There is great variability between water services authorities with respect to the level of confidence of the data supplied.

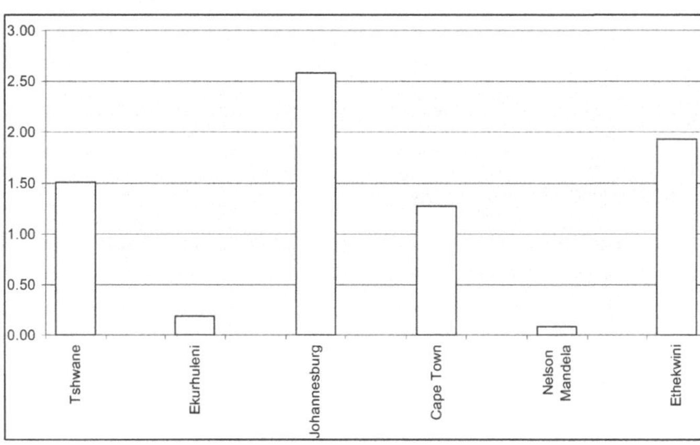

Figure 7: AVERAGE DATA CONFIDENCE ON THE METROS

3.7 Data Analysis

The NBI process ensures that the data is analysed into a report on the performance of all water service authorities in the benchmarking exercise. As yet the NBI process has noted that it cannot as yet focus on providing meaningful analysis of trends and that it is putting effort into getting the data right first and refining the indicators and benchmarking system.

The Vuna process as yet has not taken the opportunity to fully analyse performance trends. As an awards process primarily, it has been largely consumed by generating a score for municipalities based on their performance data.

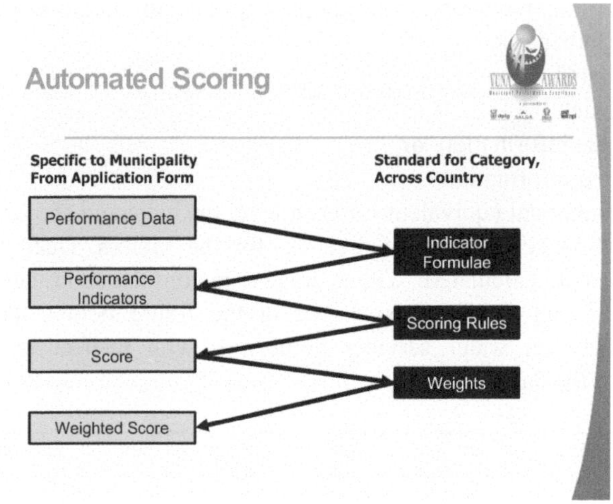

Figure 8: AUTOMATED SCORING PROCESS OF THE VUNA AWARDS

3.8 Recognition & Awards

The Vuna Awards process has been primarily a competition to provide recognition and awards in municipal performance excellence. However in many ways this placed the proverbial 'cart before the horse', in that initially it was not grounded in a credible benchmarking system to arrive at this.

The NBI process has been much more cautious and developmental, by building first a sound and credible benchmarking process, on whose foundation, awards and recognition can now be considered. In the 2006 process for the first time, it identified the best performing water service authorities in each category of municipality.

3.9 Knowledge Sharing & Process benchmarking

Both processes have had significant successes in metric benchmarking. However they can do much more to promote process benchmarking and should invest significantly in this in forthcoming years.

There are some elements of process benchmarking that they have explored.

The Vuna process has involved an appraisal of innovation and special programmes at municipalities as well as the identification of best practice projects and programmes that underlie good municipal performance. These have been documented in booklets and video / DVD for purposes of knowledge sharing.

The NBI process culminates annually in a conference which is largely aimed at both understanding benchmarking and the emerging metric benchmarks, as well as to surface the innovation and best practice that arise in leading municipalities.

For both initiatives they can develop better process benchmarking from complementing each other.

3.10 Marketing & Profile

The Vuna Awards, when compared to NBI, has been able to invest significant resources in marketing itself nationally and becoming a well recognised brand in the sector. Once again, this may have been premature without a sound and credible benchmarking process underlying it. However, it can be argued that it has been fortunate to have recovered credibility, before any irreversible damage was done to the brand and can claim presently to be a well grounded benchmarking process, with significant sector awareness and credibility.

The NBI process, in contrast, has been cautious in developing a tested and improved benchmarking system. It has operated in a understated way, in true technocratic style, with little awareness other than by the officials directly engaged with it. It has a lesser profile amongst local government councillors and the public are unlikely to be aware of it. Now that the process is working, there is a huge opportunity to improve the profile and awareness it receives.

4 CONCLUSIONS

Both these initiatives face very similar challenges. They are both voluntary benchmarking initiatives and rely on there being sufficient municipal interest in making these initiatives work. Both are highly challenged by the availability of good performance data for desired performance indicators, which in a developing country context, is likely to be one of the most inhibiting factors to good benchmarking. Both processes have relied on a foundation of metric benchmarking and are now starting to evolve a more significant element of process benchmarking.

They are very different, in terms of scope and as a consequence, in terms of depth. Because municipality-wide benchmarking is likely to cater for most municipal services and functions, the depth that can practically be attained, cannot be compared to sectoral-specific initiatives such as a Water & Sanitation benchmarking initiative. However the Water and Sanitation benchmarking initiative can be challenged to extend beyond the notion of efficient and effective service delivery to also focus on fulfillment of the institutional transformation agenda required in South Africa public institutions.

While there are differences and similarities inherent to these types of benchmarking initiatives, the approaches undertaken have also been considerably different. The Vuna Awards process can be characterised as a 'big bang' approach. It embarked initially on a high media profile process of benchmarking municipalities, initially using a set of poorly developed indicators and an inadequate process, while placing emphasis on branded awards and significant prize money as incentives for participation. While the Vuna indicators and benchmarking process was characterised by many shortcomings initially, these have evolved and have been improved significantly over recent years, restoring credibility to this process and its awards. Awareness and participation remains relatively high, although more sophisticated indicators and rigorous benchmarking processes have resulted in fewer low capacity municipalities feeling confident to participate.

In contrast, the Water & Sanitation initiative has relied on a more cautious and incremental approach. A set of good quality indicators initially used, has now evolved incrementally to a more comprehensive set and will soon be available on a web-based data collection and reporting tool. However, both media profile and awareness, at least amongst non-technical municipal officials and councillors, remains low and this, together with the lack of incentives and recognition, may be contributing to lower than ideal levels of participation.

While both these benchmarking initiatives differ significantly in scope and depth, there are important lessons that each have for the other and their relative strengths and weaknesses are complementary.

There are also important linkages that can be drawn between these two initiatives and other sectoral initiatives that may be developed for solid waste, roads and other public infrastructure services. The linkages relate to improved learning between these initiatives, the potential sharing of systems including the same software platform, common sets of indicators defined in the same way, a common framework for recognition and awards, and shared branding, media and publicity.

5 REFERENCES

Alegre, H., Hirnir, W., Baptista, J.M. and Parena, R. (2000) Performance Indicators for Water Supply Services. IWA Publishing, London, U.K.
PDG. (2006) Evaluation of the Vuna Awards for Municipal Performance Excellence
PDG. (2006) National Water Services Benchmarking Initiative, Benchmarking outcomes for 2005/6. South African Local Government Association.

Survey of performance indicators for sewer cleaning and sediments in Austria

Th. Ertl [1]

[1] *University of Natural Resources and Applied Life Sciences, Vienna, Institute of Sanitary Engineering and Water Pollution Control, Muthgasse 18, A-1190 Vienna. Email: thomas.ertl@boku.ac.at*

Keywords jet cleaning, operation, performance indicator, sediments, sewerage, survey

 Performance Assessment of Urban Infrastructure Services
Edited by Enrique Cabrera Jr and Miguel Ángel Pardo. ISBN: 9781843391913.
Published by IWA Publishing, London, UK.

ABSTRACT

A survey about jet cleaning of sewers has been carried out in Austria for the year 2005. 498 out of about 1000 potential operators participating within the so called "Neighboorhoods" of WWTPs and Sewerage (**www.kan.at**) returned filled data-sheets. Statistical analysis led to results describing the cleaning activities and sediment disposal. Furthermore an evaluation of the relevant IWA performance indicators is given considering the data availability.

1 INTRODUCTION

Although sewer systems are intended to be designed for self-cleansing, it is necessary to clean the pipes in intervals to prevent blockages, flooding and corrosion. As state of the art high-pressure-jetting is used in Austria and worldwide (as confirmed by Calomino et al., 2007; Campbell and Fairfield, 2007). To get an idea which amount of sewers is cleaned each year and how much sewer sediments are removed by cleaning, a survey has been carried out.

2 METHODS

At the yearly meeting of the speakers (coordinators) of the so called „KAN"-Neighbourhoods of WWTPs and sewer systems a questionnaire about sewer cleaning and removed sediments has been presented (see www.kan.at). Afterwards an Excel-sheet has been sent by email to all the speakers, so that they could spread it to their members and to collect them filled out and sent them back to the author.

The recipients have been 949 WWTPs and 137 sewer operators all over Austria except of Vienna. 58 sewer operators do not manage a WWTP, this leads to a basis of 962 potential survey participants. Finally 498 filled out sheets have been received for evaluation. The feedback quote results in 52 %, which is actually very satisfying.

The Questionnaire consisted of the following items:

- Name of operator, length of sewer [km] , capacity of WWTP [p.e.]
- Sewer cleaning carried out by: own cars, contractor, mixed, no cleaning in 2005 [tick if appropriate]
- Cleaned length in [m] and amount of sediments [m³], [ton]

- Analysis of the sediments: general, residue analysis according to regulation, number of analyses
- External contractor takes care of residue treatment and disposal (without documentation for operator)? [Yes, No]

Questions about treatment and temporary storage at WWTP:

- Taking over sediments from companies [Yes, No], [m^3/a]
- Treatment of sediments from own sewerage [m^3/a]
- Treatment with sediments from sand trap: [m^3/a]
- Extra treatment for sewer sediments: [m^3/a]
- Temporary storage for drainage: [m^3/a]
- Recycling as building material (filling of sewer trenches, road construction): [m^3/a]
- Disposal of sewer sediments: [Yes, No], [m^3/a]
- Certificates at hand: [Yes, No]

3 RESULTS

3.1 Statistics of sewerage lengths and design capacities of the WWTPs

When taking all the returned sheets the following sums of operated assets are included in the survey:

Total length of sewers: 21.112 km (495 values)

Total capacity of WWTPs: 10,649.824 p.e. (489 values)

When taking only data from operators knowing both the length of sewer system AND the capacity of the WWTP the following distribution can be seen. In table 1 all the participants are grouped according to the length of sewerages.

3.2 Results for cleaning of sewers

The sample for these results is 299 operators having carried out sewer cleaning in 2005.

Total length: 17.096 km (298 values)

Total p.e. capacity: 8,651.103 EW (299 values)

Total length of cleaned sewers: 3.375 km, i.e. 20% of total length

Cleaned length of sewers. Figure 2 shows the distribution of cleaned lengths of sewers.

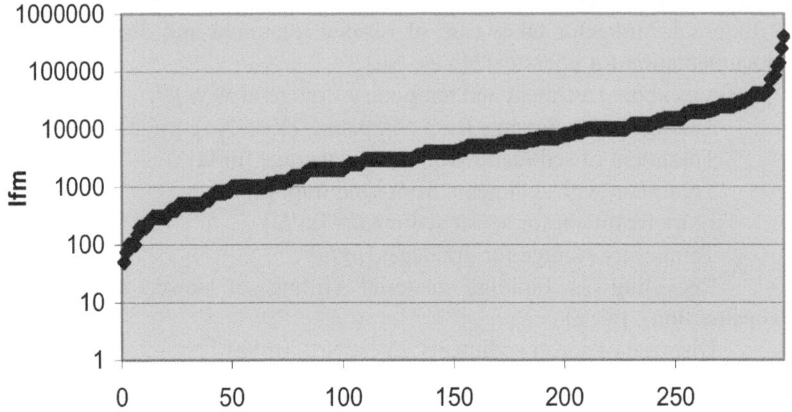

Figure 9 DISTRIBUTION OF CLEANED LENGTHS OF SEWERS

In table 1 the statistical values for the cleaned lengths are given.

Table 2 STATISTICS OF CLEANED LENGTHS OF SEWERS

Cleaned lengths	[m]	[%] of total
Average	11289	23,1
Median	4000	13,7
Min	50	0,1
Max	400000	100,0
25% Quartile	1292	6,1
75% Quartile	10000	30,3
Number of values	299	275

As Figure 1 and 2 show the range of cleaned length is enormous as well as the range of the percentage of the total length. Figure 2 shows the distribution of the cleaned length as percentage of the total sewer length.

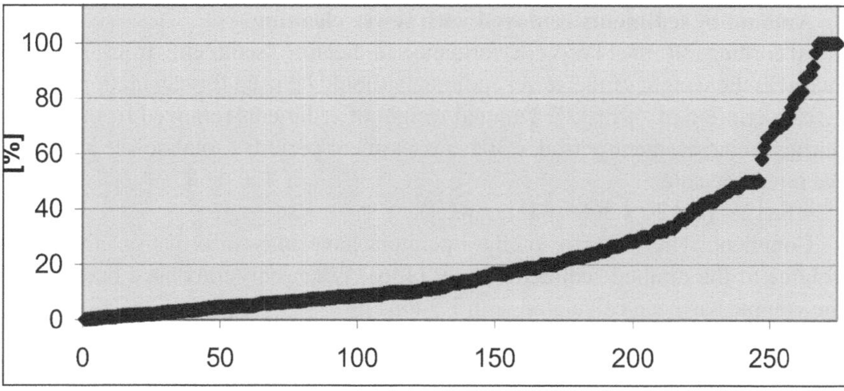

Figure 10. DISTRIBUTION OF THE CLEANED LENGTH AS PERCENTAGE OF THE TOTAL SEWER LENGTH

Figure 2 shows a nearly equally distribution from very small percentage up to the whole sewerage cleaned. A conclusion on intervals from these values should be made with care. When looking into the standards we find recommended intervals between 2 and 5 years depending on the type of system and likelihood of deposition. If the operators would clean the same amount each year the participants with percentage between 50% and 20% which is a minority would have intervals according to the standard. Milojkovic et al. (2007) calculated the IWA performance indicator (see Matos et al., 2003) "wOp2" Sewer cleaning for Belgrade Sewerage System between 10% and 40% for the years 1995 till 2005.*Comment regarding the IWA performance indicator "wOp2" Sewer cleaning (%/year).* "wOp2" should include only sewer cleaning referring to actions under a proactive management strategy. Root cutting shall be included. Curative cleaning of blockages is reflected in wOp34 "Sewer blockages". One of the main goals of the survey was to get figures of the sewer sediments brought out in total. The operators were not asked for the cause of cleaning, because blockages are poorly documented. So the results comprise both types of cleaning, proactive and reactive. This may explain the big range in percentage of cleaned length of the total systems (Figure. 2) and also the big range of sediment volumes per cleaned length (Figure. 3).

Amount of sediments removed with sewer cleaning

According to the IWA performance indicator "sediment from sewers" (wEn12) the weight of the sewer sediments should be used for calculation.

(Description of "wEn12": Drained weight of sediments removed from sewers during the assessment period x 365 / assessment period) / total sewer length at the reference date

wEn12 = (wA20 x 365 / wH1) / wC1)

Comment: The majority of the operators gave only information about the volume of the drained sediments in m^3 (136). When only tons have been given the volume has been calculated with 1.5 tons/m^3.

Sample:

- 498 feedbacks have been considered
- 199 answers with no sewer cleaning in 2005
- 299 operators with sewer cleaning in 2005
- 231 of them with a certain amount of sediments removed (77% of 299)

All the following results include data from only the 231 operators having sediments removed by sewer cleaning.

Table 3 STATISTICS OF AMOUNT OF SEDIMENTS (231 VALUES)

Amount of residues	[m^3/a]
Average	42,14
Median	8,50
Min	0,25
Max	2321,00
25% Quartile	3,00
75% Quartile	28,50

Figure 3 shows the distribution of the specific amount of sewer sediments in m^3/km resp. liter/m for all operators having removed sediments and additionally separated in those with own cars and contracted service.

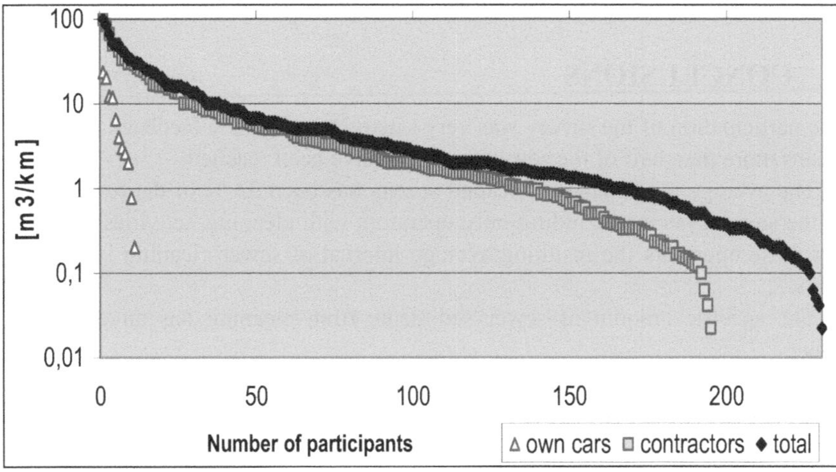

Figure 11 DISTRIBUTION OF THE SPECIFIC AMOUNTS OF SEWER SEDIMENTS RELATING TO CLEANED LENGTH

No correlation can be seen in figure 4 between the specific amount of sewer sediments and cleaning by own or contracted service. The range of both sets of values is nearly the same.

Table 4 STATISTICS OF SPECIFIC AMOUNT OF SEWER SEDIMENTS RELATING TO CLEANED LENGTH

	[m^3/km]=[ltr/m]
Weighted average	3,97
Average	6,50
Median	2,00
Min	0,02
Max	100,00
25% Quartile	0,85
75% Quartile	5,59

4 CONCLUSIONS

The participation of the survey was very satisfying with 52% feedback ratio that means more than half of the sewer operators have been reached.

The average percentage of cleaned sewers has been 23 % of the total length of the sewer systems including only operators with cleaning activities in 2005. For these operators the resulting average interval of sewer cleaning is about 4 years.

The specific amount of sewer sediments from cleaning has an enormous range.

The results of the survey cannot be directly taken for calculation of the relating IWA Performance Indicators (i.e. sewer cleaning, sewer sediments) because the data behind does not exactly correspond to the description in the IWA manual, but the operators normally cannot supply the exact data.

5 REFERENCES

Calomino F., Gabelli G., Miglio A. (2007) Water jetting for sewer cleansing. First experimental results for efficiency in non-cohesive sediment removal. Proceedings of NOVATECH 2007, the 6th international conference on sustainable techniques and strategies in urban water management. Lyon - France - June 25-28, 2007

Campbell S.A. and C.A. Fairfield (2007) An overview of the various techniques used in routine cleaning and maintenance of clay, concrete and plastic drains. Construction and Building Materials, In Press, Corrected Proof, Available online 27 September 2006

Matos, R., Cardoso, A. Ashley, R. Duarte, P., Molinari, A. and A. Schulz (2003) Performance indicators for wastewater services. Manual of Best Practice. IWA Publishing.

Milojkovic, I, et al. (2007) Preliminary Assessment of Performance Indicators for the Sewerage System in Belgrade. Proceedings of the 4th IWA Specialist Conference on Efficient Use and Management of Urban Water Supply. 21-23, May, 2007 Jeju Island, Korea

Urban water conservation based on implementing appropriate water pricing policy, network upgrading measures and public awareness activities

V. Kanakoudis[1]

[1]*Lecturer, Civil Engineering Dept.University of Thessaly, Volos, Greece. bkanakoud@civ.uth.gr, Tlf +30 6937217747*

ABSTRACT

Water conservation in urban areas under water shortage conditions is not just a prospective but the only way to sustainability. When a Water Utility manager is

 Performance Assessment of Urban Infrastructure Services
Edited by Enrique Cabrera Jr and Miguel Ángel Pardo. ISBN: 9781843391913.
Published by IWA Publishing, London, UK.

called to decide whether a water conservation project, based on implementing appropriate water pricing policy, network upgrading measures and Public awareness activities, will have the desired results being cost-effective at the same time, has to relay on the project's ex-ante evaluation. This is actually a blue print based mainly on assumptions, increasing his hesitations and second thoughts. The project's ex-post evaluation is the only way to judge whether its approval was a wise think to do or not. Ex-ante evaluation is what is available; ongoing evaluation is usually a failure-correcting process; ex-post evaluation is what is needed. But who can fast-forward the time? Experience form projects implemented in networks having similar characteristics can be used as the most-desired time machine. The paper shares this kind of experience from the ex-post evaluation of two case studies in Greece (Athens and Thessaloniki). In Athens the project included the implementation of an aggressive Water Pricing Policy, while in Thessaloniki the project included the installation of PRVs to form DMAs. Both of them included Public Awareness Campaigns.

1 INTRODUCTION

Water conservation in a rapidly developed urban area under water shortage conditions is not just a prospective but the only way to sustainability. The urban water conservation measures can be categorized as Structural, Operational, Economic and Socio-Political. Structural measures include: a) development of an accurate water use metering network; b) installation of house-kit water saving/recycling/reuse devices; c) installation of PRVs in the delivery network; d) reuse of residential waste water for public/municipal/commercial/industrial use. Operational measures include: a) leak detection; b) minimization of network's response/ resilience/ repairing of breaks time; c) water use restriction to suppress demand peaks. Economic measures include: a) establishing a water pricing policy based on an appropriate structure and level of water rates; b) adoption of economic incentives that encourage/reward water conservation; c) imposition of economic disincentives/penalties that discourage/punish water over-consumption; d) adoption of the most effective water charging cycle. Socio-Political measures include: a) public awareness campaigns for the establishment of a "water conscience"; b) public education seminars for the use of water saving devices. These measures are targeting at: a) the users (mainly residents) and especially at their mentality and the way they think and act. These measures include Public Awareness Campaigns that focus on the establishment of a water conscience; installation of home-kit water saving/reuse/recycling devices; public acceptance regarding the necessity of an appropriate (and some times unpopular) Water Pricing Policy on behalf of the Water Utility aiming at full water cost recovery (including environmental/resource costs in accordance to the WFD 2000/60/EC); b) the Water Utility especially regarding operational methods-measures. These measures include design/implementation/supervision of an effective Water Pricing Policy that will ensure water cost recovery (at which extent? This depends on the willingness and alertness of the customers); pipe replacement projects based on technico-economical analysis of all types of costs involved in a pipe break or leak; network upgrading projects based on minimizing the night time flow rate (PRVs installation).

2 THE CASE STUDY OF ATHENS

The water system of Athens, the capital of Greece, services today a population of 4,000,000 people. The local Water Utility used those days an inclining block rate to charge the water consumption (Table 1). The need for water conservation in the greater Athens area emerged for the first time in 1982, due to the fact that water demands exceeded the capacities of the water resources being used, till then (Lakes of Marathon and Yliki). This was when, for the first time ever, an integrated water conservation program was adopted.

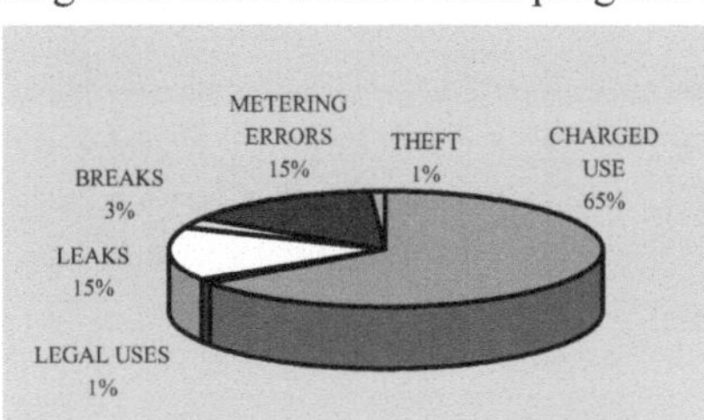

Figure 1. WATER ALLOCATION IN ATHENS

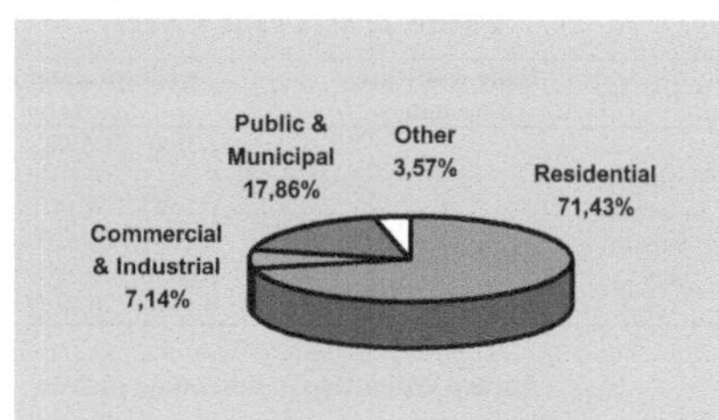

Figure 2. WATER USES IN ATHENS

Table 1. THE WATER RATES IN ATHENS

KIND OF WATER RATES	WATER RATES STRUCTURE (m^3)	WATER RATES LEVEL (euros/m^3)		
		1/1/90	1/5/90	1/7/92
Residential (per three months)	0-5	0,0851	0,1761	0,2993
	6-20	0,0851	0,2641	0,4519
	21-27	0,1966	0,5869	1,2561
	28-35	0,2348	0,8217	1,7608
	>35	0,2348	1,0271	2,2010
Industrial-Commercial (per month)	0-1000	0,1174	0,3522	0,5635
	>1000	0,1379	0,4109	0,6603
Public-Municipal (per month)	Overall	0,1467	0,4402	0,7337
Fire fighting-Harbor (per month)	Overall	0,0910	0,2847	0,4872
Charity (per month)	Overall	0,0558	0,1174	0,1966

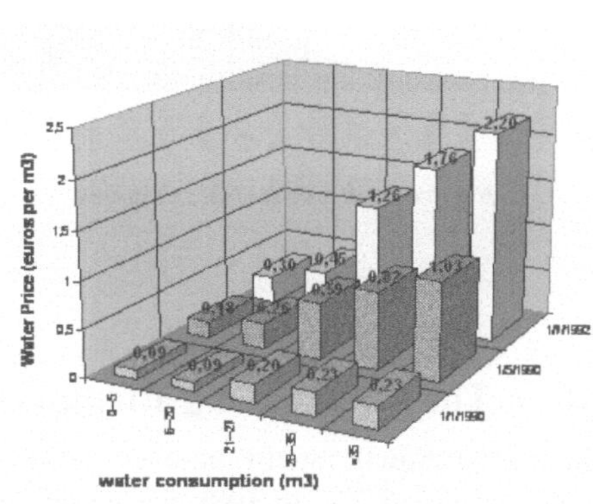

Figure 3. RESIDENTIAL USE WATER RATES

This program was based on a more aggressive pricing policy that took shape as an increase of the price levels of the water rates used. Harder and more intense water saving measures were enforced from 1990 to 1993. These measures included increase in the water prices and restriction of some particular water uses, which were thought to be a waste at that point of time. These unpopular acts seemed to be the only way to overcame the severe long drought period that the greater Athens area was experiencing since the beginning of 1988 (1988-1993) (Kanakoudis 1998). One of the main characteristics of this period was that although the decrease of the total annual precipitation was 20%, the diversity of its allocation, resulted in a dramatic decrease, of 50 to 55% of the water discharge. This was basically due to the fact that during the "used to be" rainy winter the precipitation was cut down to 30% of the normal level resulting in great values of the so called "enlargement" factor that transforms the rain to water flowing on surface.

Table 2. WATER USES

Water use	Daily water use (lt/customer)	%	Average water charge (euros/m^3)		
			1/1/90	1/5/90	1/7/92
Residential	200	71.43	0,2113	0,8070	1,6875
Industrial	20	7.14	0,1497	0,4519	0,7249
Public	50	17.86	0,1731	0,5194	0,8657
Rest	10	3.57	0,0880	0,2377	0,4050
Total	**280**	**100**	**0,1966**	**0,7102**	**1,4263**

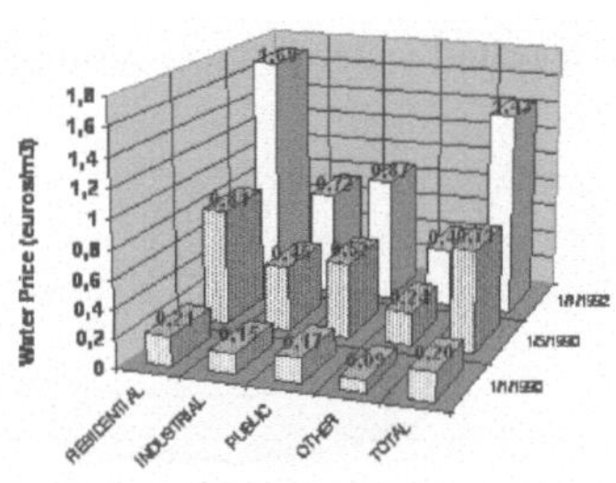

Figure 3. RESIDENTIAL USE WATER RATES

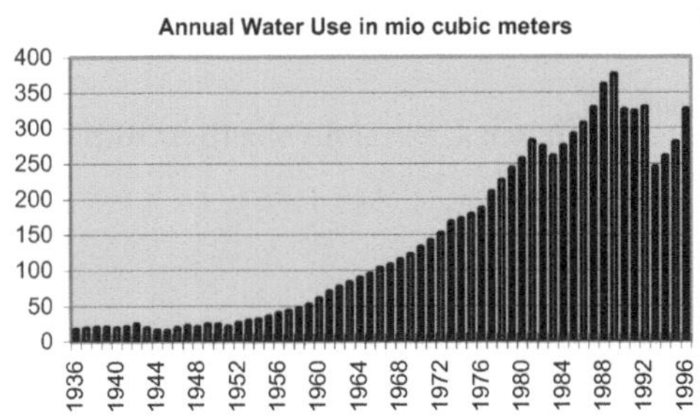

Figure 4. WATER USE (1936-96)

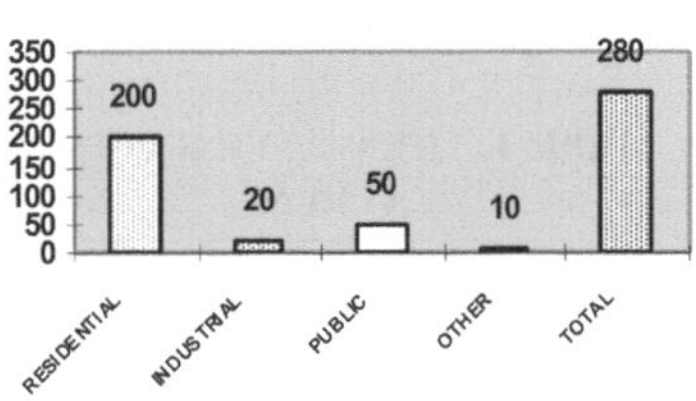

Figure 5. WATER USE

2.1 The water saving attempts of 1990

The annual growth rate of the water needs covering a 60 years period (figure 4) in Athens was 8.2%. In 1989 the water demands were almost 376 millions m^3 and till May '90 the situation was not getting any better. Considering the calculated annual growth rate, the estimated water demands would have reached the capacity of the water supply system (518 millions m^3) in 1994. In addition to this problem, the Water Utility was facing another critical situation. Since the beginning of 1988, the greater Athens area was experiencing a severe long drought period, resulting in water resources efficiency problems. The Water Utility was facing the "seemed to be" unavoidable nightmare of water shortage problems even long before the revival rains of February (Kanakoudis 1998).

Table 3. FACTORS OF THE MONTHLY CONSUMPTIONS IN ATHENS (1988)

	Factor of Monthly Consumption		Factor of Monthly Consumption		Factor of Monthly Consumption
January	0.86	May	1.04	September	1.10
February	0.79	June	1.12	October	1.05
March	0.89	July	1.21	November	0.93
April	0.91	August	1.17	December	0.93

In order to save as much water as possible, the Water Utility increased the water prices almost 300% (May 1990) and at the same time started a campaign using the media, in order to aware the public for the dramatic size of the water shortage problem and to establish a conservation consciousness. These raises on water prices yielded in saving almost 13% and 14% of the annual water demands in 1990 & 1991 respectively (Table 4). Considering the annual growth rate of the water needs (8.2%), these savings reached the levels of 20% and 26,5% respectively (Table 5). The effectiveness of the conservation program derived both from the relative increase of the water prices and the actual levels that these prices had reached. To be exact the water price levels before May '90 were extremely low, as it can be shown in Tables 1 and 2 resulting in an average monthly water charge of only 19,67 €cents/m^3. These levels were being further scaled down each year by 10% (national Rate of Inflation at that time). After the 1990 raises the average monthly water charge reached 71,02 €cents/m^3. In Athens the levels of water prices, before 1990, were not even close to the actual water cost recovering only its one third, resulting in an non-stop increase of the Water Utility's debt. The raises of 1990 just whittled down the gap between the water prices in Athens and other capitals abroad (1995 U.S.A. National Water and Wastewater Rate Survey - Jordan 1995).

2.2 The water saving attempts of 1992 and 1993

During the period that followed the raises of 1990 and till the summer of 1992, the drought became more severe, resulting in exhaustion of the water resources reserves. The situation was extremely critical as it was obvious that from June 1991 the public's sensitivity was eased off. This negative turn derived also from the unavoidable de-escalation of the 1990 raises due to the significant level of the National Inflation's Rate (10%). That's why during 1991 the water savings were only 0.8% compared with the water needs of 1990. The whole thing got out of hand during the first months of 1992, mainly on May when the water demands increased approximately 13.2% compared to the respective water demands of May '91. It was obvious that the entire water saving effort was falling apart. Facing this tragic situation, in July '92, the Water Utility doubled the 1990 water prices (Tables 1,2), resulting in an average monthly water charge of only 142,63 €cents/m^3. The new levels were the highest across the EU and almost 625% greater than the levels used before May '90.

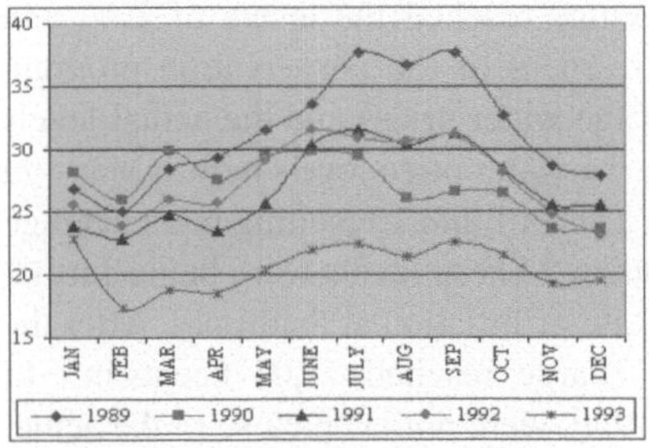

Figure 6. THE RESULTS OF THE CONSERVATION MEASURES ON THE MONTHLY WATER CONSUMPTION IN MILLIONS M^3

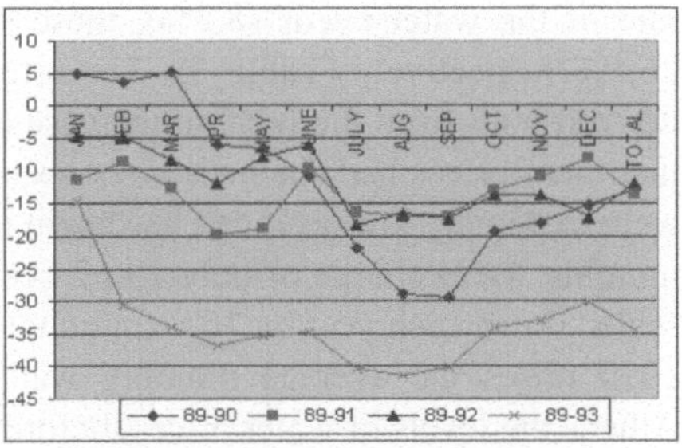

Figure 7. WATER SAVING RESULTS PER MONTH AS A PERCENTAGE OF 1989 WATER CONSUMPTION

Although the water demands of December '92 were decreased 10% since December '91, this proved not enough, considering the levels of water resources reserves. The annual water demands of 1992 were increased approximately 2% compared to those of 1991 but were reduced approximately 31% considering the annual growth rate of the water needs. At this point, for the first time ever, a program of restricting certain water uses (e.g. car and terrace washing) and punishing the water waste (water pricing penalties) was enforced to the customers. These restriction measures resulted in reducing the annual water needs of 1993 by 35% and 25% compared to the needs of 1989 and 1992 respectively. Actually the annual water demands of 1993 were reduced approximately 52% considering the annual growth rate (8.2%) of the needs.

Table 5. & Figure 8. EFFECTIVENESS OF THE WATER CONSERVATION MEASURES

	water consumption (millions of cubic meters)		
	without conservation (estimated values)	with conservation (actual values)	total reduction (%)
1989	375.9		
1990	406.8	326.5	-19.7
1991	440.3	323.8	-26.5
1992	476.5	330.3	-30.7
1993	515.7	246.4	-52.2
TOT	**1,839.3**	**1,227.0**	**-33.3**

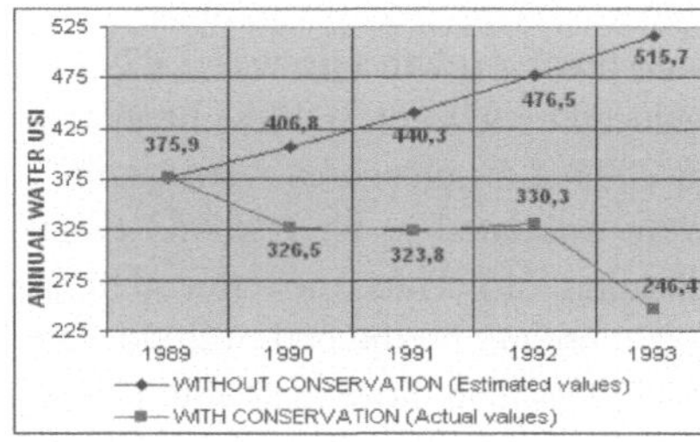

3 THE CASE STUDY OF THESSALONIKI

Thessaloniki is the second biggest city in Greece. Its network consists of 1,556km pipes, serving 1,000,000 customers. The annual water consumption is 94,000,000m3. Pipe materials include asbestos (58%, ∅80-350mm); cast-iron (16%); PVC (15%, ∅50-150mm); iron (11%). The pipe failure records being kept in the existing 8 pipe service centers were used to specify the breaks per pipe material. PVC pipes were not further examined as they experience too many annual breaks, regarding their age (installed in early 90's). This resulted from the lack of know-how and construction experience and also from the fact that some PVC pipes (connection ones) were constructed by the residents themselves (in order to satisfy their own needs) and not by trained personnel, aiming only at cost reduction instead of quality of service..

In Thessaloniki 65,000 to 83,000m3 of water are being lost every day (30% of the total water supplied). This fact forced EU to "suggest" that the Water Utility should find a way to minimize those water losses in order EU to co-finance the construction of a water supply aqueduct from Aliakmonas river to the city (studies proved that this water work will supply the city with enough water to cover its needs till 2050). Another presupposition set by EU was that a more rational use of the water should be achieved by implementing an effective water demand management policy. This forced the Water Utility to apply a huge research project based on water pressure monitoring, in order to fully understand the way that its network operates. Results revealed that in many network sections the water pressure was way more than 40m, which is considered satisfactory for an 8 floors building without a boosting device. Based on these findings, the installation of 5 PRVs in specific "nodes" of the network that took place in 2000 was considered to be the most cost effective thing to do.

The first step for the project's ex-post evaluation was to define its positive impacts on the pipe break rates. Prior to the installation of the PRVs (2000), the pipe break growth rate was 3.47%, while since 2000 breaks that occurred were reduced by 42% (1,364 less breaks) compared to those that would have occurred in 2006 if no measures were taken (Figure 9). The water losses during a break were estimated to be 1,357m3 (∅200mm pipe; water velocity: 2m/sec; break duration: 5h). Thus, the reduced breaks from 2000 to 2006 resulted in savings of 7,224,668m3 by the end of 2006, as the total breaks that actually occurred were 5,324 less than those that would have occurred since 2000.

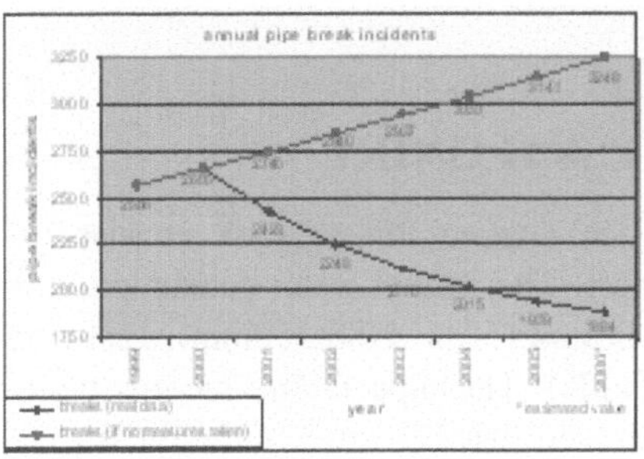

Figure 9. ANNUAL PIPE BREAK INCIDENTS

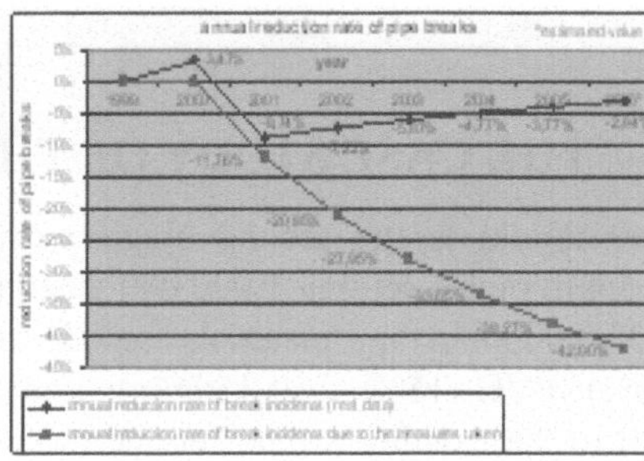

Figure 10. ANNUAL PIPE BREAK RATE

3.1 Environmental Aspect

The project had an important water saving outcome (7,224,668m^3 less water being lost by the end of 2006). Based on the Water Utility records, till 2000, although the annual growth rate of its customers was 1.5%, the respective growth rate of the water consumption was 1%. The "water savings" of 0.5% resulted from a public awareness campaign, lunched by the Water Utility. The need for immediate water conservation was urgent due to the severe water shortage problems the city is facing during the last decade. Since 2000, although the annual growth rate of the customers was also 1.5%, the total water consumption was annually reduced by 1% (0.68Mm^3/year that is 4.76Mm^3 for the period 2000-2006). This reduction was a joint result of the PRVs installation project; an Aggressive Water Pricing Policy (increased water rates, Inclining block rates); and an Aggressive Public Awareness Campaign, adopted by the Water Utility. As a result, the annual water consumption was decreased by 1% per year instead of the 1% annual increase expected. This means that since 2000, the water utility charged 1.36Mm3 less water per year (9.52Mm^3 since 2000), than it would have sold, if no water saving measures were taken. These water savings must be added to the water savings due to the reduced pipe break incidents occurred in the same period (7,224,668m^3). So the environmental importance of the entire project is strongly justified, as the total water savings since 2000 reached 16,744,668m^3.

3.2 Economical (Financial) Aspect

At first it should be noted that *the whole evaluation process "uses" the pipe break rates and has nothing to do with the (doubtful) pipe break social cost, as this cost is only used to determine the optimum replacement time of a pipe. The latter does not affect the evaluation process.* The evaluation concept is very easy as it compares the estimated costs and benefits (on monetary terms) related to the project implemented.

3.2.1 Costs vs. Benefits

Regarding the *direct costs* related to the project, the five water pressure reduction devices cost 54,514€ (purchase: 17,814€, installation 36,700€), while their annual operation/repair cost is 10,000€. The total fee of the technical consultant was 1,003,670€. There was also an annual fee of 50.000€ for its supervising services. Since 2000, the local Water Utility "sold" 1.36Mm^3 less water per year (9.52Mm^3 in total), than it would have sold, if no water saving measures were taken. Considering a mean water price of 0.60€/m^3, the *lost revenues* are estimated to be 5,712,000€. Finally the Public Awareness Campaign cost 126,000€ for 2000 and 50,000€ for each year forth. Concluding, the total cost of the water saving project was *7,556,184 €.*

Regarding the *direct benefits* related to the breaks prevented since 2000, each one costs 773€ (mean value). So, for the 5,324 breaks prevented, the Water Utility saved 4,115,452€. Water savings due to the breaks prevented since 2000 are 7,224,668m^3. Considering the relative water supply, and treatment costs (approximately 0.22€/m^3), the Water Utility saved 1.589.427€. Additionally, through the period 2000-2006, the water utility sold 1.360.000 m^3 less water per year (9,520,000 m^3 for the seven-years period), than it should have sold, if no water saving measures were taken. Considering the same water supply, and treatment costs mentioned above, the Water Utility saved 2,094,400€. So, the total benefits of the water saving project was *7,799,279* €.

3.2.2 *Cost-Benefit analysis*

The Analysis (Figure 11, Table 6) proves that the project is viable. It has a 7-years payback or return of investment period (time needed to balance the present values of benefits and costs Finally, there is the immeasurable environmental profit for the Water Utility along with the fact that it will be able to postpone the search for additional water supplies and the pack of costs related to this attempt.

Table 6. COST – BENEFIT ANALYSIS OF THE PROJECT

COSTS per year (€)	unit values	**2000**	**2001**	**2002**	**2003**	**2004**	**2005**	**2006**
water not being sold (m^3)	1,360,000	1,360,000	1,360,000	1,360,000	1,360,000	1,360,000	1,360,000	1,360,000
lost revenues (water not sold)	0.60 €/m^3	816,000	816,000	816,000	816,000	816,000	816,000	816,000
PRV's-related costs		54,514	10,000	10,000	10,000	10,000	10,000	10,000
Technical Consultant Fee		1,003,670	50,000	50,000	50,000	50,000	50,000	50,000
Public Awareness campaign cost		126,000	50,000	50,000	50,000	50,000	50,000	50,000
total annual cost		*2,000,184*	*926,000*	*926,000*	*926,000*	*926,000*	*926,000*	*926,000*
TOTAL PROJECT COST (€)		*2,000,184*	*2,926,184*	*3,852,184*	*4,778,184*	*5,704,184*	*6,630,184*	*7,556,184*
BENEFITS per year (€)								
breaks annually prevented		0	323	592	821	1,022	1,202	1,364
cost not paid for breaks' repairs	773€/break	0.00	249,679	457,616	634,633	790,006	929,146	1,054,372
water savings- less breaks (m^3)	1,357 m^3/break	0.00	438,311	803,344	1,114,097	1,386,854	1,631,114	1,850,948
money savings-less water used	0.22€/m^3	0.00	96,428.42	176,735.68	245,101.34	305,107.88	358.845.08	407,208.56
water not being sold (m^3)	1,360,000	1,360,000	1,360,000	1,360,000	1,360,000	1,360,000	1,360,000	1,360,000
money savings-less water sold	0.60 €/m^3	299,200	299,200	299,200	299,200	299,200	299,200	299,200
total annual benefit		*299,200*	*645,307.42*	*933,551.68*	*1,178,934.34*	*1,394,313.88*	*1,587,191.08*	*1,760,780.56*
TOTAL PROJECT BENEFIT (€)		*299,200*	*944,507.42*	*1,878,059.10*	*3,056,993.44*	*4,451,307.32*	*6,038,498.4*	*7,799,278.96*
NPV OF THE PROJECT (€)		*-1,700,984*	*-1,981,676.58*	*-1,974,124.90*	*-1,721,190.56*	*-1,252,876.68*	*-591,685.60*	*243,094.96*

**** all values are considered as Present Values***

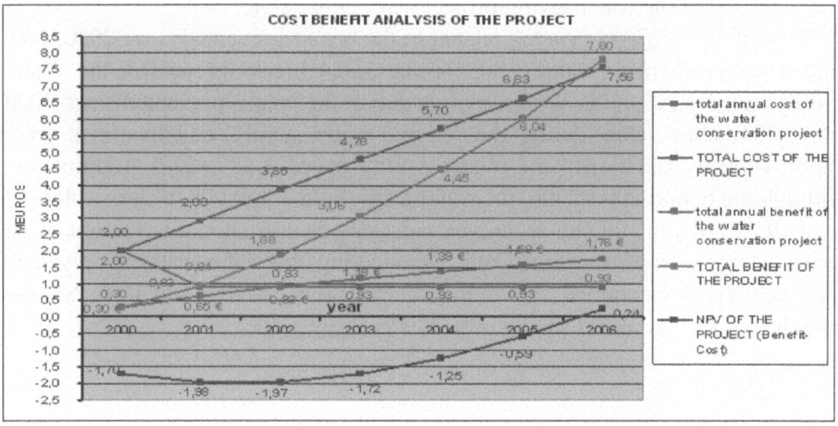

Figure 11. COST – BENEFIT ANALYSIS OF THE PROJECT FOR THE PERIOD 2000-2006

4 CONCLUSIONS

4.1 Principles of a water-pricing based conservation project

The Athens's case study resulted in establishing a number of principles for an effective water-pricing based conservation project:

- water conservation is not an objective itself, but is a part of the broader objective of total water management.
- water is a renewable, recyclable and replenishable resource, spatially and timely varied regarding its quantity/quality.
- proper water management must be holistic covering all the stages of water supply, from the exploitation (intake) of the water resources, through the several water uses and finally to water disposal and rehabilitation.
- water conservation programs must be modified according to: a) the climate conditions, b) the water uses c) the water resource supplying the system and d) the operation state of the entire water system.
- the effects on the availability and the efficiency, concerning the quality and the quantity of the water resources, resulting from the implementation of a conservation project must be estimated in advance.
- water conservation measures must be evaluated through an economic, social and environmental cost-benefit analysis taking into account their chances of achieving their goals.
- energy savings due to the conservation itself, must be estimated in advance.
- the reliability, availability and resiliency of both the water system and the water resources after achieving the goals of the water conservation program must be re-evaluated and compared to their initial levels.
- constant public education is necessary in order to establish a conservation consciousness and overcame the negative public's response towards the various measures implemented. Special education is needed to encourage the voluntary conservation and to ensure the public's compliance with the enforced but nevertheless justified measures of mandatory conservation.

4.2 Network upgrading projects can be viable

The suggestions to the Water Utility for an effective "confrontation" of water losses and their impacts can be distinguished in two types of suggestions regarding: a) the prevention of water losses and; b) the reduction of water losses or the minimization of their impacts. The pressure management project based on the installation of PRV's was the first step to control the water losses. This project should be regularly updated following the continuously changing needs of the distribution network. Another matter that should be seriously taken under consideration is the prohibition or at least the formation of requirements (setting the rules and the basic principles-standards) for the water network constructions done by the customers themselves, individually. Also, very important is the quality standards of repair/replacement works regarding the water supply mains, as the impact of their breaks may cause a severe water shortage problem for a huge part of the city. The study revealed that a lot of pipe breaks were caused by third parties (service staff of other public utilities). As the Water Utility's policy towards such failure events is extremely loose, it is advised to set a significant fine-penalty (not only for the actual repair cost, but also for the social inconvenience cost involved) so as to force the other companies' staff to be more careful. Last but not least is the better knowledge and management of the distribution network. This could be achieved through: a) installation of a S.C.A.D.A. system, providing continuous monitoring of network operation data; b) improvement of the G.I.S. system and; c) development of a detailed hydraulic simulation model. All these could be utilized to create an EDSS necessary for Crisis Management. On the other hand, in order the Water Utility to suppress the water losses or to minimize their impacts it is crucial to improve the level of alertness of its staff as well as to simplify the bureaucratic procedures followed nowadays when a private company is called for help to repair a break. The Water Utility should consider developing a mobile tool house so as to reduce the repairing time of a break and the water losses/money involved. Also important is the implementation of new techniques regarding break repairs/services. Staff capacity building will prove to be lifesaving! Finally, if the Water Utility proceeds to the adoption of this Policy, it is more than certain that it will make its customers change their mind/attitude towards the Water Utility and its actions. This is actually the most important benefit of all nowadays.

5 REFERENCES

E.Y.A.Th. records: Projects, statements and other internal records (2000-2005)

Jordan J.L. (1995). Incorporating externalities in conservation programs, American Water Works Association, 87(3), 49-56.

Kanakoudis V. (1998). The role of failure events in developing water systems pipes preventive maintenance and replacement criteria, PhD Thesis, Department of Civil Eng., Div. of Hydraulics and Environmental Eng., Aristotle University of Thessaloniki, Greece.

Kanakoudis V. and D. Tolikas (2000). The role of leaks and breaks in the management of water networks Proceedings of the 5th International Conference for the Protection and Restoration of the Environment, Thassos, Greece.

Kanakoudis V., D. Tolikas and A. Liakopoulos (2000). Leaks and breaks in water networks. Proceedings of the 8th National Conference of the Greek Hydrotechnique Association, 51-58, Athens, Greece.

Kanakoudis, V., Tolikas, D.: 2001, The role of leaks and breaks in water networks – Technical & economical solutions, J.WSRT-AQUA, IWAp;50(5), 301-311.

Kanakoudis, V., Charavgis I.: 2005, Water systems management - lifesaving interventions evaluation, EWRA 6th International Conference: Sharing a common vision for our water resources, CD-Proceedings, Menton, France.

Participatory institutional assessment for improved performance: experiences from the Swazi water sector

O. Busari[1]

[1]Development Bank of Southern Africa. olab@dbsa.org, PO Box 1234, Midrand 1685, South Africa, Tel: +27.11.313.3920, Fax: +27.11.206.3920

Keywords: Participatory-assessment, Performance-indicators, Swaziland

Edited by Enrique Cabrera Jr and Miguel Ángel Pardo. ISBN: 9781843391913.
Published by IWA Publishing, London, UK.

ABSTRACT

The identification of workable models of infrastructure service provision, including community water supply, continues to be hindered by the *project framework* within which most development agencies operate. In many parts of sub-Saharan Africa, attempts to focus on *institutions* have been largely externally-propelled, with such capacity assessment of water supply institutions frequently pointing to broad gaps, including under-funding and capacity over-stretch, inappropriate incentive systems, and inadequate administrative structures and delivery mechanisms.

In pursuit of greater and deeper understanding of archetypal *broad* institutional gaps, the policy and institutional setting at the community water services branch of the Government of Swaziland has been examined from a participatory, introspective perspective. Using performance categories (and indicators) that range from technical capability and top leadership to organizational culture and community orientation, each of the Branch's units, including those at the district level, quantified its own level of performance and practices and suggested strategies for attaining higher levels. While certain service delivery obstacles – such as poor informal communication, low inter-unit project input and non-supportive external policies – were considered as common to the entire institution, staff members were able to clearly identify the strengths and weaknesses of other units, providing an effective springboard for programmed and holistic capacity building.

1 INTRODUCTION

The UNEP Finance Initiative (2004) observes that, in many developing countries, effective water governance is jeopardized by weak capacity – technical, administrative and financial – as well as by fragmented and shifting institutional responsibilities for water allocation, supply and management. Almost a decade earlier, in its assessment of the impacts of its policies and operational tools and practices on capacity building in Africa, the World Bank (1996) attributed the weaknesses in the continent's public and private sector institutions to ineffective management techniques, inefficient procedures and practices, and poor communication systems. In the case of the public sector, such weaknesses were seen as being reinforced by the dearth of professionals with vital skills, sub-optimal allocation of resources and poor logistic support.

First, it is true that insufficient incentive systems, weak administrative structures and delivery mechanisms, and poor leadership in its broad sense account for much of the problems in public institutions. In many countries, bad governance and political interference have eroded public service independence, accountability and overall professionalism, and left in their wake low motivation, commitment and productivity.

But, second, it is also true that most of the initiatives for responding to the foregoing challenges of capacity building and institutional development have emanated from external support agencies, including UN agencies and the World Bank, and have been frequently led by them. Gaps in institutional performance are often identified by external agencies in a broad capacity assessment, with the strategies for addressing capacity problems also prescribed and driven by such organizations. Further, external support to institutional and policy improvements is usually made available within the framework of *packaged* projects. With the proliferation of poorly coordinated projects in virtually all sectors, policy overload and institutional distortions easily result.

In an independent evaluation of the water resources strategy of the World Bank, Pitman (2002) reports that the Bank was drawn too much to the demands of influential voices to improve infrastructure services, with several projects simply addressing immediate water supply problems in major cities, irrespective of other important demands. The study argued that the development effectiveness and sustainability of operations was often hindered by the adoption of such easy options, instead of exercising the leverage to promote critical sector reforms. In particular, Pitman (2002) noted that, in Colombia and Honduras, while Bank loans achieved their central objective of ensuring better access to drinking water, they failed in their *institutional objectives*, primarily because 'the project scope exceeded the capacity of sector institutions that suffered from frequent managerial changes and political interference' (p. 34). Similar project challenges were recorded in Africa, especially Ghana and Algeria, with poorly defined performance objectives co-existing with poor governance.

To bring a meaningful long-term perspective to national strategy formulation with respect to capacity building, institutions in developing countries, including those for water services, need to first assume ownership of the prevailing institutional environment and to take responsibility for the inherent challenges. The institutional assessment case study presented in this paper – involving the Community Water Supply Branch of the Government of Swaziland – goes that full distance. By quantifying their own levels of performance and identifying their relative strengths and weaknesses, the units in the Branch (as well as the Branch as a whole) simply call for interventions that would complement the commitment already shown in starting out on the capacity building pathway.

2 INSTITUTIONAL AND POLICY SETTING

The Community Water Supply Branch, through a total of nine units comprising five central sections and four regional depots, is mandated to oversee the design, construction and maintenance of community water supply schemes in Swaziland. While the agency assumed that responsibility on an *ad hoc* basis since the inception of serious sector business in the mid-1970s, official authorization to supervise and coordinate all sector activities only emerged in 1995. At that point, however, the planning and implementation of community water supply was undertaken in an uncoordinated fashion by several organizations, each with its own priorities and approaches. External support institutions were not different (Busari et al., 1996), with their interests and agendas divergent at times.

Within the foregoing context, the Branch easily recognized the urgent need to take full responsibility for proactive, coherent and effective sector development and management: identifying development agents, pinpointing goals, formulating plans, establishing priorities and spearheading policy reforms.

Consequent upon an introspective assessment of its own track record, the Branch has indeed dedicated itself to joining hands with principal role players as full and equal partners in the sector development process. Constructive partnership is evidenced, *inter alia*, by the joint formulation of sector policy guidelines from 1998, especially those for design and construction, community organization and participation, and linkage of water supply, sanitation and hygiene education.

With a national policy keyed largely to the attainment of universal coverage by the year 2020, success in policy implementation hinges on the hope that political commitment would be obtained at the highest level, translating into increased government funding and the motivation of households to contribute their quota (Busari, 2002). More specifically, government needs to facilitate the enabling political climate for development agents to operationalize appropriate cost-sharing, cost-recovery and financing mechanisms for equity and sustainability.

3 PARTICIPATORY INSTITUTIONAL ASSESSMENT

Efforts to diagnose institutional problems as a first step in the formulation of improvement interventions are common to many sectors and countries. But the self-driven and participatory assessment by the community water supply institution in Swaziland, to generate forward-looking lessons for its future service delivery programmes and to assure accountability, is unique for the following reasons:

- staff members themselves initiated the assessment and mapped out the entire exercise;
- staff members identified the issues and indicators to be addressed, planned the retreat for addressing them, and gathered the information necessary for informed discussions and performance rating;
- staff discussed the issues in self-lead unit-based teams and proceeded to present their findings and indicator ratings in plenary, freely challenging the findings of one another; and
- the facilitator was challenged to keep the process on track, even if occasionally chaotic: assisting with filling in gaps in information and data, capturing the core proceedings in a balanced manner, and isolating prominent institutional weaknesses and how they inform the strategy for capacity building.

The assessment process incorporated six performance categories – sets of related skills, procedures and capabilities defining particular institutional performance areas (USAID, 1988). The categories, for which the degree of successful unit performance was rated on a scale of 1 (low) to 10 (high) using the indicators summarized in Table 1, are as follows:

- Top Leadership;
- Management and Administrative Systems;
- Technical Capability;
- Community Orientation;
- Organizational Culture; and
- External Relations.

Tables 1 and 2 show that, irrespective of whether they are located at the headquarters or at the district level in the four administrative regions of the country, institutional units consider themselves as having varying degrees of skills and practices in each of the performance categories (Government of Swaziland, 1996). However, arriving at what appear to be simple rankings was not an easy task, with the journey bedevilled by:

- the initial difficulty of enlisting the participation of an appropriate number of district level staff members, active stakeholders who interface frequently with s
- the sustained sceptical attitude of lower-level officers, as to whether it was safe to freely critique the role and performance of senior staff and headquarters-based units;
- the embarrassment and discomfort which accompanied the expression of frank opinions and concerns about the performance of one another; and
- the extra but critical effort by the facilitator not to push up his own views in discussions, with the understanding that every officer and unit would certainly note other participants' perspective of their performance and overall point of view, even whilst, expectedly, everyone did not agree to everything.

Notwithstanding these challenges and the rethinking associated with them in the course of the exercise, both the assessment process and its outcomes have been useful and instructive. In general, the mean category ranking ranges from 4.7 to a widely disputed 10, although clouding indicator ratings as low as 2.0 in some cases. Apart from an anticipated criticism and low ranking of external factors (central ministries and departments and the wider government policies and procedures), the most interesting and lowest rating is presented in the case of organizational culture.

There was consensus that poor internal awareness (and performance) of some of the Branch's projects was being sustained by the related factors of poor informal communication and low inter-unit programme support. First, informal personal contacts were considered a powerful tool for directly dealing with the largely traditional community structures vital to sustainable water system management. But then, the use of that tool where it matters most calls for the devolution of certain decision-making powers to district level. Second, constituting specific task forces across headquarters and regional units was believed to be an effective input coordination mechanism for fostering an inclusive commitment to the success of every project and programme, and for engineering team spirit over the long term.

Table 1. PERFORMANCE RATING BY REGIONAL UNITS (LOW 1 TO HIGH 10)

Performance Category and Indicators	Hhohho Region	Lubombo Region	Manzini Region	Shiselweni Region
TOP LEADERSHIP (averaged for: sense of mission, role-modelling, operational knowledge, technical guidance, hardwork, competence, listening capacity, personal integrity and decision-making)	6.7	6.7	7.3	4.8
MANAGEMENT & ADMIN SYSTEMS (averaged for: understanding of mission & responsibilities, work planning, goal setting & appropriateness, regularity of follow-through, functionality of admin systems, & transparency & accountability)	6.4	4.9	6.2	5.0
TECHNICAL CAPABILITY (averaged for: soundness of technical decisions, timeliness & economy of completion, quality control, adequacy of skilled staff, interest in learning & info sharing)	5.5	6.2	5.2	4.7
COMMUNITY ORIENTATION (averaged for: promotion of the Branch's positive image, commitment & responsiveness to communities, effectiveness of interaction & interfacing, community participation)	6.4	6.0	6.6	4.9

ORGANIZATIONAL CULTURE (averaged for: team spirit & team work, pride in work, informal communication & programme awareness, & value placed on assets)	5.6	5.4	5.2	5.2
EXTERNAL RELATIONS (averaged for: government & public support, leadership in civil society activities, & knowledge/relevance of government policies & procedures)	5.8	5.0	5.3	5.0

Table 2. **PERFORMANCE RATING BY HEADQUARTERS UNITS (LOW 1 TO HIGH 10)**

Performance Category	Design	Finance	Public Health	Planning & Constru ction	Community Liaison
TOP LEADERSHIP	7.9	6.5	9.0	9.0	6.6
MANAGEMENT & ADMIN SYSTEMS	8.4	7.1	7.4	8.9	7.2
TECHNICAL CAPABILITY	9.3	5.8	8.5	8.5	5.8
COMMUNITY ORIENTATION	8.0	6.6	5.1	9.0	10.0
ORGANIZATIONAL CULTURE	8.4	5.6	4.8	8.8	10.0
EXTERNAL RELATIONS	7.8	7.5	7.5	8.8	10.0

4 CONCLUSIONS

Irrespective of the pinpointed challenges to participatory institutional assessment and the rethinking associated with them in the course of this particular case study, both the participatory process and its outcomes have been useful and instructive. The criticism and low ranking of factors external to the institution, especially the role of supervisory and other ministries and the impact of overarching government policies and procedures, were anticipated. But it is instructive that staff members returned the lowest rating for their organizational culture. Specifically, there was consensus that poor internal awareness and performance of some projects was being sustained by the related factors of poor informal communication and low inter-unit programme support.

In the light of the difficulties with supporting institutional development within their own long-term framework, external agencies need to increasingly seek coherent entry opportunities pointed to in each country's long-term development strategy for the sector, providing capacity support that couples their external perspective to an equally important internal perception of gaps and weaknesses.

5 ACKNOWLEDGEMENT

The assistance provided by Stephen Dlamini, Chief Water Engineer in Swaziland's Ministry of Natural Resources and Energy is gratefully acknowledged.

6 REFERENCES

Busari, O. (2002). Millennial policies and strategies for promoting household water security: A Southern African Example. *Canadian Water Resources Journal* **27**(3), 273-291.

Busari, O., Dlamini, S. and Mayisela, M. (1996). Coordination: coverage catalyst in Swaziland. In *Proceedings of the 22nd WEDC Conference*, New Delhi, pp. 27-29.

Government of Swaziland (1996). Towards improved programme delivery: as you see it. Workshop Facilitation Report, Ministry of Natural Resources and Energy, Mbabane.

Pitman, G.K. (2002). Bridging troubled waters: assessing the World Bank water resources strategy. Operations Evaluation Department, World Bank, Washington, DC.

UNEP Finance Initiative (2004). Challenges of water scarcity: a business case for financial institutions. United Nations Environment Programme – Finance Initiative, Geneva.

USAID (1988). Guidelines for institutional assessment: water and wastewater institutions. WASH Technical Report No. 37. United States Agency for International Development, Washington DC.

World Bank (1996). Partnership for capacity building in Africa: strategy and program of action. Report on the Impact of Bank Policies, Instruments and Operational Practices on Capacity Building in Africa. World Bank, Washington, DC.

Measuring the Performance of Portuguese Solid Waste Services

P. Simões[1] *and R.C. Marques*[2]

[1] *Center for Urban and Regional Systems, DECivil-IST, Technical University of Lisbon.* psimoes@civil.ist.utl.pt, Av. Rovisco Pais, 1049-001 Lisbon, Portugal, *Tlf: +351 218418316, Fax: +351 218409884*
[2] *Center for Urban and Regional Systems, DECivil-IST, Technical University of Lisbon.* rcmar@civil.ist.utl.pt, Av. Rovisco Pais, 1049-001 Lisbon, Portugal, *Tlf: +351 218418319, Fax: +351218409884*

Keywords: Solid Waste services; Performance; Benchmarking; DEA

Edited by Enrique Cabrera Jr and Miguel Ángel Pardo. ISBN: 9781843391913.
Published by IWA Publishing, London, UK.

ABSTRACT

The current paper presents a research that measures the performance of the Portuguese solid waste disposal services. It begins by an introduction to the solid waste sector, highlighting the importance of performance evaluation in this sector. Then, it reviews the literature and systematizes the major studies about this issue. After, it uses the non-parametric frontier benchmarking technique of data envelopment analysis (DEA) to compute the efficiency of 29 solid waste disposal services which encompass the whole population of Continental Portugal. The study refers to the year 2005. Several models were employed including the integration of the operational environment. The outcomes have pointed to significant inefficiency. In the constant returns to scale model the potential savings attain 124 million Euros if the solid waste disposal services works efficiently. The GDP per capita was the only statistically significant explanatory factor and no differences were found between regulated and non-regulated services or between private and public owned services.

1 INTRODUCTION

Nowadays, one of the most serious problems faced by modern societies is related to the production of solid waste in connection to the economic development. Solid waste management transformed itself in one of the main local public services with the same or higher importance than the drinking water supply or sewerage services (Simões, 2007). The municipalities are responsible for solid waste collection and one of their main goals consists in keeping the areas under their jurisdiction in adequate aesthetic and hygienic conditions.

The European Union (EU) requirements as well as the difficulties of the municipalities to handle the complexity of the solid waste services have brought about reforms in this sector all over Europe in the past few years (Sanchez, 2006). In Portugal this issue has only been taken into account after the 1990s. This was caused by the continuous growth of the solid waste production and the inadequate treatment facilities (uncontrolled landfills). At this time the Portuguese government implemented a significant set of reforms and measures which were translated into significant progress on environmental issues. Portugal has become an innovative and pioneer country in several fields, for example by creating an economic regulator for the sector.

The need for a larger control of the public accounts and external pressures from the EU or from the users themselves have played an important role in raising the awareness of public institutions to assess and develop the provision of the solid waste service in accordance with the principles of economic efficiency and cost effectiveness. There has been an increase in the number of studies aiming at evaluating the public sector efficiency either due to budget restrictions or to the growing importance of this sector in the economy of every country.

The evaluation of performance and, particularly, the use of benchmarking can play a major role to challenge the problems faced by the urban solid waste services which usually act as natural monopolies and thus are characterized by reduced incentives towards efficiency and innovation. This study which is based on a recent academic research (Simões, 2007) applies benchmarking to measure the performance of the Portuguese solid waste sector. It makes use of data envelopment analysis (DEA), a mathematical programming technique used for assessing the relative performance of organizational units. DEA deals easily with multiple inputs and outputs and allows for the identification of a group of efficient organizations to each inefficient organization with a similar combination of inputs and outputs.

This paper is organized as follows. Next to this introduction section 2 presents the literature review on the efficiency studies in this sector and in section 3 a performance assessment of the solid waste "wholesale" operators in Portugal is carried out. Section 4 concludes the paper.

2. REVIEWING THE LITERATURE

The efficiency studies developed in the scope of the solid waste sector are mostly directed to its performance evaluation, to the comparison between public and private management and to determine the optimal market structure, particularly with regard to the effect of the reforms concerning the sector's unbundling or horizontal integration. Many studies have focused on ownership, most of them leading to opposite conclusions depending on the sample and on the technique applied.

The techniques adopted to determine the efficiency are often classified into parametric and non-parametric. Sometimes they are further divided into frontier and non frontier according to whether they compute the best practices or the average adjustment and into stochastic or deterministic if they consider or not the random error (Marques, 2005).

The solid waste efficiency studies collected from the literature amount to a total number of 55, from which 18 use non-parametric methods and 37 draw on parametric methods (Marques and Simões, 2007). Concerning the DEA technique and the stochastic frontiers, despite the larger number of works under development by academics, there are also some regulators and governmental organisms that apply this type of methods. Many studies from the literature regard the performance of the municipalities as organizations, including also the activities related to the urban solid waste provision.

The increasing importance of the private sector within the solid waste collection has lead to a larger number of studies on efficiency measurement in countries where the service provision is carried out by private entities. The strong presence of the private sector, namely in Spain, creates an environment favourable to the development of this type of studies (see, for example, Bel and Costas, 2006).

In Portugal, due to the lack of information and to the fact that the service is still mostly provided by public entities, few studies exist about the efficiency of solid waste sector. Gaiola (2002) and Marques *et al.* (2006) are exceptions. The literature has revealed studies in 13 countries, namely the USA, Canada, Switzerland, Spain, UK, Ireland, Denmark, the Netherlands, Sweden, Australia, Portugal, Belgium and Finland (Simões, 2007).

Underlying the activity of solid waste services provision, more importance is often observed with regard to particular inputs and outputs. Consequently some of them are more used than others by different authors. For the non-parametric methods adopted, the main variables for the inputs were the number of employees and the total cost of the service (OPEX plus CAPEX). The most adopted outputs were the quantity of collected waste and the number of service users. Regarding the explanatory factors the distance to the landfills and the population density were the ones most often used.

From the 18 studies that employ the DEA technique mentioned in the literature, 3 of them try to study the influence of scale efficiency. From the remaining studies 10 considered only CRS and the other 5 VRS (Marques and Simões, 2007). 12 out of the 18 studies considered oriented models and 6 were non-oriented highlighting a balance between the type of orientation (input minimization or output maximization).

3 EFFICIENCY OF PORTUGUESE SOLID WASTE DISPOSAL SERVICES

3.1 Methodology

The non-parametric frontier technique of DEA is based on mathematical programming to measure the relative efficiency of organizational units which present a homogeneous set of inputs and outputs. Its large use is explained by its several advantages (Marques, 2005), specifically, the identification of best practices that can be followed by other operators, the calculation of the operators optimal size, the estimation of the potential efficiency earnings, the accomplishment of marginal rates of substitution between production factors, the determination of productivity change through time of each operator and the computation of the most efficient organizational structure, such as private or public entities.

The DEA technique enables the determination of the production frontier and consequently the relative efficiencies of each operator. One of the main advantages is related to the fact that unlike other benchmarking (parametric) techniques it does not need a previous specification to each input/output weight and does not require judgements about the production function. Technical efficiency (TE), obtained by DEA shows the capability each operator has to spend a minimum number of inputs for a given level of outputs or vice-versa. TE can assume values between 0 and 1. The latter indicates that the operator is efficient and located on the frontier.

In 1978 the so-called CCR model was developed based on CRS and strong disposability of inputs (Charnes *et al.*, 1978). Considering the minimisation of inputs we obtain the following formulation where W_k corresponds to the efficiency of each operator that produces the outputs y_i, i=1, ..., I, from the inputs x_j, j=1, ..., J, with the respective weights a_i and b_j and M is the total number of operators.

$$
\begin{aligned}
& Max: W_k = \sum_{i=1}^{I} a_i y_{im} \\
& s.a. \\
& \sum_{j=1}^{J} b_j x_{jm} \leq \sum_{i=1}^{I} a_i y_{im} && m = 1, \ldots, k, \ldots, M \\
& \sum_{j=1}^{J} b_j x_{jk} = 1 \\
& a_i, b_j > 0 && i = 1, \ldots, I;\ j, \ldots, J \qquad (1)
\end{aligned}
$$

Afterwards, there was the development of a model considering VRS (BCC model) by adding to the previous model the sum of the weights restrictions equal to 1 (Banker *et al.*, 1984). If the TE obtained by CRS (CCR model) are related to those of VRS (BCC model), they can be decomposed into two parts, one due to scale efficiency (SE) and the other due to pure technical efficiency (PTE). This result can be obtained by the DEA models application assuming CRS and VRS. If there is a difference between the two TE for the same operator, it means that the operator reveals scale inefficiency and that it can be computed by means of the results reached with the different methods. The SE accounts for the degree of savings attained if the operator was operating at an optimal scale.

3.2 Case-study

3.2.1 Sample and data

The current research aims at evaluating the solid waste services efficiency in Portugal. It included the 29 operators that act in the secondary market (solid waste treatment), encompassing the whole Continental Portugal population. The year under analysis was 2005 but data available for the year 2004 was always used by comparison to check some possible divergences.

The data collected was predominantly provided by the operators. However, occasionally the data quality was not very good and it was necessary to do some simplifications. It was also necessary to ask information to the regulator (Institute for the Regulation of Water and Solid Waste - IRAR) in order to compare information between the available documents always there was some doubt.

3.2.2 Specification of the model

The model applied for the solid waste services comprises the inputs operational expenses (OPEX) and capital costs (CAPEX) whereas the outputs are the treated solid waste and the recycled waste. Concerning the inputs the values for the CAPEX were the ones present in the account reports regarding depreciation and interest expenses. For the OPEX the model included in the analysis the internal manpower costs, external services, energy, chemicals, other consumables and materials for maintenance and repair and other operating costs. The outputs incorporated all the treated solid waste in the system (including non-hazardous industrial waste) and the recycled waste.

The model considered the population density, the GDP *per capita* (by region) and the average distance to treatment facilities (landfills, composting and incineration plants) as explanatory factors. Table 1 shows the main features of the sample.

Table 1. FEATURES OF THE SAMPLE ADOPTED

	Minimum	Maximum	Average	St. deviation	Median
Output					
Treated solid waste (t)	13.323	789.583	178.143	190.796	90.650
Recycled waste (t)	835	48.601	9.546	12.548	3.607
Input					
CAPEX (€)	199.375	18.299.907	3.012.942	4.148.434	1.504.729
OPEX (€)	742.471	29.301.336	5.303.628	6.781.132	2.228.297
Explanatory factors					
Population density	15	1.993	353	561	88
GDP *per capita* (€)	7.522	23.422	11.838	3.961	10.732
Distance to landfill (km)	4,2	37,5	14,1	8,6	11,4

The model is oriented towards input minimisation as all the operators have to serve all the urban solid waste users and because this service follows a demand side management policy. It would not be sensible to choose an input maximisation oriented model and no justification would be found for the additional effort of using non-oriented models. For this reason, the option was for an input-minimising orientation where more importance was given to the maintenance or improvement of the level of service provided with the minimal use possible of inputs.

Several models were computed admitting CRS (CCR model) and VRS (BCC model) along with other constraints aiming at a better description of the production technology. Adjusting for environment was also comprised in the analysis by means of a second stage where Tobit regression was applied.

3.2.3 Results

Table 2 displays the summary of the main results obtained for the Portuguese sold waste disposal services with a GAMS code developed by the authors. Figures 1 show the operators TE and PTE sorted by their scores classes.

Table 2. RESULTS FOR THE PORTUGUESE SOLID WASTE SERVICES

Results	Values obtained
Average TE	0,590
Average PTE	0,798
Average SE	0,739
Minimum TE	0,205
Minimum PTE	0,291
Minimum SE	0,325
Number of efficient operators (TE)	2
Number of efficient operators (PTE)	12

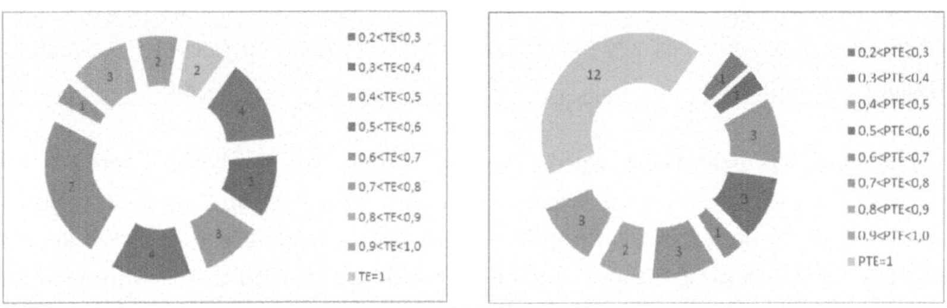

Figure 1. SOLID WASTE SERVICES BY TE AND PTE CLASSES

3.2.4 Results analysis

The inefficiency levels of the Portuguese solid waste disposal services for the year 2005 were meaningful. The average TE value was 0,590. In Portugal, the solid waste services show potential savings of the average TE of 41,0%. This means that, on average, each operator can reduce the inputs consumed by 41,0%, i.e., they can reduce their OPEX and CAPEX and produce the same quantity of outputs (treated and recycled solid waste), corresponding to 124 million Euros of savings for the CRS model. For the VRS model the savings would be about 40 millions of Euros.

The solid waste services would consume less 26,1% of inputs for the same quantity of produced outputs if there was an optimal scale (CRS). From the 29 operators, 15 of them present increasing returns to scale (IVRS), 12 decreasing returns to scale (DVRS) and 2 CRS. It is estimated that the optimal size of the Portuguese solid waste disposal services would comprise about 300 000 inhabitants.

For the Portuguese solid waste services (disposal market) and for the VRS model, the operators 6 and 27 are more often used as peers (best practices). Operator 6 is used 17 times as peer whereas utility 27 is used 10 times. Table 3 illustrates the targets of CAPEX and OPEX for the operators 2, 15 and 19 and their peers for CRS model.

Table 3. TARGETS AND PEERS OF OPERATORS 2, 15 AND 19 FOR CRS MODEL

	Operator 2		Operator 15		Operator 19	
Input	Value	Target	Value	Target	Value	Target
OPEX (€)	8.013.853	5.217.482	1.442.213	1.196.795	1.082.216	711.692
CAPEX (€)	7.228.766	687.068	2.948.532	849.779	866.779	487.907
Peers	—	6	—	6, 10	—	6, 10

3.2.5 Adjusting for environment

The capability of transforming the resources consumed into desired products and results depends not only on the operators TE, but also on the operational environment that characterises them. Hence, DEA models have a limited value if the operational environment where the operators work is not deemed (Fried *et al.*, 1999).

Here, a two-stage methodology was considered. In a first stage, this technique solves the DEA algorithm without the explanatory factors and in a second stage there is regression application to the results, where the dependent variable is the efficiency obtained in the first stage and the independent variables are the explanatory factors.

As explanatory factors the research adopted the population density, the GDP *per capita* in the area under the operators' jurisdiction and the average distance to the treatment facilities. The efficiencies of the operators are expected to increase if there is a higher population density and a higher GDP *per capita*. On the other hand, the operators' efficiency will be lower when the distance to treatment facility is longer.

We concluded that the regression coefficients are positive both for the population density and for the GDP *per capita* and negative for the distance to the landfill. However, only the GDP *per capita* is statistically significant at more than 95 % and just for the VRS model. Table 4 presents these results.

Table 4. EXPLANATORY FACTORS STATISTICAL SIGNIFICANCE

Model	CRS			VRS		
Variables	Population density	GDP *per capita*	Distance to landfill	Population density	GDP *per capita*	Distance to landfill
Coeffic.	0.00002	0.00001	-0.00369	0.00017	0.00003	-0.00473
Error	0.00013	0.00002	0.00579	0.00013	0.00002	0.00537
Stat. T	0,13	9.86	-0.64	1.33	1.97	-0.88
Proba. P	0.898	0.389	0.524	0.185	**0.049**	0.378

3.2.6 Some policy considerations

Two exercises were made in view of the reforms carried out in the solid waste sector during the last years. Although these reforms are very recent, we tried to study their influence on performance. Firstly, we compared the efficiency between regulated and non-regulated solid waste disposal services in Portugal. Secondly, we compared the efficiency of private capital ownership services with the public owned ones. The statistical test analyses were performed through the Wilcoxon-Mann-Whitney and T tests and led to the conclusion that no statistical significance between the different groups exists. Although the private sector seems to have higher TE than public owned services, PTE points to the opposite effect. Nonetheless, the results are not statistically significant. The regulated solid waste disposal services seem, a priori, more efficient than the non-regulated ones both with regard to TE and PTE. Similarly, the statistical tests show indifference between them.

4. CONCLUDING REMARKS

The current research used the non-parametric benchmarking technique of DEA to compute the efficiency of Portuguese solid waste disposal services. The models applied, admitting CRS and VRS, pointed to the conclusion the Portuguese operators' inefficiency levels for the year 2005 were somewhat significant. If the utilities were efficiently operated a cost reduction of about 124 million Euros would take place considering the CRS model and of 40 million Euros in the VRS model. This study has proven that the systems have an optimal size if they comprise about 300 000 inhabitants.

An analysis to the influence of the operational environment was also undertaken. In terms of the explanatory factors included in the model, particularly the population density, the GDP *per capita* and the distance to the treatment facilities (the ones most adopted in literature), we noticed that only the

GDP *per capita* was statistically significant at 95 % and for the VRS model. Furthermore, unsurprisingly, the operational environment proved to have a significant influence on the operators' efficiency.

Finally, we did not detect influence on efficiency of the recent reforms carried out in Portugal concerning the increase of the private sector presence and the effect of regulation.

5 REFERENCES

Banker, R., Charnes, A. and Cooper, W. (1984) Some models for estimating technical and scale inefficiencies in data envelopment analysis. *Management Science.* **30**(9), 1078–1092.

Bel, G. and Costas, A. (2006) Do public sector reforms get rusty? Local privatisation in Spain. *The Journal of Policy Reform.* **9**(1), 1–24.

Charnes, A., Clark, T., Cooper, W. and Golany, B. (1985) A development study of data envelopment analysis in measuring the efficiency of maintenance units in the U.S. air forces. *Annals of Operations Research.* **2**, 95–112.

Charnes, A., Cooper, W. and Rhodes, E. (1978) Measuring the efficiency of decision making units. *European Journal of Operational Research.* **2**(6), 429–444.

Fried, H., Schmidt, S. and Yaisawarng, S. (1999) Incorporating the operating environment into a nonparametric measure of technical efficiency. *Journal of Productivity Analysis.* **12**(3), 249–267.

Gaiola, A. (2002) Efficiency evaluation in the urban solid waste systems of Portugal using data envelopment analysis. *Symposium at a Glance.* Moscow.

Marques, R. (2005) *Regulação de Serviços Públicos.* Sílabo Press, Lda., 1st Edition.

Marques, R., Levy, J. and Pinela, A. (2006) Medição da eficiência dos serviços de resíduos sólidos urbanos em Portugal. In *IO 2006*, 12 Congress of the Portuguese Association for Operations Research, Lisbon, October.

Marques, R. and Simões, P. (2007) Incentive regulation and performance measurement of the Portuguese solid waste disposal services. Mimeo.

Simões, P. (2007). Eficiência dos serviços de resíduos sólidos em Portugal. M.Sc Dissertation. Instituto Superior Técnico, Technical University of Lisbon, Lisbon, Portugal.

Sanchez, I. (2006) Efficiency of solid waste collection in Spain. *Waste Management and the Environment III.* WIT press, Third International Conference on Waste Management and the Environment, 593–601.

Development and Application of the Decision-Making Support System for Planning Sewer Rehabilitation

J. D. Son[1], *S. Kim*[1], *W. Kang*[1], *K. Jin*[1] *and H. Hwang*[1]

[1] *Dept. of Water Supply Project Design, Kwater(Korea Water Resources Corporation), 6-2 Yeonchuk-dong, Taeduck-gu, Taejeon city, South Korea, disohn@kwater.or.kr, seonghyo@kwater.or.kr, wanghee@kwater.or.kr jinkh@kwater.or.kr, nutria86@pangaea.co.kr, T. +82-19-9476-7476*

 Performance Assessment of Urban Infrastructure Services
Edited by Enrique Cabrera Jr and Miguel Ángel Pardo. ISBN: 9781843391913.
Published by IWA Publishing, London, UK.

ABSTRACT

The goal of this study is in developing a decision-making support system, which can be utilized efficiently in the processes of planning sewer rehabilitation. To achieve the goal, tasks associated with sewer planning and rehabilitation were analyzed. The collected information on these tasks was integrated to build a decision-making support database, an essential tool for the decision-making support system consisting of 4 modules: selection of target pipes, selection of rehabilitation method, budget estimation, and determination of project priority. The system was verified using the sewer data of Sangdang-gu Chongju city, Korea.

1 INTRODUCTION

As important elements of urban infrastructure, sewer systems transport storm water and wastewater efficiently. For the construction of a sewer system, a long-term investment is required. Recently in Korea, large-scale built-transfer-and-lease (BTL) sewer projects based on the funds of private sectors have been coordinated by local governments for stable money flow. To efficiently implement sewer rehabilitation projects, setting up of a comprehensive plan meeting the purposes of the project is essential by evaluating existing pipes, selecting target pipes and rehabilitation methods, estimating budget, and determining project priority.

Planning of sewer systems consists of a series of decision-making processes. For the efficient execution of these processes, collection of information on in-depth background database, and applicable standards and criteria are considered to be preceded. Recently, as supporting tools for these processes, various analytical computer-aided tools and decision-making support system become available.

The purpose of this study is in developing a decision-making support system to support complicated tasks associated with planning sewer rehabilitation. These tasks were carefully analyzed to build a decision-making support database, an essential tool for the decision-making support system consisting of 4 modules: selection of target pipes, selection of rehabilitation method, budget estimation, and determination of project priority. The developed decision-making support system was verified using the data of Sangdang-gu Chongju city, Korea.

2 PRINCIPLES

In many countries in Europe, there has been increasing demand for sewer rehabilitation because of outdated sewer systems. Engineers faced with multiple difficulties in the decision-making processes of sewer rehabilitation projects including inconsistency in applicable criteria or standards and inaccuracy in evaluating sewers, selecting target pipes, estimating budget, and determining project priority. To minimize these limitations, Computer Aided Rehabilitation of Sewer and Stormwater Networks (CARE-S) was developed by standardizing the decision-making processes and applicable criteria (Saegrov, 2006. As illustrated in Figure 1, the system consisted of hydraulic reliability module, structural condition module and deterioration module. For each module, GIS-based integrated database was provided for computation. The hydraulic reliability module compatible with existing modeling software such as MOUSE, InfoWorks, SWMM, and Fluent, was used for evaluating the stability of structures and the capacity of sewer pipes. For the evaluation of the performance level of pipes, performance indicator (PI) in 5 different fields such as environmental impact, physical condition, service level, and budget capacity, was set to evaluate performance level of sewer pipes. The PIs were used in the deterioration module. Multi-criteria decision model was applied for determining rehabilitation of pipes on the basis of a line management concept and for selecting optimum rehabilitation methods. By integrating quality factors in PIs, scientific background in decision-making processes was increased significantly.

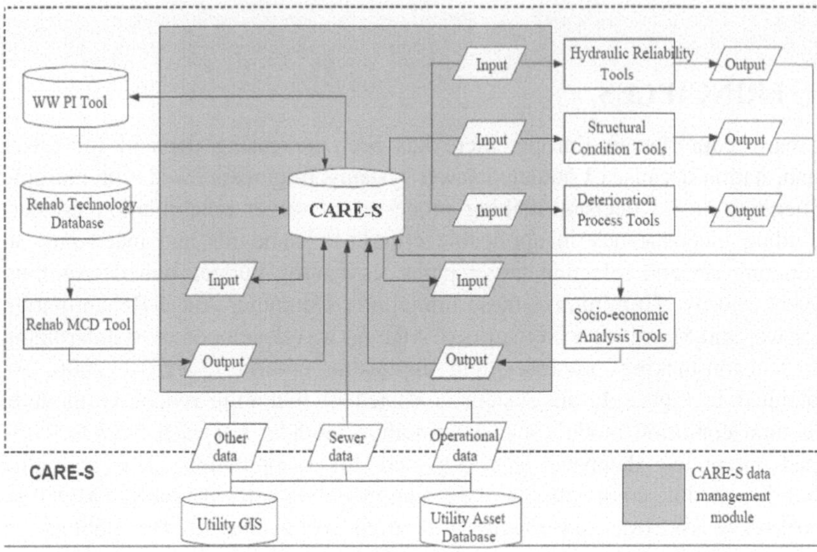

Figure 1. CONCEPTUAL DIAGRAM OF THE STRUCTURE OF CARE-S

Fenner, R.A. and Sweeting, L. (1999) devised a decision-making support system for the operation and maintenance (O&M) of non-critical sewer systems. In U.K., during the past 15 years, complete rehabilitation of critical sewers, consisting 20% of target sewers, had been executed. However, as for the non-critical sewers, consisting 80% of target sewers, no actions had been made. In this study, the model for determining the priority of rehabilitation works was suggested by sectioning the target area in grids and using property information in the GIS database and management history. After sectioning the area, probability of pipe damage of pipelines were analyzed through Bayesian statistical evaluation. According to the calculated probability, the priority for rehabilitation works was determined. Though this model could not be an absolute tool for estimating sewer behaviour, it was proved that the model could be efficient when used for the purpose of sewer O&M.

Plenker (2002) developed a support system for the selection of rehabilitation methods for defected pipes using spatial and property information. As the features and costs of individual rehabilitation methods were analyzed in depth, the system suggested the optimum methods for the pipes with wide range of defects. Multiple standards decision making method, eliminating undesirable alternatives according to weighing and ranking procedure using knock-out criteria, was adopted for the system.

These preceding studies became bench mark studies in developing CARE-S. Further, this led into studies in international standard decision-making methodology (Saegrove et al., 2006). However, to implement these theories, massive-scale database need to be prepared in addition to standardized standards and criteria. Due to lack of required information and data, it is difficult to adopt these methodologies in Korea.

In Korea, many of sewer rehabilitation projects have been coordinated by estimating budget and evaluating project priority after selecting target pipes based on the assessment results. For the series of decision-making processes associated with these tasks, score systems for the 5 to 6 criteria including the ratio of the pipes of low capacity, amount of infiltration and inflow (I/I) per unit area, amount of I/I per unit pipe length, number of defects per unit pipe length, budget allocation rate, and repair costs per unit I/I were used. However, considerations on important factors such as pollution loads were not included. Moreover, the influence of personal preference on the final decision could not be ignored. Recently, efforts to minimize these problems have been made by numbers of experts.

Chung et al. (2001) suggested application of genetic algorithm in determining the defect rate of sewer pipes in their studies on optimizing sewer maintenance methods in a small-size catchment. The project priority was determined by using a function designed to approach to the minimum rehabilitation costs when required rehabilitation costs and potential loss were given. Through this method, effects-costs analysis was achieved by converting results of rehabilitation into effects. The results could be changed by user-specified budget scale, effects-costs factor, and availability of technologies. For the limitations of this method, lack of criteria for pipe defect rate and pipe rehabilitation method was raised. In calculating the amount of I/I, the difference between the amount of estimated sewer volume and measured sewer volume was evenly distributed for each defect criteria. The weight factors allocated to each criterion were set to be 2 times of corresponding score. Though trials for incorporating social costs in the model were considered to be outstanding, there was no definite scientific background to support the chosen social cost factor.

Kim (2004) suggested the minimization of I/I is the most important factor in sewer rehabilitation. A decision-making support system was built to determine project priority under the criteria of the amount of decreased I/I per costs. Using genetic algorithm, the amount of I/I before the rehabilitation was calculated and distributed to each pipe according to the results of defect rate evaluation. The degree of effects caused by defect on the amount of I/I was calculated using analytic hierarchy process (AHP) and allocated on each pipe. The required costs were determined by selecting a function and limiting factors depending on sewer treatment costs and restrictions in budget implementation. The project priority was figured out based on the capacity results acquired in SWMM. The logical limitation in this study was considered to be in counting the I/I factor as the most important factor in sewer rehabilitation. Due to lack of scientific background for determining the target pipes, it was difficult to estimate accurate budget for different rehabilitation methods.

3 SYSTEM DEVELOPMENT

3.1 Design

By analyzing the tasks associated with planning sewer rehabilitation, core tasks were determined. Those selected core tasks include collection of sewer information, development of database, selection of target pipes, selection of rehabilitation methods, budget estimation and economic assessment, and determination of project priority. For each task, related information was collected and an integrated decision-making support database was prepared.

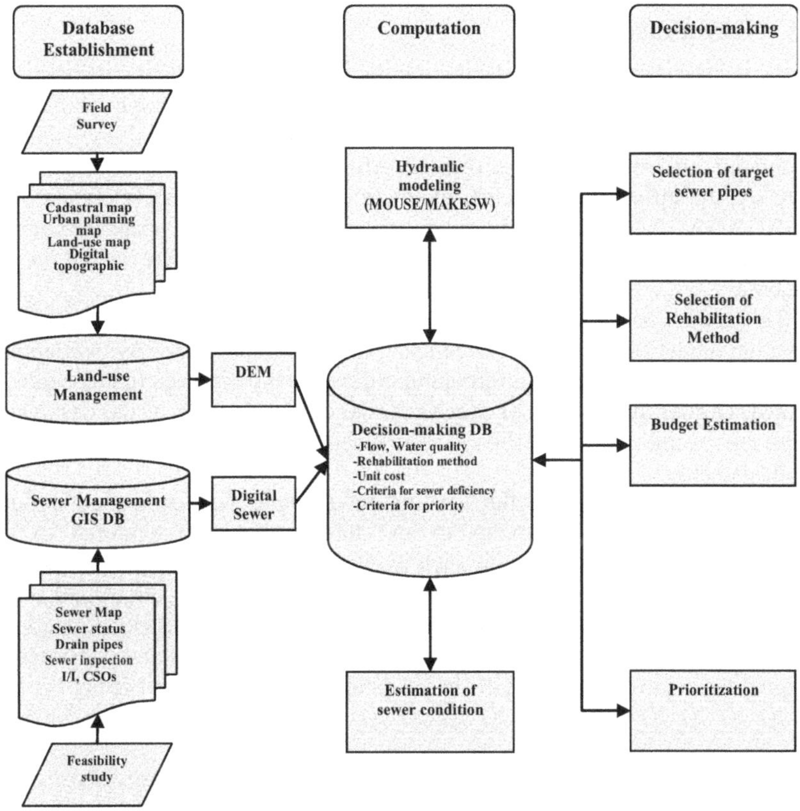

Figure 2 . STRUCTURE OF THE DECISION-MAKING SUPPORT SYSTEM

The decision-making support system is designed to have the flow of conventional decision-making processes starting from database setup followed by computation to decision-making as shown in Figure 2. In the database setup stage, a basic decision-making support database is created using raw and processed data. External modeling results acquired from MOUSE or MAKESW in addition to the on-site inspection results are integrated in the database in the computation stage. In the decision-making stage, target pipes, optimum rehabilitation methods, costs estimation, and determination of project priority are determined.

3.2 Database Setup

In the database setup stage, a database built on basic information was prepared. A GIS reference map was made using the digital maps (topography, buildings, roads, streams, land utilization, etc.) of target areas provided by National Geographic Information System (NGIS). After the boundaries of the target areas were set, the information on each pipe was digitized for the development of a digital sewer map compatible with existing CAD sewer maps. Extracted information on the nodes (manholes) and links (pipes) were added in the database.

To acquire basic information required for computer modeling, a digital elevation model (DEM) was created using the topography map. By overlapping land utilization (land cover) information, information on surface runoff could be created. During these series of processes, the catchment area of the designated nodes and overall runoff coefficient were calculated automatically and recorded on the database.

Structural evaluation on the existing sewer pipes was conducted through MOUSE and MAKESW modeling and the results were exported to the database. Additional information such as water quality and flow rate at major points acquired through on-site survey (CCTV or visual inspection) and sewer management documents were added in the database. The structure of all the information needed for the decision-making stage was defined and converted as a digital format to construct a GIS-based database with high consistency.

3.3 Computation

In the computation stage, the modeling results and on-site inspection results were summarized and incorporated in the decision-making database. The information on capacity, flow rate, and slope acquired through MOUSE or MAKESW were reviewed to extract the pipes, which did not meet the criteria, and further to create background data, which was to be used in the following decision-making stage. The on-site inspection results were used to identify the types of defects and to count the number of defects. The creation of the database was completed in the computation stage by adding the values of calculated total defect rate and defect rate per pipe length in the computation stage.

3.4 Decision -Making

There are 3 modules under the decision-making stage including the modules for selection of target pipes, selection of rehabilitation method, budget estimation, and determination of project priority. Each module is designed to produce interim and final results during and after the analysis. Those results may be modified by a user depending on the principal purposes and conditions of the specific project.

In the decision-making stage, the priority of projects is determined after all the results produced by other modules. The criteria used in determining the project priority are chosen after reviewing criteria used widely in Korea for the decision-making processes of sewer rehabilitation projects. Those criteria include pollution load, I/I rate, pipe defect rate, ratio of pipes of low capacity, pipe density, and ratio of pipes of low flow rate. Table 1 shows the definition and the selection background of each criterion.

Figure 3 shows the flow diagram of the module for the determination of project priority. In many rehabilitation projects, for the 5 criteria except for the ratio of pipes of low flow rate, similar importance was allocated. However, imerging new demands such as increase in flowrate requires flexibility in selecting the criteria and allocating different importance level for each criteria. To meet recent demands, this system is designed to adopt flexible criteria and the importance level allocated on each criteria. Depending on specific features of the target area, the weight factors can be adjusted using the standard AHP questionaires. With the introduction of flexible scoring system depending on specific conditions, absolute evaluation factors can be adopted.

Table 1. DEFINITION AND SELECTION BACKGROUND OF 6 CRITERIA USED IN DETERMINING THE PROJECT PRIORITY

Criteria	Definition	Selection Background
Pollution load	Design pollution load (kg BOD/d) and measured or computed pollution load (kg BOD/d) in the designated area	Important factor determining the treatment efficiency of sewer treatment facilities. Using pollution load, transportation of pollution load by pipes can be evaluated.
I/I rate	Measured or estimated amount of I/I per pipe length or the area of the designated area. There are I/I rate per average pipe diameter and pipe length (m3/d/km/mm) or I/I rate per area (m3/d/ha)	I/I rate, proportional to pipe defect rate, is important in determining dilution factors of pollution load in pipes. Minimization of I/I rate is often counted as major goal of sewer rehabiltation projects.
Pipe defect rate	The value calculated by dividing the total defect score by pipe length (score/km). The total defect score is the sum of the number of defects multiplied by defect score.	The defree of defects observed during on-site inspection can be used for evalutation of pipe condition.
Ratio of pipes of low capacity	The ratio of the length of pipes of low capacity to the total pipe length (%)	The capacity of storm water transportation can be estimated. Using the ratio of pipes of low capacity, selection of target pipes and target areas becomes easier.
Pipe density	The ratio of pipe length to population in the designated area. There are pipe density per population (km/1,000 people) and pipe density per area (km/ha).	Connection public sewer ratio and service quality can be evaluated using pipe density. Usually, high pipe density is required in urbanized zones.
Ratio of pipes of low flowrate	The ratio of the length of pipes of low flowrate to the total pipe length (%)	Low flowrate causes settlement, odors and decomposition within pipes.

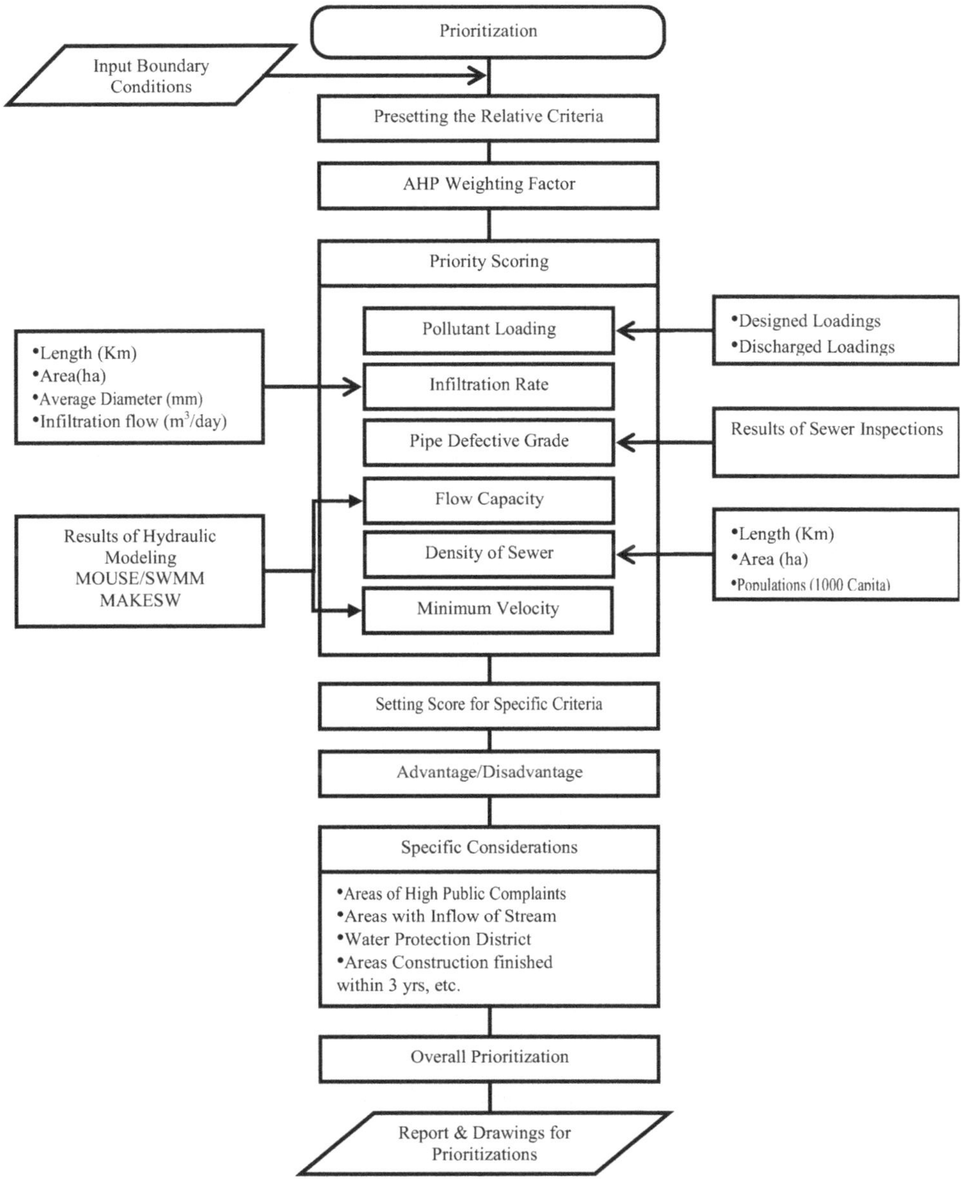

Figure 3. FLOW DIAGRAM OF THE MODULE FOR THE DETERMINATION OF PROJECT PRIORITY

3.5 Weight factors according to AHP

For the determination of weight factors of the criteria summarized in Table 1, surveys and interviews were coordinated for a specific project. According to AHP, a popular method used as one of multiple-criteria decision analysis, a standard questionnaire was prepared and distributed to sewer experts for the determination of weight factors for each criterion.

In this system, as illustrated in Figure 4, a user may directly evaluate relative importance or utilize AHP module to integrate suggestions by experts for the evaluation of relative importance.

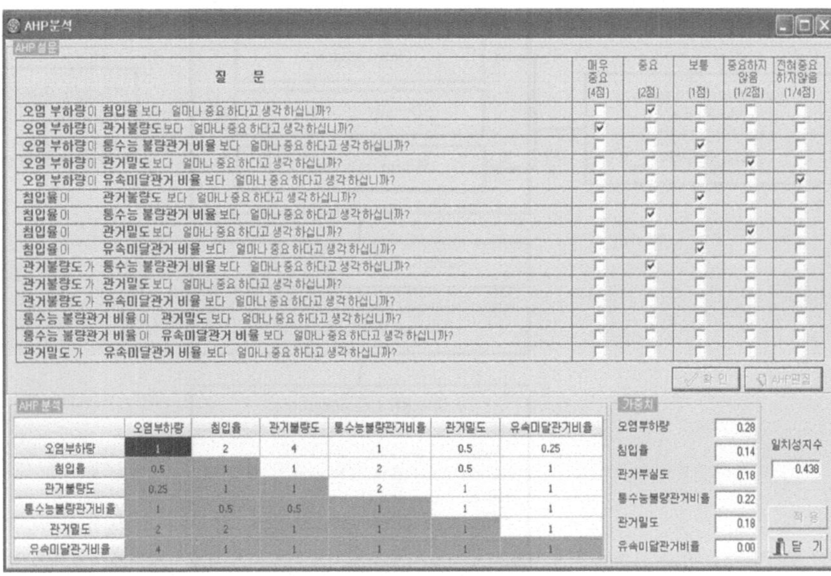

Figure 4. AHP MODULE USED TO DETERMINE WEIGHT FACTORS FOR RELATIVE EVALUATION

For this study, 30 sewer experts, who had extensive experiences in designing sewer networks and planning sewer rehabilitation, were selected. The AHP questionnaire containing the questions shown in Figure 4 was distributed to the experts. Based on the survey results, geometric means of weight factors were calculated to create supermatrixes. Table 2 shows the normalized results of those matrixes. The relative importance was highest for I/I rate (0.241), defect rate (0.204), ratio of pipes of low capacity (0.182), and pollution load (0.178) in order. Low relative importance was observed in pipe density (0.104) and ratio of pipes of low flowrate (0.091). The results showed the major issues associated with rehabilitation of sewers in Korea to be minimization of I/I, separation of storm water, and transport of pollution load. As this relative importance can be changed significantly depending on the projects, the AHP analysis module needs to be utilized for different projects to modify weight factors.

Table 2. RELATIVE IMPORTANCE ANALYSIS RESULTS OF SEWER PIPE EVALUATION CRITERIA

Item	Pollution load	I/I rate	Defect rate	Ratio of pipes of low capacity	Pipe density	Ratio of pipes of low flowrate	Relative importance (weight factor)
Pollution load	1.000	0.831	0.776	0.891	1.701	2.245	0.178
I/I rate	1.203	1.000	1.382	1.382	2.406	2.351	0.241
Defect rate	1.289	0.724	1.000	1.072	2.194	2.245	0.204
Ratio of pipes of low capacity	1.122	0.724	0.933	1.000	1.823	1.701	0.182
Pipe density	0.588	0.416	0.456	0.548	1.000	1.350	0.104
Ratio of pipes of low flowrate	0.445	0.425	0.445	0.588	0.741	1.000	0.091

4 VERIFICATION

For the verification of the system, utilization of sewer data of Sangdang-gu and Hungduk-gu in Chongju city, Korea was considered. As there was an on-going sewer rehabilitation project in Sangdang-gu, Sangdang-gu (area: 69.0km2, the length of pipes: 319km) was determined to be the target area. Major reference materials include Plan on Sewer Maintenance in Chongju city (March, 2001), Assessment of Chongju city Sewer Maintenance Project (December, 2002), and Feasibility Study on BTL projects and Plan on Sewer Systems (May 2006).

4.1 Target areas and database setup

There are old downtown areas developed along the Moosimchon, a stream bisecting Chongju city. The target area consists of 6 sewage treatment districts. The area of each treatment district is relatively large, and pipe networks are well developed in the target area. For convenience, after dividing the Moosimchon treatment district into 3 sub-districts, Moosim 1, Moosim 2, and Moosim 3, information on a total of 8 districts was collected. Table 3 summarizes property information of each district including pipe length, area, average pipe diameter, population and the amount of I/I. In Table 4, the factors required for the determination of project priority are summarized for each treatment district. As the factor for the ratio of pipes of low flowrate is not included in the criteria provided in Plan on Sewer Maintenance in Chongju city (March, 2001), the ratio of pipes of low flowrate is excluded in this study

Table 3. PROPERTY INFORMATION OF THE SEWAGE TREATMENT DISTRICTS IN SANGDANG-GU, CHONGJU CITY

Treatment district	Pipe length (km)	Area (ha)	Average pipe diameter (mm)	Population	Amount of I/I (m^3/d)
Balsan	10.97	347	625	9,300	767
Yulyang	51.26	321	569	94,700	7,806
Moosim 1	65.95	244	645	12,970	1,069
Moosim 2	69.24	332	684	16,720	1,378
Moosim 3	97.21	414	640	19,010	1,567
Myongam	57.17	317	584	50,600	4,171
Youngwoon	54.74	261	564	64,100	5,284
Wolwoon	9.28	218	689	1,700	140
Total	415.82	2,452	-	269,100	22,182

Table 4. SUMMARY OF FACTORS REQUIRED FOR DETERMINING THE PROJECT PRIORITY

Treatment district	Pollution load		I/I rate		Pipe defect rate (score/km)	Ratio of pipes of low capacity	Pipe density	
	Design (kg/d)	Measured (kg/d)	Per length (m3/d/km/mm)	Per area (m3/d/ha)			Per population (km/thousand people)	Per area (km/ha)
Balsan	489	545	111.869	2210.37	238	79.12	1.18	0.03
Yulyang	4976	5002	267.632	24317.76	57	25.50	0.54	0.16
Moosim 1	682	456	25.131	4381.15	43	14.72	5.08	0.27
Moosim 2	879	136	29.096	4150.60	39	2.44	4.14	0.21
Moosim 3	999	1885	25.187	3785.02	30	8.27	5.11	0.23
Myongam	2659	2760	124.928	13157.73	59	29.12	1.13	0.18
Youngwoon	3368	3717	171.151	20245.21	51	7.22	0.85	0.21
Wolwoon	89	99	21.896	642.20	303	37.72	5.46	0.04

Detail property and spatial information on the digital maps, sewer maps, and topography maps acquired from Plan on Sewer Maintenance in Chongju city (March, 2001), and Assessment of Chongju city Sewer Maintenance Project (December, 2002) was modified. As shown in Figure 5, after the target areas were defined on the digital maps at a scale of 1:1,000, sewer maps were modified. These maps were utilized as basic maps and property information for the GIS-based decision-making support database. The MOUSE and MAKESW modeling results were incorporated. The on-site inspection results were evaluated and used for the calculation of defect rate for each pipe and treatment district.

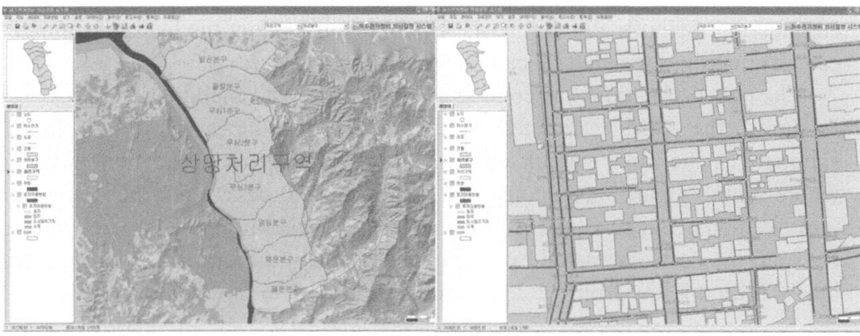

Figure 5. DIGITAL MAP OF THE TARGET AREA (LEFT) AND SEWER MAP (RIGHT)

4.2 System Verification

4.2.1 Selection of Target Pipes

Capacity, flowrate, and pipe gradient acquired from MOUSE and MAKESW modeling were reviewed to extract the pipes, which did not meet design standards. For rehabilitation of the pipes with defect rate higher than 20% and less than 20%, non-excavation whole-scale repair, and partial repair were assigned respectively. All the information was marked on the map for the selection of target pipes. Depending on the direction of a specific rehabilitation project, a user could include or exclude the pipes of low flowrate in the category of target pipes. Additionally, based on the user's need, the pipes requiring whole-scale repair can be converted into the pipes requiring replacement.

4.2.2 Selection of Rehabilitation Methods and Costs Estimation

In Korea, recommendation on optimum rehabilitation methods for target pipes requiring improvement or repair was not made in numbers of reports on sewer rehabilitation. In some countries, there are systems, which can suggest details of optimum rehabilitation methods depending on the types and degrees of pipe defect (Knolmar et al., 2003, and Saegrove, 2006). Expecting potential changes in engineering methods and the standards on defect evaluation, the engineering methods under the sub-major classification suggested by Korea Water and

Wastewater Works Association (2006) were adopted in the system. The conventional cost estimation method was improved from calculating standard costs for pipe length to figuring out the project costs in consideration of specific conditions and types of defect based on the information on classified engineering methods and unit costs. Upon the changes in the factors required for cost estimation, the database can be modified by a user.

The cost estimation was conducted for 2 different rehabilitation cases. Table 5 shows the costs estimated automatically by the system and Table 6 summarizes the costs calculated when all the pipes requiring non-excavation whole-scale repair are converted into the pipes requiring excavation improvement. Because the costs for non-excavation methods are generally cheaper, the costs estimated by changing the rehabilitation method are increased by 1.4 billion KRW from 72.2 billion KRW to 73.6 billion KRW.

Table 5. COSTS ESTIMATED WHEN NON-EXCAVATION WHOLE-SCALE REPAIR IS APPLIED

District	Sum (million KRW)	Improvement		Repair		
		Sum	Combined	Sum	Whole-scale	Partial
Total	73,631	71,635	71,635	1,996	1,851	145
Myongam	7,472	7,286	7,286	187	184	3
Moosim 1	13,108	12,787	12,787	322	301	21
Moosim 2	14,484	14,380	14,380	104	100	3
Moosim 3	24,973	24,831	24,831	142	137	5
Balsan	1,733	1,248	1,248	485	471	14
Youngwoon	4,552	4,241	4,241	311	291	20
Wolwoon	1,913	1,662	1,662	251	209	43
Yulyang	5,395	5,201	5,201	195	159	36

Table 6. COSTS ESTIMATED WHEN EXCAVATION IMPROVEMENT METHODS ARE APPLIED

District	Sum (million KRW)	Improvement		Repair		
		Sum	Combined	Sum	Whole-scale	Partial
Total	72,230	72,085	72,085	145	-	145
Myongam	7,332	7,329	7,329	3	-	3
Moosim 1	12,880	12,859	12,859	21	-	21
Moosim 2	14,407	14,404	14,404	3	-	3
Moosim 3	24,868	24,863	24,863	5	-	5
Balsan	1,387	1,372	1,372	14	-	14
Youngwoon	4,326	4,305	4,305	20	-	20
Wolwoon	1,757	1,714	1,714	43	-	43
Yulyang	5,275	5,239	5,239	36	-	36

In this study, the module was specified to select the lower costs after estimating costs for the cases of non-excavation whole-scale repair and excavation improvement. A user may accept the result or alter some conditions to get different results if needed.

4.2.3 Determination of Project Priority

The project priority for each treatment district was determined by varying the weight factors and importance of 5 criteria: pollution load, ratio of pipes of low capacity, defect rate, I/I rate, and pipe density. Depending on specific features of the target areas and the direction of the rehabilitation project, the weight factors were adjusted based on AHP. Through this, a user may establish a systematic rehabilitation plans for different scenarios. In this study, 3 scenarios were created for the determination of project priority for the treatment districts in Sangdang-gu Chongju city.

The 1st scenario consisted of the target pipes determined based on the MAKESW modeling results, the costs estimated automatically by the system, and the 5 criteria with the same weight factors. The project priority is listed in Table 7. In this scenario, the highest priority was allocated on the most crowded districts, the 3 Mooshim districts. Less crowded districts got lower priority.

The 2nd scenario is different from the 1st scenario because it is assumed that the pipes of poor flow rate were not included in the category of target pipes. As shown in the table, the estimated costs were decreased by 28.4%. The higher priority was given to the Wolwoon district, where variable of pipe network density, than the Moosim 2 district.

In the 3rd scenario, the weight factors were adjusted based on AHP. Higher weight factors were given to ratio of pipes of low capacity and pollution load than other factors. When compared with the 1st scenario, changes in the priority of the Wolwoon district were observed. The reason for the changes in the priority was considered to be relatively new and less defective sewer networks in the district. Consequently, depending on the purposes and styles of a project, changes in project priority could be observed.

Table 7. PROJECT PRIORITY DETERMINED USING THE MODELING RESULTS WITHOUT MODIFICATION

District	Pollution load (20 points)	I/I rate (20points)		Pipe defect rate (20 points)	Ratio of pipes of low capacity (20 points)	Pipe density (20 points)		Specific condition	Sum	Priority	Costs (million KRW)
		Per unit length (10 points)	Per unit area (10 points)			Per capita (10 points)	Per unit area (10 points)				
Moosim 1	20	10	10	4	20	10	4	10	88	1	13,108
Moosim 3	20	10	10	4	20	10	4	8	86	2	24,973
Moosim 2	20	10	10	4	20	8	4	5	81	3	14,484
Wolwoon	16	10	10	8	20	10	2	3	79	4	1,913
Youngwoon	20	10	10	4	20	2	4	6	76	5	4,552
Yulyang	20	10	10	4	20	2	2	6	74	6	5,395
Balsan	20	10	10	4	20	2	2	6	74	7	1,733
Myoungam	20	10	10	4	20	2	2	-10	58	8	7,472

Table 8:PROJECT PRIORITY DETERMINED WHEN PIPES OF POOR FLOWRATE WAS EXCLUDED FROM CATEGORY OF TARGET PIPES

District	Pollution load (20 points)	I/I rate (20points)		Pipe defect rate (20 points)	Ratio of pipes of low capacity (20 points)	Pipe density (20 points)		Specific condition	Sum	Priority	Costs (million KRW)
		Per unit length (10 points)	Per unit area (10 points)			Per capita (10 points)	Per unit area (10 points)				
Moosim 1	20	10	10	4	20	10	4	10	88	1	9,132
Moosim 3	20	10	10	4	20	10	4	8	86	2	18,042
Wolwoon	16	10	10	12	20	10	2	3	83	3	1,523
Moosim2	20	10	10	4	20	8	4	5	81	4	11,342
Youngwoon	20	10	10	4	20	2	4	6	76	5	3,829
Yulyang	20	10	10	4	20	2	2	6	74	6	4,926
Balsan	20	10	10	4	20	2	2	6	74	7	1,328
Myoungam	20	10	10	4	20	2	2	-10	58	8	5,238

Table 9: PROJECT PRIORITY WHEN WEIGHT FACTORS OF THE CRITERIA WERE ADJUSTED BASED ON AHP.

District	Pollution load (22 points)	I/I rate (20points)		Pipe defect rate (18 points)	Ratio of pipes of low capacity (24 points)	Pipe density (20 points)		Specific condition	Sum	Priority	Costs (million KRW)
		Per unit length (8 points)	Per unit area (8 points)			Per capita (5 points)	Per unit area (5 points)				
Moosim 1	22	8	8	4	24	5	2	10	83	1	13,108
Moosim 3	22	8	8	4	24	5	2	8	81	2	24,973
Moosim 2	22	8	8	4	24	4	2	5	77	3	14,484
Youngwoon	22	8	8	4	24	1	2	6	75	4	4,552
Yulyang	22	8	8	4	24	1	1	6	74	5	5,395
Wolwoon	18	8	8	7	24	5	1	3	74	6	1,913
Balsan	22	8	8	4	24	1	1	6	74	7	1,733
Myoungam	22	8	8	4	24	1	1	-10	58	8	7,472

5 CONCLUSION

Through this study, the decision-making support system to support core tasks including cost estimation and determination of project priority of sewer rehabilitation works was developed. The system was verified and modified using the data of Sangdang-gu Chongju city, where there was an on-going sewer rehabilitation project. Throughout the courses of this study, following conclusion was made.

1. Because the system was designed and developed to be compatible with standardized digital information, digital maps, sewer inventory, and GIS maps of NGIS can be efficiently utilized. Additionally, the decision-making support system can be applied in numbers of sewer-related projects efficiently due to its systematic structure and vast amount of imbedded information.

2. In addition to design performance of the rehabilitation project, an integrated evaluation results on the basis of specific features and conditions of a project can be presented. Upon changes in the project, the system provides fast and accurate alternatives by minimizing repeated computation and simplifying complicated decision-making processes.

3. Considering the new trend of sewer projects, the system is designed to be compatible with international standards. The criteria and multiple elements of the system can be modified and specified by a user upon the changes in the purposes of a project or applicable laws and standards. In the future, through modification, the system could be used in conjunction with sewer management systems for higher efficiency

6 REFERENCES:

Kim T. (2004). The Development of Sewer Optimal Rehabilitation Decision Making System by Considering Inflow/Infiltration. Master's degree thesis. Korea University

Chung C., Park K., and Chung Y. (2001). A Multiple Objective Mixed Integer Programming Modeling for Sewer Rehabilitation Planning. Journal of Korean Society of Civil Engineers, Vol. 21, pp. 295-304.

Korea Water and Wastewater Works Association (2006) Revised Manual for Sewer Construction

Ministry of Environment (1997) Study on the Maintenance of Urban Sewer Pipes

Fenner R. A. and Sweeting, L. (1999). A Decision Support Model for the Rehabilitation of "Non-Critical" Sewers. Water Science and Technology, Vol. 39, No. 9, pp. 193-200.

Knolmar, M. and Szabo, G. C. (Ed.) (2003). Classification Systems Based on Visual Inspection. CARE-S D3 Report, WP2-Structural Condition.

Plenker, T. (2002). Computer Aided Decision Support on Choosing the Right Technology for Sewer Rehabilitation. Water Science and Technology, Vol. 46, No. 6-7, pp. 403-410.

Saegrov S. (Ed.) (2006). CARE-S: Computer Aided Rehabilitation of Sewer and Storm Water Networks. European Water Research Report, EVK1-CT-2002-00106. IWA Publishing, UK.

Performance indicators implemented in the city of Larissa water distribution network

S. Tsitsifli and V. Kanakoudis

ABSTRACT

The present paper is trying to implement the IWA standard international water balance for the case of the distribution network of the city of Larissa. Some of the performance indicators are also calculated and analysed. The goals of the study are to develop the IWA standard international water balance, calculate performance indicators and assess them in relation to other national and international data given by the bibliography. The final aim is to propose

 Performance Assessment of Urban Infrastrucutre Services
Edited by Enrique Cabrera Jr and Miguel Ángel Pardo. ISBN: 9781843391913.
Published by IWA Publishing, London, UK.

techniques (PRVs etc.) introduced to other Water Utilities internationally in order to minimize water losses.

1 INTRODUCTION

An urgent need for a common international terminology is apparent the last few years since there is a wide diversity of formats and definitions used for water balance calculations internationally or even within the same country. IWA Task Forces on Water Losses and Performance Indicators produced an international "best practice" standard approach for water balance calculations using the best practice from many countries (Farley et al., 2003).

This IWA standard international water balance and the performance indicators are being implemented for the city of Larissa distribution network. Specifically the research team is estimating the billed and unbilled authorized water consumption, the apparent and the real losses in order to finally estimate the non-revenue water and its cost. The time period analyzed is from 2001 to 2006. Many performance indicators are being estimated. The difficulties experienced in order to collect and analyze the data are being introduced. Finally the four basic methods of managing real losses, that is: a) pressure management; b) speed and quality of repairs; c) active leakage control; d) pipeline and assets management, are being reviewed and examined in order to evaluate their impact in reducing water losses of the system.

2 IWA WATER BALANCE INTERNATIONAL STANDARD AND PERFORMANCE INDICATORS (PIS)

Based on the IWA "best practice" standard water balance certain definitions are involved (Farley et al., 2003):

- System input volume is the annual volume input to that part of the water supply system.
- Authorised consumption is the annual volume of metered and/or non-metered water taken by registered customers, the water supplier and others who are implicitly or explicitly authorised to do so. It includes water exported and leaks and overflows after the point of customer metering.
- Unbilled authorised consumption includes items such as fire fighting and training, flushing of mains and sewers, cleaning of suppliers' storage tanks, filling of water tankers, water taken from hydrants, street cleaning, watering of municipal gardens, public fountains, frost protection, building water etc. Unbilled authorised consumption should

normally be only a small component of the water balance (less than 1% of system input volume).

- Non-revenue water (NRW) is the difference between system input volume and billed authorised consumption. NRW consists of: a) unbilled authorised consumption and b) water losses.
- Water losses are the difference between system input volume and authorised consumption and consists of apparent losses and real losses.
- Apparent losses consist of unauthorised consumption and all types of metering inaccuracies.
- Unauthorised consumption (theft and illegal use) occurs to a greater or lesser extent in most systems worldwide, but in reasonably well managed systems it should not exceed 1% of the system input volume.
- Customer metering errors include: a) random errors due to accounting procedures (misread meters, incorrect estimates fro stopped meters, improper calculations etc.) b) systematic errors due to under-registration or over-registration of customer meters.
- Real losses are the annual volumes lost through all types of leaks, bursts and overflows on mains, service reservoirs and service connections, up to the point of customer metering.

The final goal of any performance indicators system is to provide information. The performance indicators measure the efficiency and effectiveness of the delivery of the services by an undertaking that results from the combination of several variables. In the present study a number of operational indicators are calculated (Alegre et al., 2006):

- Instrumentation calibration PI: Meter replacement (Op8)
- Rehabilitation PI: Mains replacement (Op18)
- Operational water losses PI : Water losses per connection (Op23), Water losses per mains length (Op24), Apparent losses (Op25), Apparent losses per system volume (Op26), Real losses per connection (op27), Real losses per mains length (Op28) and Infrastructure leakage index (Op29).

Also an economic and financial indicator is calculated:

- Economic water losses PI: Non-revenue water by volume (Fi46).

The mathematical formulas of the indicators are given in table 1 (Alegre et al., 2006):

Table 1. PERFORMANCE INDICATORS AND THEIR MATHEMATICAL FORMULAS

	Indicators	Formula
Op8	Meter replacement	(D45*365/H1)/E6
Op18	Mains replacement	[(D22*365/H1)/C8]*100
Op23	Water losses per connection	(A15*365/H1)/C24
Op24	Water losses per mains length	(A15/H1)/C8
Op25	Apparent losses	[A18/(A3-A5-A7)]*100
Op26	Apparent losses per system volume	A18/A3
Op27	Real losses per connection	(A19*1000)/(C24*H2/24)
Op28	Real losses per mains length	(A19*1000)/(C8*H2/24)
Op29	Infrastructure leakage index	Op27/[18*C8/(C24+0,8+0,025*C25)/(D34/10)]
Fi46	Non-revenue water by volume	(A21/A3)*100

Where, D45 is the number of the meters replaced, H1 is the assessment period (days), E6 is the number of customer meters, D22 are the mains replaced (Km), C8 is the mains length (Km), A15 are the water losses (m^3), C24 is the number of service connections, A18 are the apparent losses (m^3), A3 is the system input volume (m^3), A5 is the exported raw water (m^3), A7 is the exported treated water (m^3), A19 are the real losses (m^3), H2 is the time system is pressurized (hours), C25 is the average service connection length (m), D34 is the average operating pressure (KPa), A21 is the non-revenue water (m^3).

Real losses cannot be eliminated totally. The lowest technically achievable annual volume of real losses for well-maintained and well-managed systems is the unavoidable annual real losses (UARL). The difference between UARL and the current annual real losses (CARL) is the potentially recoverable real losses. The ratio of the CARL to the UARL is the Infrastructure Leakage Index (ILI). The ILI measures how effectively the infrastructure activities (repairs, active leakage control and pipeline/assets management) are being managed at current operating pressure.

3 THE CASE STUDY NETWORK

The case study network is the water distribution network of the city of Larissa, in Greece. The city is situated in the eastern part of the Greek mainland, being the capital city of the Thessaly area. Its population is about 192.000 inhabitants. Its water distribution network consists of 627.300 km of pipes of different materials and it serves 72.000 connections (data for the year 2006). The pipes are made of cast iron, asbestos cement, steel and PVC (Fig. 1).

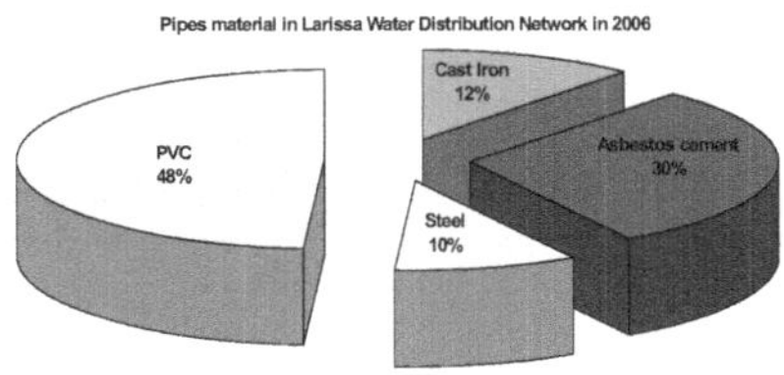

Figure 1. PIPES MATERIAL

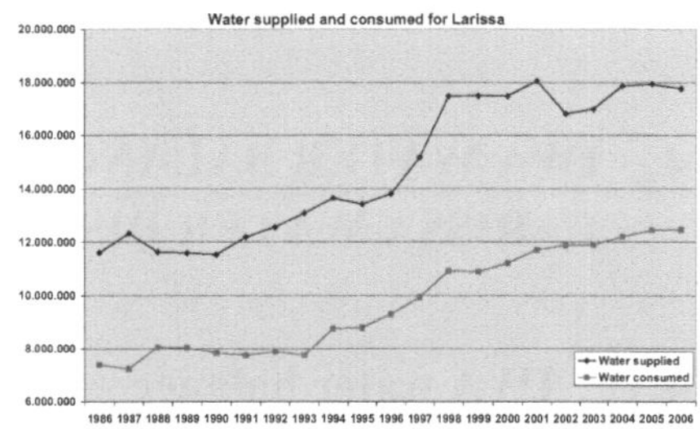

Figure 2. WATER SUPPLY AND CONSUMPTION 1986-2006

The water distribution system is operated by the municipal Utility called DEYAL. The city is being supplied with water taken from three groups of water bores. The water is being concentrated in two water tanks from where it is being supplied by gravity in the two zones that the distribution network is divided.

DEYAL is keeping extensive data of the water supplied, consumed, metered and unmetered since 2001. Figure 2 shows the water supplied and consumed since 1986.

4 IWA WATER BALANCE AND PIS IMPLEMENTED IN LARISSA WATER DISTRIBUTION NETWORK

4.1 IWA water balance

Based on DEYAL's data the IWA water balance was created for 6 years (Fig. 3).

IWA WATER BALANCE FOR LARISSA WATER DISTRIBUTION NETWORK FOR 2001

Annual System Input Volume 18.059.662 m3/year	Authorised Consumption 11.705.246 m3/year	Billed Authorised Consumption 10.977.845 m3/year	Billed Metered Consumption 8.802.563 m3/year	Revenue Water 10.977.845 m3/year
			Billed Unmetered Consumption 2.175.282 m3/year	
		Unbilled Authorised Consumption 727.401 m3/year	Unbilled Metered Consumption 727.401 m3/year	Non - Revenue Water 7.081.817 m3/year
			Unbilled Unmetered Consumption 0 m3/year	
	Water Losses 6.354.416 m3/year	Apparent Losses 1.602.809 m3/year	Unauthorised Consumption 173.314 m3/year	
			Customer Meter Inaccuracies and Data Handling Errors 1.429.495 m3/year	
		Real Losses 4.751.607 m3/year		

IWA WATER BALANCE FOR LARISSA WATER DISTRIBUTION NETWORK FOR 2006

Annual System Input Volume 17.770.139 m3/year	Authorised Consumption 12.458.122 m3/year	Billed Authorised Consumption 11.687.062 m3/year	Billed Metered Consumption 9.095.465 m3/year	Revenue Water 11.682.062 m3/year
			Billed Unmetered Consumption 2.591.597 m3/year	
		Unbilled Authorised Consumption 771.060 m3/year	Unbilled Metered Consumption 771.060 m3/year	Non - Revenue Water 6.083.077 m3/year
			Unbilled Unmetered Consumption 0 m3/year	
	Water Losses 5.312.017 m3/year	Apparent Losses 1.656.764 m3/year	Unauthorised Consumption 176.786 m3/year	
			Customer Meter Inaccuracies and Data Handling Errors 1.479.979 m3/year	
		Real Losses 3.655.253 m3/year		

Figure 3. THE IWA WATER BALANCE FOR 2001 AND 2006 FOR LARISSA

The WB-Easy Calc (Liemberger, 2006) was used in order to calculate the water balance. Certain assumptions have been made in the implementation of the IWA water balance. The Utility is measuring the unbilled metered consumption used for fire services, parks, municipal buildings, etc. The unbilled authorised consumption is the unbilled metered consumption, meaning that there is no unbilled unmetered consumption. The unauthorised consumption is calculated assuming that the illegal connections and meter bypasses etc. represent about the 1% of the water input based on national data (Kanakoudis, 1998). The metering errors are estimated to be 15% of the metered bulk supply. Also the water tariff was put in the model (table 2) in order to calculate the value of the non-revenue water. From the IWA water balance for the 6 years the water losses volume and the water losses value are calculated (figure 4).

Table 2. WATER TARIFF FOR LARISSA WATER NETWORK

Year	2001	2002	2003	2004	2005	2006
Water tariff (Euro/m^3)	0,41	0,41	0,41	0,43	0,43	0,43

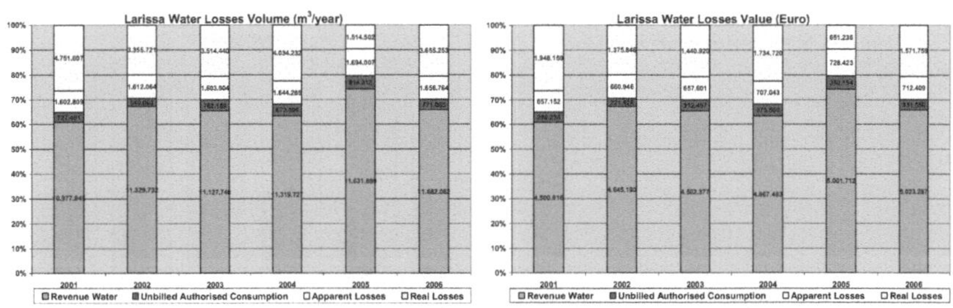

Figure 4. WATER LOSSES VOLUME AND WATER LOSSES VALUE

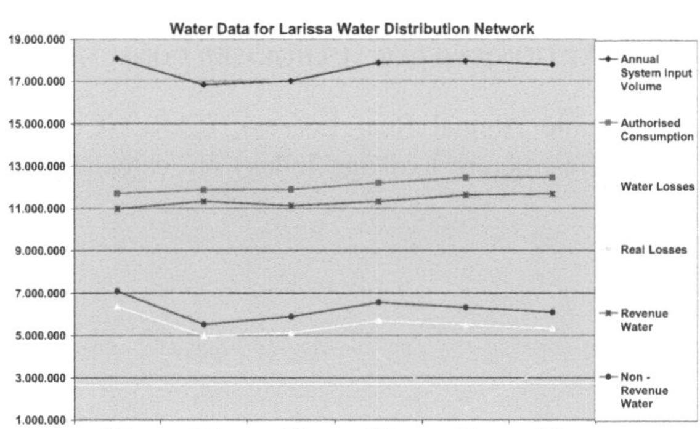

Figure 5. WATER DATA FOR LARISSA WATER SUPPLY NETWORK

The water distribution data for the system (annual system input volume, authorised consumption, water losses, real losses, revenue water, non-revenue water) is presented in figure 5 for 6 years.

4.2 IWA PIs

The Performance Indicators (PIs) analyzed in section 3 (Op8, 18, 23, 24, 25, 26, 27 and 28) are calculated for Larissa distribution network (figure 6).

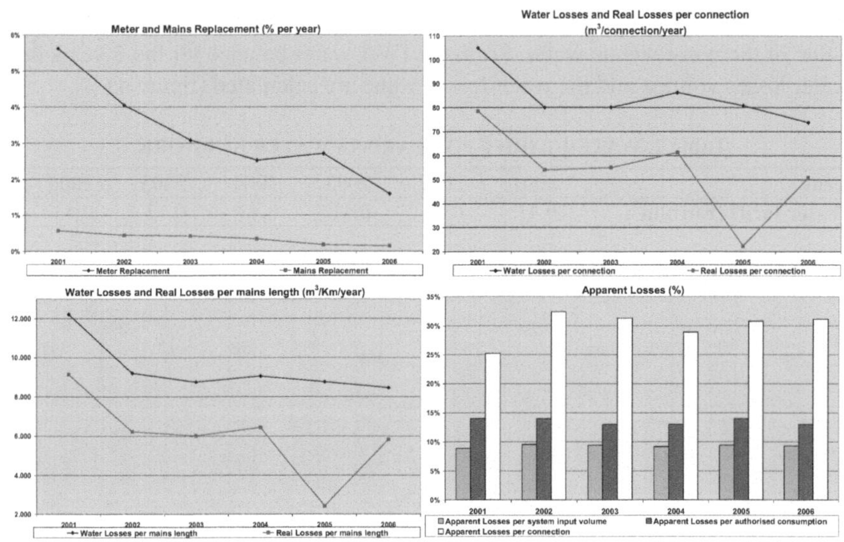

Figure 6. **METER AND MAINS REPLACEMENT; WATER AND REAL LOSSES PER CONNECTION AND PER MAINS LENGTH; APPARENT LOSSES PER SYSTEM INPUT VOLUME, PER CONNECTION AND PER AUTHORISED CONSUMPTION**

UARL (Unavoidable Annual Real Losses), CARL (Current Annual Real Losses) and ILI (Infrastructure Leakage Index) are calculated for the network (figure 7).

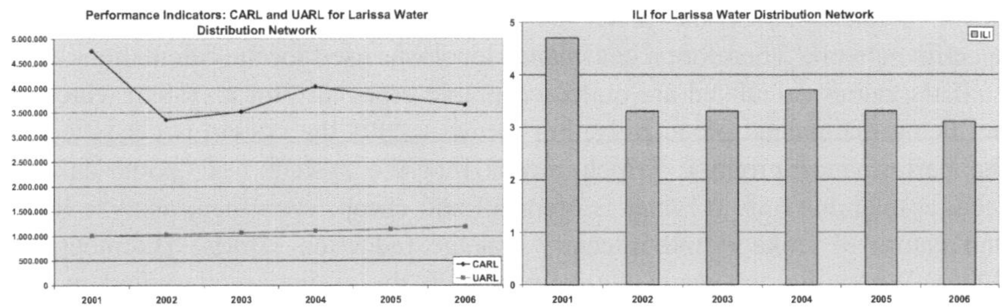

Figure 7. UARL, CARL AND ILI FOR THE LARISSA WATER SUPPLY NETWORK

ILI's minimum and maximum value for Larissa for the 6-year period compared to other randomly selected countries are presented in figure 8 (Liemberger, 2006).

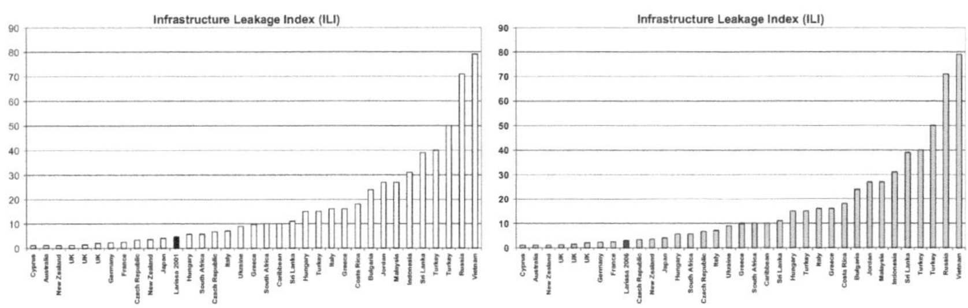

Figure 8. LARISSA'S ILI COMPARED TO OTHER RANDOMLY SELECTED COUNTRIES

5 CONCLUSIONS

The IWA standard water balance is a diagnostic approach, that is giving to the Water Utility the chance to evaluate correctly its system and develop a water loss strategy; The volume of losses are being quantified and the real losses have been measured revealing a different and more correct approach than just deduct the billed consumption from the supply input. But during the application to a certain Utility some problems arise.

The main problems faced in the Larissa distribution network were related to the difficulties to get the data from the Utility. In most of the cases the Utilities are not keeping exact data except of the water supplied and billed to the customers. The Utility has never implemented a project in order to calculate the illegal connections or the meter under-registration and malfunction for the specific network. Therefore a confidence level was used for the calculations.

ILI's values calculated are quite acceptable especially for a system which is not being maintained proactively. For the year 2001 the system is classified in the performance group C which means that the system has "poor leakage recode; tolerable only if water is plentiful and cheap; even then, analyze level and nature of leakage and intensify leakage reduction efforts" (Liemberger, 2006). Where for the years 2002 until 2006 the system is classified in the performance group B where there is "potential for marked improvements; consider pressure management, better active leakage control practices; and better network maintenance" (Liemberger, 2006). Generally ILI's values are falling. This is probably due to the efforts from the Utility to manage water supply better and the installations of PRVs (pressure reduction valves).

As mentioned before, real losses cannot be eliminated totally. The lowest technically achievable annual volume of real losses for well-maintained and well-managed systems is the UARL, represented by the smaller rectangle in figure 9. The ILI measures how effectively the infrastructure activities (repairs, active leakage control and pipeline/assets management) are being managed at current operating pressure. For each of the four activities, there is some economic level of investment and activity, which needs to be calculated or assessed, depending upon the marginal value placed on the real losses. Depending upon local circumstances and practice, the marginal value placed on real losses may be low or high, and this profoundly influences the economic management policies for controlling real losses.

In figure 9, suppose that the area of the large rectangle represents the current annual real losses (CARL). As the system ages, there is a tendency for a natural rate of rise of real losses through new leaks and bursts, some of which will not be reported to the utility. This tendency is controlled and managed by some combination of the four primary components of real losses management, namely:

- Pipeline and assets management
- Pressure management (which may mean increases or decreases of pressure)
- Speed and quality of repairs
- Active leakage control, to locate unreported leaks

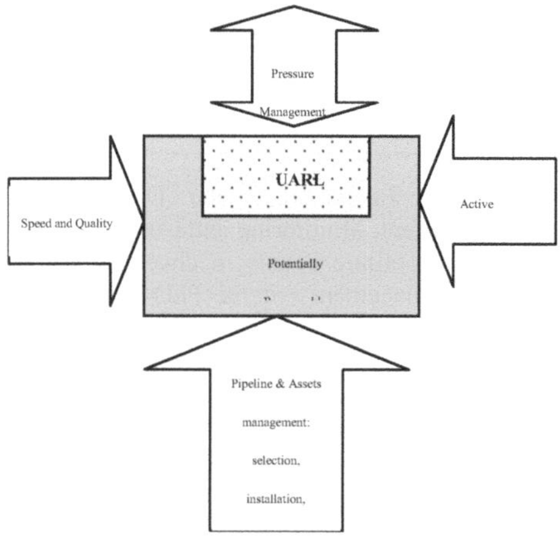

Figure 9. THE FOUR BASIC METHODS OF MANAGING REAL LOSSES (FARLEY ET AL., 2003).

The number of new leaks arising each year is influenced primarily by long-term pipeline management. Pressure management can influence the frequency of new leaks, and the flow rates of all leaks and bursts, The average duration of the leaks is limited by the speed and quality of repairs, and the active leakage control strategy controls how long unreported leaks run before they are located. The extent to which each of these four activities is carried out will determine whether the volume of annual real losses increases, decreases or remains constant.

The recommendations for the Larissa water distribution network are firstly to manage pressure in a better way by starting a pressure management project containing the installation of more PRVs. This project should be regularly updated following the continuously changing needs of the distribution network. In the case study of Thessaloniki network where PRVs were installed the pipe break incidents were reduced up to 33,8% for the year 2004 (Kanakoudis et al., 2005). Another matter is the repair or replacement of the water supply mains, as the impact of their breaks may result the entire city to a water shortage problem. This can be achieved by determining the optimum replacement time of the pipes using a technico-economic analysis (Kanakoudis, 1998). Last but not least is the better knowledge and management of the distribution network which could be achieved through: a) installation of a S.C.A.D.A. system, providing continuous monitoring of pressure and flow data; b) installation of a G.I.S. system and; c) development of a detailed hydraulic simulation model.

6 REFERENCES

Alegre, H., Baptista, J.M., Cabrera, E. Jr., Cubillo, F., Duarte, P., Hirner, W., Merkel, W. and Parena, R. (2006) Performance Indicators for Water Supply Services, Second Edition, IWA Publishing , UK

Farley, M. and Trow, S. (2003) Losses in Water Distribution Networks – A practinioner's Guide to Assessment, Monitoring and Control, IWA Publishing, UK

Kanakoudis, V. (1998) The role of failure events in developing water systems pipes preventive maintenance and replacement criteria. PhD Thesis, Aristotle University of Thessaloniki, Greece

Kanakoudis, V. and Charavgis, I. (2005) Water systems management – Lifesaving interventions evaluation, *e-Proceedings of the 6th International Conference of EWRA,* Menton, 7-10 September

Lambert, A.O., Brown, T.G., Takizawa, M. and Weimer, D. (1999) A Review of Perfrormance Indicators for Real Losses from Water Supply Systems. *Aqua, Journal of Water Services Research and Technology*, **48**(2), 227-237.

Lambert, A. and Hirner, W. (2000) Losses from Water Supply Systems: Standard Terminology and Recommended Performance Measures, The Blue Pages, IWA

Liemberger, R. and partners (2006) WB-Easy Calc software

Performance Indicators for analysing the water supply and wastewater services: The case of Portugal

P. Freixial[1]

[1] *Instituto Regulador de Águas e Resíduos. paula.freixial@irar.pt, Centro Empresarial Torres de Lisboa, Rua Tomás da Fonseca, Torre G – 8º, 1600-209 LISBOA, PORTUGAL, Tlf +351 210 052 200*
Fax: +351 210 052 529

Keywords: Regulation model, quality of service assessment, performance indicators

ABSTRACT

The main objectives of this paper are to provide an overview of the Portuguese Regulator's (IRAR) activity concerning the project of the performance indicators (PI) systems to be used as basis for comparison of the utilities performance. Three groups of PI were defined, related to: 1) protection of the user interests; 2) sustainability of the utility; and 3) environmental sustainability. Twenty PI have been set for every type of service: water supply, wastewater and solid wastes. These PI systems are a fundamental tool for IRAR to implement the quality of service regulation, as an essential component of its regulatory model. The quality of service regulation frames the utility performance as far as the quality of service provided to users is concerned and shall not be dissociated from the economic regulation. At present, the consolidation of regulation is crucial envisaging the enhancement of the current quality of service to the users. IRAR aims at having a leverage effect in the transition of the country from the current stage of infrastructural investment in new assets to a new stage of stability and high quality of service delivery.

1 INTRODUCTION

The consolidation of a clear and effective regulatory model for the public water supply and urban wastewater services to the population is one of the current priorities of the Portuguese Institute for the Regulation of Water and Solid Wastes (IRAR - Instituto Regulador de Águas e Resíduos). This paper aims to explain the Institute's main strategic guidelines, as well as the approach adopted for the quality of service assessment.

According to IRAR's bylaws legislation, its scope covers water supply, wastewater and solid wastes services. Such activities are a structural part of public service, being vital to the general welfare, public health and overall safety of the population, as well as to the economic activities and the environment protection.

Therefore, it deals with natural monopolies, either at a local or regional level, in particular as far as public water supply and urban wastewater treatment and disposal are concerned; such a feature is probably more evident here than in any other public service markets. Natural monopolies occur whenever the cost structure is characterized by the production average marginal costs reduction as the productive system develops, due to the existence of scale economies. In such cases, the whole production cost for a given demand decreases whenever there is a single service supplier.

Constraints to competition exist at a natural monopolist market. In particular, natural barriers to the entry into the market of new utilities arise, independently of the level of privatization. Naturally, such a fact does not induce a continuous incentive to increased efficiency.

Such services allow for scale economies, as the investment and production unit costs tend to decrease as far as demand increases, to a given extent. On the other hand, there are also significant economies range, where the production unit costs are likely to decrease as far as similar activities are integrated (e.g., whenever managing together the water supply and the wastewater services), due to the synergies in human resources , equipment and facilities management. Such activities are also characterized by high value assets, representing an intensive capital sector. Additionally, these are assets built up for a life period of many decades and depreciation periods, designed for peak situations and generating idle capabilities which are not used for long periods. The infrastructures show a high degree of immobilization, as they are intended for a specific purpose for which their transaction is difficult. Finally, this market shows a strong correlation between the asset values and the revenues generated and low demand-price elasticity, as they are essential services.

Such features contribute to the limitation of competition among the sector. In practice, only the existence of a single utilities for a given geographic area is feasible. This originates a monopoly, as in such a context the user cannot choose either an utility or the most convenient price-quality ratio.

The regulatory model for the water and solid wastes sector in Portugal is giving its first steps. It is IRAR's priority to consolidate a clear and effective regulation policy, taking into account the regulation practice in other fields in Portugal and internationally. Such is vital for every agent at this sector, mainly the utiliies, so that they know in advance the regulation model rules and may safely decide upon their attitude.

IRAR's strategy follows two major action plans, the first being at the structural regulation level of the sector, and the second at the utilities level (BAPTISTA, J.M. et al, 2003).

The direct intervention by IRAR at the utilities level includes economic and quality of service regulation, in order to promote adequate performance of the utilities.

In association to the economic regulation, the quality of service regulation is an effective way of promoting an adequate balance between price and quality of the service provided to the users.

The project carried out by IRAR, in association with LNEC, aimed the definition of the performance indicators (PI) system to be used as basis for comparison of utilities performance. Three groups of PI were defined, related to: 1) protection of the user interests; 2) sustainability of the utility; and 3) environmental sustainability. Twenty PI have been set for every type of service: water supply, wastewater and solid wastes. These PI systems are a fundamental tool for IRAR to implement the quality of service regulation, as an essential component of its regulatory model. The quality of service regulation frames the utilities performance as far as the quality of service provided to users is concerned and shall not be dissociated from the economic regulation. At present, the consolidation of regulation is crucial envisaging the enhancement of the current quality of service to the users. IRAR aims at having a leverage effect in the transition of the country from the current stage of infrastructural investment in new assets to a new stage of stability and high quality of service delivery.

2 OBJECTIVES

The IRAR's regulation model is complemented by using tools from the utilities self-monitoring performance. The results thereof will be validated by the regulatory body using assessment tools, and further compared with the results of similar utilities at different geographic areas (benchmarking), adopting an educational and added value attitude. For that purpose, the regulatory body must obtain from the utilities the information needed to calculate performance indicators previously set and, upon validation, to carry out the benchmarking based on the utilities self historical record, in order to know their own management evolution over time, and to compare with other similar utilities, aiming to settle references which enable the setting up of new efficiency targets in a realistic way.

The results of such a comparison will be publicly explained, as this procedure is a user's fundamental right. On the other hand, utilities do not wish to be poorly graded and public visibility of their performance promotes efficiency. The purpose is also to consolidate a true information culture, concise, credible and easily understood.

An efficient regulation of natural monopolies recommends the joint application of both sector's structural regulation and performance regulation. This is why utilities regulation must incorporate the following major issues: structural regulation of the sector, utilities' economical regulation, and utilities quality of service regulation, naturally completed by benchmarking and public dissemination of the results. At the water supply, wastewater treatment and disposal, and solid wastes collection, recovery, treatment and disposal sectors – once stated that they are quite static, due to the slow changes on the market conditions and technologies (less perceptible in the current domestic context, still unstable, but tending to develop towards consolidation) - it is foreseeable that the most suitable will be the prevalence of the regulation of the utilities performance over the structural regulation of the sector. In this case, the role of regulation would be, above all, to facilitate the performance regulation. On the contrary, at other service sectors with a more dynamic technology, prevalence of the sector's structural regulation over the performance regulation might be advisable.

IRAR's duties, according to its present ruling law, are limited to the multi-municipal and municipal concessions for public water supply, urban wastewater and solid wastes management. The local administration bodies in charge of the systems direct management are not liable to IRAR, except as far as monitoring of water quality for human consumption is concerned. From the assessment of its duties resulting from its current ruling – the revision of which should be shortly carried out - IRAR is mainly empowered to "proposing, issuing recommending, pronouncing itself, requesting information, dissemination, promoting assessment, calling attention, examining, cooperating, assessing, auditing, informing and promoting" actions. Without prejudice to its duty as an "influence magisterial" through a constructive interaction with the remaining implicated actors, IRAR must use its capabilities as well as allowed by law "issuing binding instructions to settle detected irregularities". Its action may also be strengthened by using the capability to request information to the utilities, to its validation and to disclose the levels of service and comparative summaries. This enables IRAR to employ both an indirect tool to promote merit by disclosing real cases that may be considered as quality references, and a penalty tool that calls the attention of the community to the occurrence of abnormal situations.

3 SECTOR STRUCTURE AND CONTEXT INFORMATION

The performance indicator (PI) system was applied in 2006 exclusively to the multi-municipal and municipal concessions for public water supply and urban wastewater utilities, according to IRAR's present ruling law.

This system was applied to 43 utilities, in which 36 have the water supply service and 32 have wastewater service, according to the following:

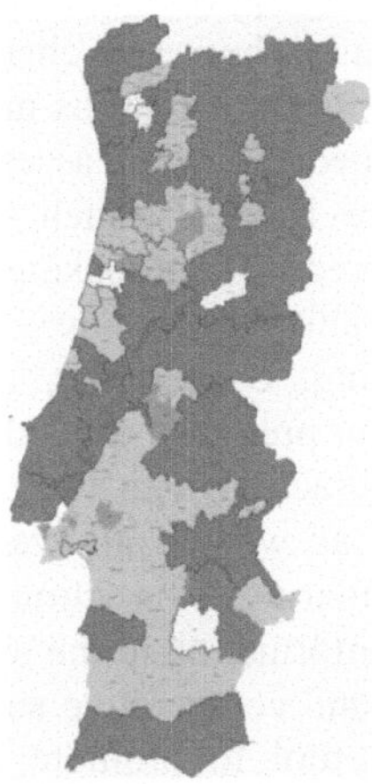

Multi-municipal concessions with water supply service (2006)			
Utility size group	Number of participants	Average water volume supplied	Average network length
Mio m³/year	-	m³/(km year)	km
<0,5	-	-	-
<1,5	7	63.516	13
<10,0	3	27.822	189
<20,0	1	51.892	328
<50,0	1	121.079	249
<100,0	1	158.036	397
<200,0	1	281.788	403
<300,0	1	328.605	744

Figure 1. MULTI-MUNICIPAL CONCESSIONS WITH PUBLIC WATER SUPPLY SERVICE (2006): 11 UTILITIES (DARK BLUE).

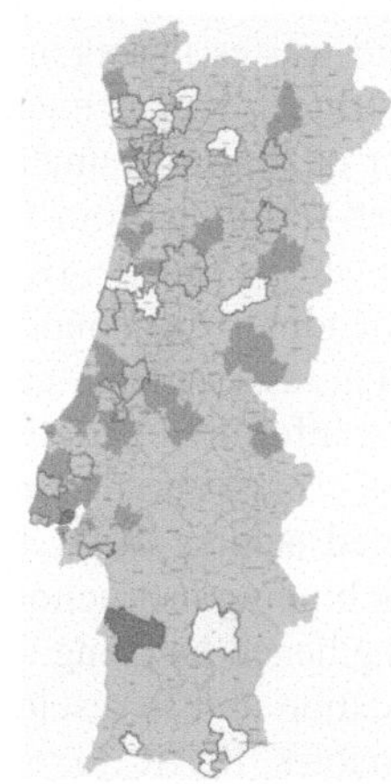

Municipal concessions with water supply service (2006)			
Utility size group	Number of participants	Average water volume supplied	Average network length
Mio m³/year	-	m³/(km year)	km
<0,5	-	-	-
<1,5	5	3.579	278
<10,0	14	6.680	564
<20,0	2	15.693	758
<50,0	1	17.432	1.310
<100,0	1	59.011	1.409
<200,0	-	-	-
<300,0	-	-	-

Figure 2. MUNICIPAL CONCESSIONS WITH PUBLIC WATER SUPPLY (2006): 19 UTILITIES (PALE BLUE).

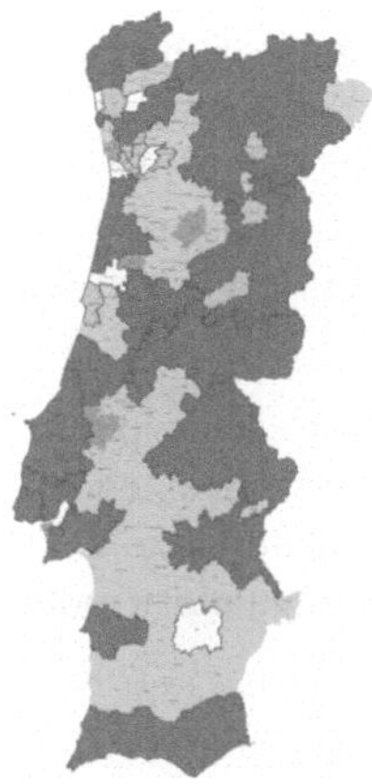

Multi-municipal concessions with wastewater service (2006)			
Utility size group	Number of participants	Average wastewater volume collected	Average network length
Mio m³/year	-	m³/(km year)	km
<0,5	-	-	-
<1,5	1	-	-
<10,0	7	69.363	72
<20,0	4	87.023	150
<50,0	4	784	34.715
<100,0	2	448.094	166
<200,0	-	-	-
<300,0	-	-	-

Figure 3. MULTI-MUNICIPAL CONCESSIONS WITH WASTEWATER (2006): 12 UTILITIES (DARK GREEN).

Municipal concessions with wastewater service (2006)			
Utility size group	Number of participants	Average wastewater volume collected	Average netw length
Mio m³/year	-	m³/(km year)	km
<0,5	2	2722,485294	102
<1,5	4	7.087	1
<10,0	8	21.744	2
<20,0	1	22.716	7
<50,0	-	-	-
<100,0	-	-	-
<200,0	-	-	-
<300,0	-	-	-

Figure 4. MUNICIPAL CONCESSIONS WITH WASTEWATER (2006): 11 UTILITIES (PALE GREEN).

4 ESTABLISHMENT OF THE PI SYSTEM

As far as the quality of the utilities service to be assessed by IRAR is concerned, and having in mind the complexity of the matter, it shall be essential to use performance indicators, which are a measure of the utilities effectiveness and efficiency related to specific issues of the activity developed or to the system performance. The indicators translate the performance levels actually obtained and make clear and transparent the comparison between the management objectives and the obtained results, by making easier a situation that would be quite complex otherwise. With this objective in view, IRAR developed, in association with the National Laboratory of Civil Engineering (LNEC), the following three groups of performance indicators:

a) Protection of the user interests: the purpose being to assess in which degree the user interests are protected, as far as the service accessibility and the service quality are concerned.

b) Sustainability of the opera utility: the purpose being to assess in which degree the technical and economical sustainability of the utility, as well as its legitimate interests, are protected, under the economic and financial, infrastructural, operational and human resources' points of view.
c) Environmental sustainability: the purpose being to assess in which degree the environmental aspects associated with the utility's activities are being considered.

Twenty performance indicators have been set for each type of service (see tables 1, 2 and 3). These are currently being analysed in an extended way. In this way IRAR shall further own a fundamental tool to materialize the regulation of the quality of service, an essential component of the regulatory model.

Table 1. PERFORMANCE INDICATORS FOR PUBLIC WATER SUPPLY SERVICES REGULATION

.Protection of the user interests

User service accessibility

- AA 01 - Service coverage (%)
- AA 02 - Average water charges (€/m^3)

Quality of service supplied to users

- AA 03 - Service interruptions (nr./ delivery point or /1000 serviced connection)
- AA 04 - Water tests performed (%)
- AA 05 - Quality of supplied water (%)
- AA 06 - Response to written complaints (%)

Sustainability of the operator

Operator's economical and financial sustainability

- AA 07 - Operating cost coverage ratio (-)
- AA 08 - Unit running costs (€/m^3)
- AA 09 - Debt equity ratio (-)
- AA 10 - Non-revenue water (%)

Operator's infrastructural sustainability

- AA 11 - Fulfilment of the water intake licensing (%)
- AA 12 - Treatment utilisation (%)
- AA 13 - Transmission and distribution storage capacity (days)
- AA 14 - Mains rehabilitation (%)
- AA 15 - Service connection rehabilitation (%)

Operator's operational sustainability

- AA 16 - Mains failures (nr./100 km)

Operator's human resources sustainability

- AA 17 - Personnel (nr./1000 service connection or / m^3)

Environmental sustainability

- AA 18 - Inefficiency of use of water resources (%)
- AA 19 - Standardised energy consumption (kWh/ m^3/100 m)
- AA 20 - Disposal of sludge from the water treatment (%)

Table 2. PERFORMANCE INDICATORS FOR URBAN WASTEWATER SERVICES REGULATION

Protection of the user interests

User service accessibility

- AR 01 - Service coverage (%)
- AR 02 - Average water charges (€/10^3)

Quality of service supplied to users

- AR 03 – Flooding occurrence (nr/100 km of sewers or m^2/100 km of sewers)
- AR 04 – Response to written complaints (%)

Sustainability of the operator

Operator's economical and financial sustainability

- AR 05 - Operating cost coverage ratio (-)
- AR 06 - Unit running costs (€/10^3)
- AR 07 – Debt equity ratio (-)

Operator's infrastructural sustainability

- AR 08 - Treatment utilisation (%)
- AR 09 - Treatment of collected wastewater (%)
- AR 10 - Wastewater pumping capacity (%)
- AR 11 - Sewer rehabilitation (%)
- AR 12 - Service connection rehabilitation (%)

Operator's operational sustainability

- AR 13 - Sewer blockages (nr./100 km of sewers)
- AR 14 – Pump failures (hours/pump)
- AR 15 - Sewer collapses (nr./100 km of sewers)

Operator's human resources sustainability

- AR 16 – Personnel (nr./100 km of sewers)

Environmental sustainability

- AR 17 - Wastewater tests performed (%)
- AR 18 - Fulfilment of the wastewater discharge parameters (%)
- AR 19 – Energy resources efficiency use (kWh/ m^3/ m)
- AR 20 – Sludge disposal (%)

In addition, the use of such performance assessment tools enables comparisons of the results between similar utilities (benchmarking); such results will be reported and made public on an yearly basis, as above mentioned, as this encourages utilities to provide an improved service as they aim to achieve an advantageous position, and at the same time to fulfil a fundamental right of the users.

A reference document has been edited by IRAR and LNEC (BAPTISTA, J.M. et al, 2004), as well as additional technical documents (ALEGRE, H. e al, 2000; MATOS, M.R. et al, 2003) which are the Portuguese versions of the International Water Association manuals of best practice on performance indicators for water supply and wastewater services.

These documents contain clear procedures for the implementation of the assessment system, particularly on the definition of the data to be supplied by the utilities, on the data validation and auditing by IRAR, on the data processing, and on results interpretation anchored on explanatory factors, according to Figure 5.

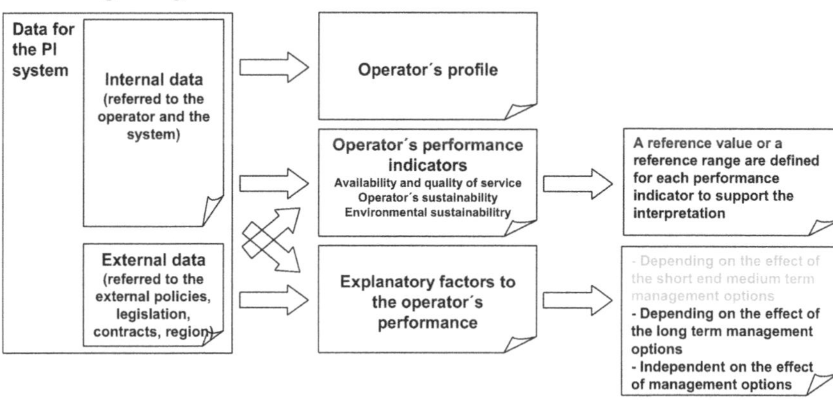

Figure 5. THE IRAR PERFORMANCE INDICATORS SYSTEM

An Extranet application for easier data input by the utilities has been developed and is now available. A forum (Extranet application) is being used to answer to "frequently asked questions" and to promote discussions among utilities previously registered (http://portal.irar.pt).

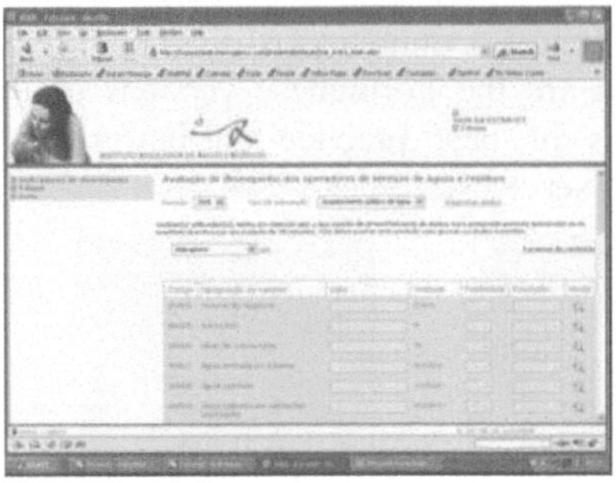

5 RESULTS AND DETAILS OF ANALYSIS, SYNTHESIS AND INTERPRETATION METHODS

The third year performance indicators (PI) system results are presented in “Relatório Anual do Sector de Águas e Resíduos em Portugal 2006”, 3th volume titled “Avaliação da qualidade do serviço prestado aos utilizadores”.[6].

[6] “Annual Report of Water and Waste Sector in Portugal 2006” 3th volume titled “Quality of service provided to user’s”.

In this document is presented the institutional context, the methodology adopted and the results interpretation. The utilities analyses are presented in a chapter where is described each entity and his own system, followed by a global assessment of the quality service with remarks aiming the quality service increase as in the following example:

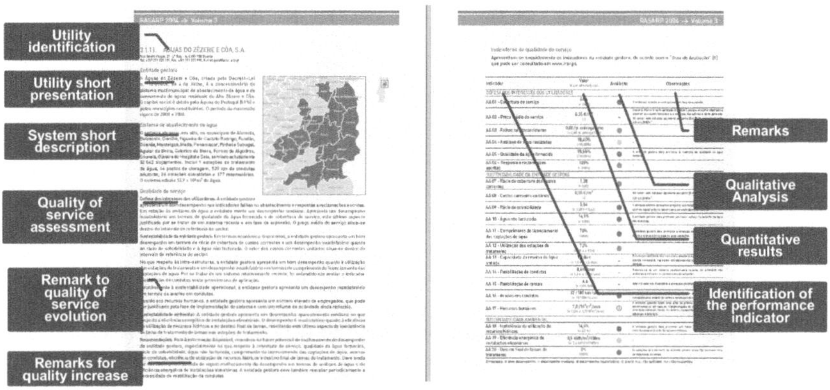

The detailed analysis by performance indicator (PI) is presented in a separate chapter that contains the description of each indicator, values and intervals of reference considered (when applicable), such as the following example:

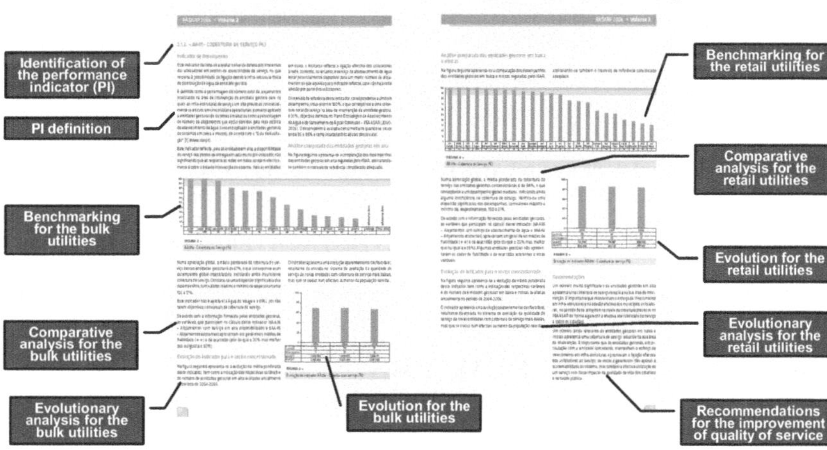

In the final chapter is presented a general quantitative analysis of performance indicators without evaluation, and by assessment good (green), medium quality (yellow) and insufficient (red) according to results presented in the following graphics.

Further on, are presented per activity and per bulk/ retail utilities the main conclusions, with special attention to the necessary improvement of the data collection in general and the improvement in the performance of some indicators.

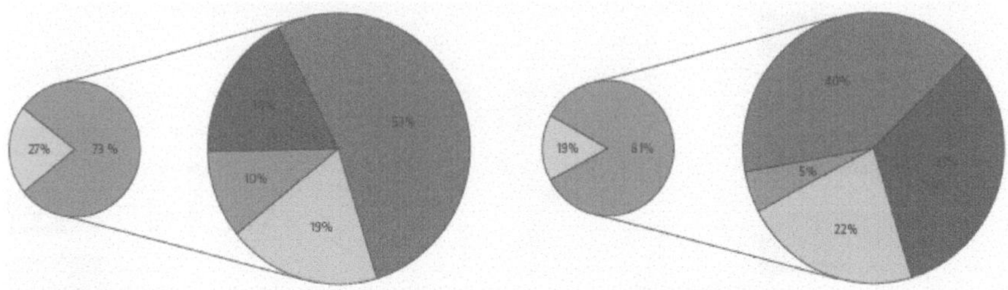

Figure 6 and 7. PERCENTAGE OF UTILITIES (BULK WATER SUPPLY SERVICE – FIG. 6 AND RETAIL WATER SUPPLY FIG. 7) BY ASSESSMENT - GOOD QUALITY (GREEN), MEDIUM QUALITY (YELLOW), INSUFFICIENT QUALITY (RED), WITHOUT EVALUATION OR DATA (GREY) IN EACH PERFORMANCE INDICATOR

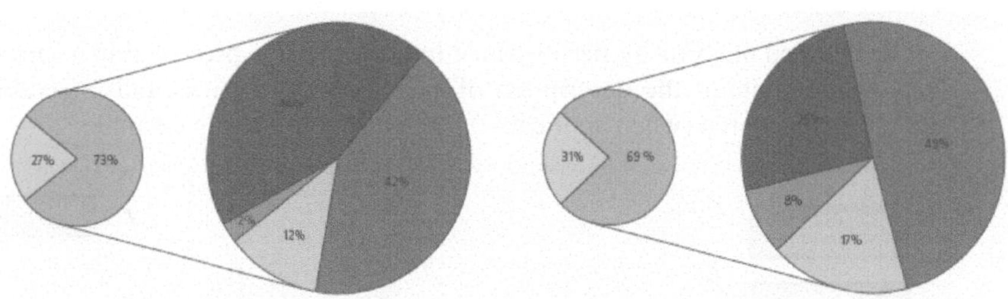

Figure 8 and 9. PERCENTAGE OF UTILITIES (BULK WASTE WATER SERVICE FIG. 8 AND RETAIL WASTE WATER SERVICE FIG. 9) BY ASSESSMENT - GOOD QUALITY (GREEN), MEDIUM QUALITY (YELLOW), INSUFFICIENT QUALITY (RED), WITHOUT EVALUATION OR DATA (GREY) IN EACH PERFORMANCE INDICATOR

6 CONCLUDING REMARKS

The publicly explained of assessment results is a user's fundamental right. On the other hand, utilities do not wish to be poorly graded and public visibility of their performance promotes efficiency. The purpose is also to consolidate a true information culture, concise, credible and easily understood.

The consolidation of regulation envisages a well-balanced growth of these sectors in Portugal, and will assist in the transition from the current stage of infrastructural investments in new assets to a new stage we all aim to be characterized by stability and high quality of services delivery. It is our belief that a clear gain may result from regulation for the services users, for the sector in general, and for the country at large.

7 REFERENCES

Baptista, J.M. *et al, 2005,* Presentation in "10th International Conference on Urban Drainage", Copenhagen, Danmark, 21 a 26 de Agosto de 2005

Baptista, J.M. *et al*, 2004, Guia de avaliação da qualidade de serviço no sector de águas e dos resíduos - IRAR's Technical Guide

Alegre, H. *et al*, 2000; Portuguese version of the International Water Association manual of best practice on performance indicators for water supply service – IRAR's Technical Guide n.º1

Matos, M.R. *et al*, 2003, Portuguese version of the international water association manual of best practice on performance indicators for wastewater service - IRAR's Technical Guide n.º2

IRAR, 2005, RASARP 2004 - Relatório Anual do Sector de Águas e Resíduos em Portugal.

IRAR, 2006, RASARP 2005 - Relatório Anual do Sector de Águas e Resíduos em Portugal.

IRAR, 2007, RASARP 2006 - Relatório Anual do Sector de Águas e Resíduos em Portugal.

IWA Publishing's authorised EU representative for General Product Safety Regulations is Diane D'Arras, 15 rue Duret, 75116 Paris, France, e-mail: safety@iwap.co.uk.

Printed and bound by CPI Group (UK) Ltd, Croydon, CR0 4YY
16/06/2026
02140381-0001